Luger, Modern X-Ray Analysis on Single Crystals

Peter Luger

Modern X-Ray Analysis on Single Crystals

W DE G

Walter de Gruyter · Berlin · New York 1980

Dr. rer. nat. Peter Luger
Professor of Crystallography
Department of Chemistry
Freie Universität Berlin
Takustraße 6
D 1000-Berlin 33

CIP-Kurztitelaufnahme der Deutschen Bibliothek

Luger, Peter:
Modern X-ray analysis on single crystals / Peter Luger. – Berlin, New York · de Gruyter, 1980. –
ISBN 3-11-006830-3

Library of Congress Cataloging in Publication Data

Luger Peter, 1943-
Modern X-ray analysis on single crystals.

Includes Index.
1. X-ray crystallography. I. Title.
QD945.L77 548'.83 80-12115
ISBN 3-110-068303-7

 Typesetting: Tutte Druckerei GmbH, Salzweg-Passau. – Printing: Karl Gerike, Berlin. – Binding: Lüderitz & Bauer, Buchgewerbe GmbH, Berlin. – Printed in Germany.

To Katrin

Preface

Seventeen years after the discovery of X-rays by Wilhelm Conrad Röntgen in 1895, von Laue and his collaborators Friedrich and Knipping found that this novel type of radiation showed the property of diffraction when passed through a crystal lattice. In their classic experiment of 1912 they proved that X-rays, like all other electromagnetic waves, interact with the electron sheath when exposed to a sample of matter, thus causing a diffraction process.

This experiment can be regarded as the foundation of X-ray analysis, a new method for determining the structure of solid matter. However, in the first half-century after its invention this method could seldom be applied, and then only under restrictive conditions. The execution of one structure determination frequently took several months, in some cases even a few years. Moreover the results were of limited accuracy and not always unambiguous.

This situation has changed totally in the last few years. In a dynamic development modern X-ray analysis has become an instrument of structure determination which yields the most detailed, safe and precise information on molecular and crystal geometry available.

Two major reasons can be given for this decisive progress. First, it is a consequence of the development of the so called "Direct Methods" of phase determination in the last 20 years. From this it is now possible to work on a large number of compounds, especially of organic chemistry, which could never have been treated before. Second, the possibility of executing the extensive numerical calculations with the help of ever faster and larger computers has reduced the time needed for a structure determination or has even made its execution possible in the case of larger structures.

Today it can be stated that an X-ray analysis can be performed on any crystalline compound within a reasonable amount of time if its molecular weight is not too large.

Being now comparable in speed and expense to other methods and superior in results, the application of X-ray analysis is increasing not only in all chemical laboratories but also in biological and biochemical as well as in physical research projects; the number of scientists using this method is becoming larger from year to year.

Today, by means of highly sophisticated computer programs controlling fairly automatically the measuring and structure determination process, an X-ray analysis can be processed with little effort on the part of the user. In general, extensive previous knowledge of theoretical crystallography is unnecessary; instead, much practical experience is more helpful for the experimenter to continue his investigation. However, in spite of automation several sources of error remain for the user, each capable of preventing a succesful solution to a structural problem.

It seemed therefore appropriate to have a guide for practical work in X-ray analysis directed at those who are not highly experienced in crystallography but who need structure determination as a method for solving some of their problems. In this book the fundamentals of crystallography are presented together with those topics that are helpful for the execution of a structure analysis. The contents were selected with respect to practical applicability; most questions arising in the course of practical work are treated. This book is addressed to graduate students intending to use this method in any part of an examination as well as to scientists in any research or industrial laboratory, hence to all people concerned with a structural problem which might be solved by the method of single crystal analysis.

In the first part mainly theoretical aspects are presented. Note that no effort has been made to derive all results of diffraction theory. This is not the aim of this book since we are more interested in practical problems. In the second and subsequent parts we describe the process of an X-ray structure determination in all details, starting with the diffraction experiments, then dealing with the phase determination, the refinement and finally with the representation and documentation of results. The presentation of three structures as examples supports the orientation of this book toward practical work.

I have tried to give as modern a formulation as possible of the mathematical aspects that figure so largely in X-ray analysis because the modern mathematical language seems to be the most appropriate for a clear understanding, despite its somewhat abstract nature.

It is the aim of this book to serve as a guide and to enable the reader to solve his structural problems almost without further preparation. It is desired that this book will be a contribution to the further dissemination of X-ray analysis as a modern method of structure determination to an ever increasing number of scientists.

Acknowledgement

I am deeply indebted to Professor G. A. Jeffrey, University of Pittsburgh, U.S.A., for having accepted the task of a linguistic revision of my English manuscript. I also owe him thanks for important suggestions aimed to a scientific improvement of the manuscript.

Many of my collegues of the Institut für Kristallographie, Freie Universität Berlin, and of the Institut für Anorganische und Analytische Chemie, Technische Universität Berlin, supported my work on this book. Their various contributions are gratefully accepted.

I am obliged to Miss U. Ahlers, Mrs. H. Bombosch, Mrs. G. Malchow, Miss E. Müller and Miss S. Schiedlausky for typewriting the manuscript and drawing most of the graphical representations and to Mr. K. Wichmann, who took most of the photographs.

Finally I thank the de Gruyter Verlag, Berlin, for the cooperation in preparing and publishing this book.

Berlin, May 1979 *Peter Luger*

Contents

1 Theoretical Basis

1.1 Matrices, Vectors

1.1.1 *Introduction*

The first part of this chapter is concerned with some mathematics which will be used in the later chapters of this book. We assume that the fundamentals of arithmetic and of integral and differential calculus are well-known to the reader, but students of chemistry often have difficulties with the theory of vector and matrix algebra. Since we will make frequent use of these mathematical formalisms, the most important properties of vectors and matrices are briefly discussed.

1.1.2 *Matrices, Determinants, Linear Equations*

A rectangular array, A, arranged in the form

$$A = \begin{pmatrix} a_{11} & a_{12} & a_{13} & \dots & a_{1n} \\ a_{21} & a_{22} & a_{23} & \dots & a_{2n} \\ a_{31} & a_{32} & a_{33} & \dots & a_{3n} \\ \cdot & & & & \cdot \\ a_{m1} & a_{m2} & a_{m3} & \dots & a_{mn} \end{pmatrix}$$

is called a matrix. The elements a_{ik} can be arbitrary numbers. If the number of rows is m and the number of columns is n, the matrix is said to be of the order $m \times n$. If $m = n$ the matrix is called a square matrix of order n. The index i of the element a_{ik} indicates its row and the index k the corresponding column. As will be shown in the next chapter, the matrix formalism is a very convenient way to describe vector operations, vector transformations and it provides a very elegant method for solving linear equations.

The introduction of matrices requires a knowledge of matrix algebra. First, we define the basic arithmetic operations of matrices.

The equality of matrices. Two matrices are said to be of equal type if their numbers of rows and columns are equal. Two matrices are equal, if they are of equal type and if all elements in corresponding rows and columns are equal.

Example:

(a) The matrices

$$A = \begin{pmatrix} 4 & 0 \\ 0 & 1 \end{pmatrix} \quad \text{and} \quad B = \begin{pmatrix} 4 & 0 \\ 1 & 0 \end{pmatrix}$$

are of the same type. They are square matrices of order 2. But they are not equal. For instance, the element in the second row and first column a_{21} is equal to zero and is not equal to $b_{21} = 1$.

(b) The matrices

$$\begin{pmatrix} 4 & 0 \\ 0 & 1 \end{pmatrix} \quad \text{and} \quad \begin{pmatrix} 2^2 & 0 \\ 0 & e^0 \end{pmatrix} \quad \text{are equal.}$$

Now we can proceed to the algebraic operations. If $A = (a_{ik})$, $B = (b_{ik})$ are matrices of the same type, the matrix $C = (c_{ik})$ is called the sum (difference) of A and B,

$$C = A \pm B$$

if

$$c_{ik} = a_{ik} \pm b_{ik} \tag{1.1}$$

It can be clearly seen that C must be of the same type as A and B, and that the calculation of a sum or difference is impossible if A and B are of different types.

The product λA of a matrix $A = (a_{ik})$ with a single factor λ is defined by

$$\lambda A = \begin{pmatrix} \lambda a_{11} & \lambda a_{12} & \dots & \lambda a_{1n} \\ \lambda a_{21} & \lambda a_{22} & \dots & \lambda a_{2n} \\ \vdots & & & \\ \lambda a_{m1} & \lambda a_{m2} & \dots & \lambda a_{mn} \end{pmatrix}$$

Example:

(a) Given

$$A = \begin{pmatrix} 2 & 3 \\ -3 & 2 \end{pmatrix} \quad B = \begin{pmatrix} 1 & -1 & 1 \\ -1 & 1 & -1 \end{pmatrix},$$

calculate $A - (1/2)B$.

Answer: This calculation is impossible, because A and B are of different type. If we change

$$B \text{ to } B' = \begin{pmatrix} 1 & -1 \\ -1 & 1 \end{pmatrix}$$

we obtain

$$A - (1/2)B' = \begin{pmatrix} (2 - 1/2) & (3 + 1/2) \\ -(3 - 1/2) & (2 - 1/2) \end{pmatrix}.$$

(b) Let

$$A = \begin{pmatrix} a & b \\ c & d \end{pmatrix} \quad E = \begin{pmatrix} 1 & 0 \\ 0 & 1 \end{pmatrix}$$

and λ an arbitrary constant. Calculate $B = A - \lambda E$.

Solution:

$$B = \begin{pmatrix} (a-\lambda) & b \\ c & (d-\lambda) \end{pmatrix}.$$

The product of two matrices is not as simple an operation as that of the sum and difference. Let $A = (a_{ik})$ be a matrix of order $m \times n$, $B = (b_{ik})$ a matrix of order $n \times p$. The $m \times p$ matrix, $C = (c_{ik})$ is said to be the product of A and B,

$$C = AB$$

if

$$c_{ik} = \sum_{\varrho=1}^{n} a_{i\varrho} b_{\varrho k}$$

for

$$i = 1, \ldots, m \quad \text{and} \quad k = 1, \ldots, p \tag{1.2}$$

The definition of the matrix product is not self-evident and at this stage no simple reason for this extraordinary operation can be given. It can only be stated that a large number of important operations can be expressed by way of matrix products in a very clear and simple way.

Let us give an explanation of (1.2) which may be somewhat clearer. The element c_{ik} of the product matrix C is obtained by multiplication of the elements of the i-th row of A by the corresponding elements of the k-th column of B followed by summation of the n products. A stringent requirement for this procedure is that the length of rows of the first matrix is equal to the length of columns of the second. That is, if A has n columns, B must have n rows, otherwise the product is not defined.

Problem:

(a) Let $P(x_0, y_0)$ be an arbitrary point in a plane, with (x_0, y_0) its cartesian coordinates. Show that the rotation of P about an angle φ can be expressed by a special matrix.

(b) Show that the rotation about φ followed by a rotation about ω can be expressed by a matrix product.

Answer:

(a) As shown in Fig. 1.1, we solve the problem by rotating the $x - y$ system about the same angle φ. In the new $x_\varphi - y_\varphi$ system the point P′ has the coordinates $x_\varphi = \overline{OC}$ and $y_\varphi = \overline{P'C}$ which are of course equal to x_0 and y_0. Then we have (see Fig. 1.1)

$$x' = \overline{OA} = \overline{OB} - \overline{AB} = \overline{OB} - \overline{DC} = x_0 \cos\varphi - y_0 \sin\varphi$$

$$y' = \overline{AP'} = \overline{AD} + \overline{DP'} = \overline{BC} + \overline{DP'} = x_0 \sin\varphi + y_0 \cos\varphi$$

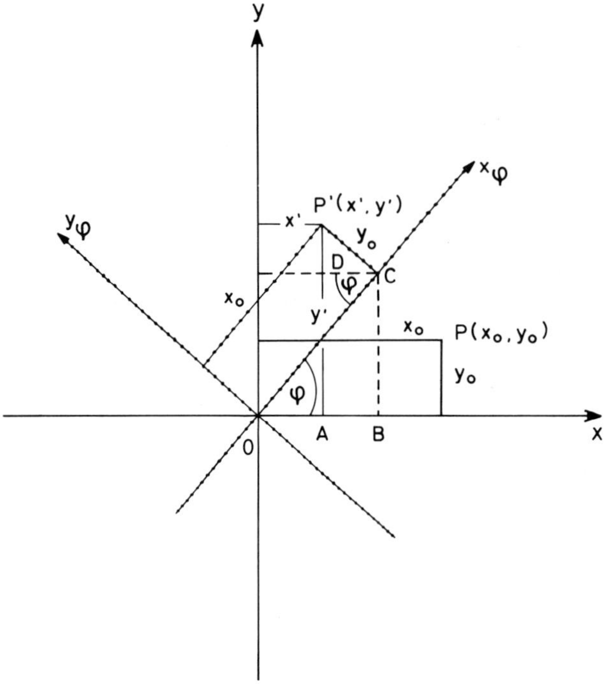

Fig. 1.1. Rotation of a point P about an angle φ.

If we now write (x', y') and (x_0, y_0) as 2×1 matrices, we get

$$\begin{pmatrix} x' \\ y' \end{pmatrix} = \begin{pmatrix} \cos\varphi & -\sin\varphi \\ \sin\varphi & \cos\varphi \end{pmatrix} \begin{pmatrix} x_0 \\ y_0 \end{pmatrix} \tag{1.3}$$

and the matrix

$$A(\varphi) = \begin{pmatrix} \cos\varphi & -\sin\varphi \\ \sin\varphi & \cos\varphi \end{pmatrix}$$

represents the rotation about the angle φ.

(b) From the rotation of P' by the angle ω we get P'' (x'', y'') and (x'', y'') are obtained by

$$x'' = x' \cos\omega - y' \sin\omega$$
$$y'' = x' \sin\omega + y' \cos\omega .$$

With the expressions of (a) for x' and y' we have then

$$x'' = (x_0 \cos\varphi - y_0 \sin\varphi) \cos\omega - (x_0 \sin\varphi + y_0 \cos\varphi) \sin\omega$$
$$y'' = (x_0 \cos\varphi - y_0 \sin\varphi) \sin\omega + (x_0 \sin\varphi + y_0 \cos\varphi) \cos\omega$$

or

$$x'' = (\cos\omega\cos\varphi - \sin\omega\sin\varphi)\,x_0 + (-\cos\omega\sin\varphi - \sin\omega\cos\varphi)\,y_0$$
$$y'' = (\sin\omega\cos\varphi + \cos\omega\sin\varphi)\,x_0 + (-\sin\omega\sin\varphi + \cos\omega\cos\varphi)\,y_0$$

The last two equations become quite simple when written as matrix product:

$$\begin{pmatrix} x'' \\ y'' \end{pmatrix} = \begin{pmatrix} \cos\omega & -\sin\omega \\ \sin\omega & \cos\omega \end{pmatrix} \begin{pmatrix} \cos\varphi & -\sin\varphi \\ \sin\varphi & \cos\varphi \end{pmatrix} \begin{pmatrix} x_0 \\ y_0 \end{pmatrix}$$

By evaluating the last matrix product you get the expressions developed above.

On the other hand we get from multiple angle formulae

$$\begin{aligned} \cos\omega\cos\varphi - \sin\omega\sin\varphi &= \cos(\omega+\varphi) \\ -(\cos\omega\sin\varphi + \sin\omega\cos\varphi) &= -\sin(\omega+\varphi) \\ \sin\omega\cos\varphi + \cos\omega\sin\varphi &= \sin(\omega+\varphi) \\ -\sin\omega\sin\varphi + \cos\omega\cos\varphi &= \cos(\omega+\varphi) \end{aligned}$$

Hence we get

$$\begin{pmatrix} x'' \\ y'' \end{pmatrix} = A(\omega+\varphi)\begin{pmatrix} x_0 \\ y_0 \end{pmatrix} \quad \text{or } A(\omega+\varphi) = A(\omega)\,A(\varphi).$$

Some properties of the arithmetic operations of matrices are as follows:

(1) $A \pm B = B \pm A$

(2) $(A+B)+C = A+(B+C)$

(3) $\lambda(A+B) = \lambda A + \lambda B$

(4) $\mu(\lambda A) = (\mu\lambda)A$

(5) $(\mu+\lambda)A = \mu A + \lambda A$

(6) $(AB)C = A(BC)$

(7) $A(B+C) = AB + AC$

(8) In general, the matrix product is not cummutative, that is, $AB \neq BA$ (the reader may show this by an example).

Special matrices are as follows:

(a) A matrix O with all elements being equal to zero is called a null matrix. It has the property $A + O = O + A = A$ for all matrices A of the same type.

(b) A square matrix $E = (e_{ik})$ with $e_{ik} = \Delta_{ik}$ (Kronecker symbol) is called a unit matrix. It has the property $EA = AE = A$ for all square matrices of the same order.

(c) For a given matrix $A = (a_{ik})$, its transposition matrix A' is defined by

$$A' = (a_{ki}) = \begin{pmatrix} a_{11} & a_{21} & \dots & a_{m1} \\ a_{12} & a_{22} & \dots & a_{m2} \\ \vdots & & & \vdots \\ a_{1n} & a_{2n} & \dots & a_{mn} \end{pmatrix}$$

A′ is obtained by reflection of the elements of A across the principal diagonal, which is defined by the elements $a_{11}, a_{22}, \ldots a_{ii} \ldots$ etc.

(d) For a given square matrix A the matrix B is called the inverse matrix of A, if BA = E. B is then denoted by $B = A^{-1}$.

The problem of how to find the inverse matrix of A is very important, but its solution is not trivial. Let us demonstrate the importance of the inverse matrix by an example:

A system of n linear equations is of the form

$$\begin{aligned} a_{11}x_1 + a_{12}x_2 + \ldots a_{1n}x_n &= b_1 \\ a_{21}x_1 + a_{22}x_2 + \ldots a_{2n}x_n &= b_2 \\ \vdots \qquad\qquad\qquad \vdots \quad &\ \ \vdots \\ a_{n1}x_1 + a_{n2}x_2 + \ldots a_{nn}x_n &= b_n \end{aligned} \tag{1.4}$$

In matrix notation, with

$$A = (a_{ik}),\ b = \begin{pmatrix} b_1 \\ \vdots \\ b_n \end{pmatrix}, \quad x = \begin{pmatrix} x_1 \\ \vdots \\ x_n \end{pmatrix},$$

we obtain

$$Ax = b.$$

If we were able to calculate the inverse matrix, we could have immediately the solution for x,

$$x = A^{-1}b.$$

The example shows that there is a close connection between the solution of n linear equations and the calculation of the inverse matrix and we shall see that the procedure of solution is the same for both problems. Before starting this, we must define a determinant.

Let $A = (a_{ik})$ be a square matrix of order n. Let us denote by A_{ik} the matrix obtained from A by deleting the i-th row and k-th column:

$$A_{ik} = \left(\begin{array}{ccc|ccc} a_{11} & \cdots & a_{1,k-1} & a_{1,k+1} & \cdots & a_{1n} \\ \vdots & & & & & \\ a_{i-1,1} & \cdots & a_{i-1,k-1} & a_{i-1,k+1} & \cdots & a_{i-1,n} \\ \hline a_{i+1,1} & \cdots & a_{i+1,k-1} & a_{i+1,k+1} & \cdots & a_{i+1,n} \\ \vdots & & & & & \\ a_{n1} & \cdots & a_{n,k-1} & a_{n,k+1} & \cdots & a_{nn} \end{array}\right)$$

A_{ik} is then of order n − 1.

Let

$$A = \begin{pmatrix} a_{11} & a_{12} \\ a_{21} & a_{22} \end{pmatrix},$$

a square matrix of order 2. Its determinant $|A|$ is a number obtained from A by

$$|A| = a_{11}a_{22} - a_{21}a_{12}$$

The determinant of a square matrix of higher order is defined iteratively:

Let $A = (a_{ik})$ be a square matrix of order n. Its determinant $|A|$ is defined by

$$|A| = \sum_{k=1}^{n} (-1)^{1+k} a_{1k} |A_{1k}|$$

This expansion is done with respect to the first row. It can be shown that the expansion is independent of our choice of the row; furthermore it can be done referring to a column. This property is expressed by the *Theorem of Laplace*:

The determinant $|A|$ of a square matrix $A = (a_{ik})$ is given by

$$|A| = \sum_{k=1}^{n} (-1)^{i+k} a_{ik} |A_{ik}| \text{ for arbitrary } i = 1, \ldots, n$$

or (1.5)

$$|A| = \sum_{i=1}^{n} (-1)^{i+k} a_{ik} |A_{ik}| \text{ for arbitrary } k = 1, \ldots, n$$

For convenience, the expression $(-1)^{i+k}|A_{ik}|$ shall have its own name. It is called the minor of the matrix element a_{ik} and designated α_{ik}.

With the notation of minors, Laplace theorem reads

$$|A| = \sum_{k=1}^{n} a_{ik}\alpha_{ik} \text{ for } i = 1, \ldots, n$$

or

$$|A| = \sum_{i=1}^{n} a_{ik}\alpha_{ik} \text{ for } k = 1, \ldots, n.$$

In practice the calculation of a determinant of higher order is a joyless task. There is some help from the rules for the arithmetic of determinants, but we shall not describe them because in the progress of a structure analysis the calculation of the determinants is done by the computer and at most we ourselves calculate determinants of order two or three.

The application of all these rules for determinant arithmetic leads to an important generalization of the Laplace theorem:

Generalized Expansion Theorem: Let A be a square matrix of order n. Then for its determinant the following equations hold:

$$|A|\Delta_{ik} = \sum_{s=1}^{n} a_{is}\alpha_{ks}$$

$$|A|\Delta_{ik} = \sum_{s=1}^{n} a_{si}\alpha_{sk}$$

(Δ_{ik} = Kronecker-Symbol)

For i = k we get the Laplace theorem. Its generalization largely contains the solution of the inverse matrix problem.

Let us regard a matrix, denoted by A^A, containing the minors α_{ki} as elements. Note that, in A^A the usual order of subscripts is exchanged. In $A = (a_{ik})$ the element a_{ik} is positioned in the i-th row and k-th column, while in A^A the corresponding minor α_{ik} is in the k-th row and i-th column. The minor matrix A^A has the important property, that $AA^A = |A|E$ holds.

Proof: From the definition of matrix product we get

$$\begin{pmatrix} a_{11} & \cdots & a_{1n} \\ \vdots & & \ddots \vdots \\ a_{n1} & & a_{nn} \end{pmatrix} \cdot \begin{pmatrix} \alpha_{11} & \cdots & \alpha_{n1} \\ \vdots & & \vdots \\ \alpha_{1n} & & \alpha_{nn} \end{pmatrix} = \left(\sum_{s=1}^{n} a_{is} \alpha_{ks} \right)$$

The sum on the right side is equal to $|A| \Delta_{ik}$, from generalized expansion theorem. So we get

$$AA^A = \begin{pmatrix} |A| & & 0 \\ & |A| & \\ 0 & & |A| \end{pmatrix} = |A|E$$

Now it is easy to see how to calculate the inverse matrix A^{-1}. It is simply

$$A^{-1} = (1/|A|)A^A \quad \text{if } |A| \neq 0 \tag{1.6}$$

For a matrix A with $|A| = 0$, A^{-1} does not exist!

Problem:
Given

$$A = \begin{pmatrix} -1 & 0 & 1 \\ 2 & 1 & -1 \\ 3 & -2 & 0 \end{pmatrix},$$

calculate A^{-1}.

Solution: The determinant $|A|$, calculated by expansion with respect to the third column is

$$|A| = 1 \begin{vmatrix} 2 & 1 \\ 3 & -2 \end{vmatrix} - (-1) \begin{vmatrix} -1 & 0 \\ 3 & -2 \end{vmatrix} = -5.$$

Since $|A| \neq 0$, A^{-1} exists. The minors α_{ik} are then

$$\alpha_{11} = \begin{vmatrix} 1 & -1 \\ -2 & 0 \end{vmatrix} = -2 \qquad \alpha_{12} = -\begin{vmatrix} 2 & -1 \\ 3 & 0 \end{vmatrix} = -3$$

$$\alpha_{13} = \begin{vmatrix} 2 & 1 \\ 3 & -2 \end{vmatrix} = -7 \cdot \qquad \alpha_{21} = -\begin{vmatrix} 0 & 1 \\ -2 & 0 \end{vmatrix} = -2$$

$$\alpha_{22} = \begin{vmatrix} -1 & 1 \\ 3 & 0 \end{vmatrix} = -3 \qquad \alpha_{23} = -\begin{vmatrix} -1 & 0 \\ 3 & -2 \end{vmatrix} = -2$$

$$\alpha_{31} = \begin{vmatrix} 0 & 1 \\ 1 & -1 \end{vmatrix} = -1 \qquad \alpha_{32} = -\begin{vmatrix} -1 & 1 \\ 2 & -1 \end{vmatrix} = 1$$

$$\alpha_{33} = \begin{vmatrix} -1 & 0 \\ 2 & 1 \end{vmatrix} = -1$$

It follows that

$$A^{-1} = (1/|A|)A^A = -1/5 \begin{pmatrix} -2 & -2 & -1 \\ -3 & -3 & 1 \\ -7 & -2 & -1 \end{pmatrix}$$

To check whether we have made any mistakes, let us calculate

$$A^{-1}A = -1/5 \begin{pmatrix} -2 & -2 & -1 \\ -3 & -3 & 1 \\ -7 & -2 & -1 \end{pmatrix} \begin{pmatrix} -1 & 0 & 1 \\ 2 & 1 & -1 \\ 3 & -2 & 0 \end{pmatrix}$$

$$= -1/5 \begin{pmatrix} -5 & 0 & 0 \\ 0 & -5 & 0 \\ 0 & 0 & -5 \end{pmatrix} = E$$

Since the result is equal to E, our calculation was correct.

As well as deriving the inverse matrix from the generalized expansion theorem, we get the solution of a system of n linear equations from the same proposition:

Cramer's rule: Let $Ax = b$ be a system of n linear equations of n unknowns with $|A| \neq 0$. Let A_k be the matrix obtained by replacing the k-th column by the column b. Then the unknown x_k is given by

$$x_k = \frac{|A_k|}{|A|} \quad \text{for all } k = 1, \ldots, n \tag{1.7}$$

Proof: Let us take a fixed k and multiply all equations by α_{ik}. We get

$$\begin{aligned} &a_{11}\alpha_{1k}x_1 + \ldots a_{1k}\alpha_{1k}x_k + \ldots a_{1n}\alpha_{1k}x_n = \alpha_{1k}b_1 \\ &\vdots \qquad\qquad\qquad\qquad\qquad\qquad \vdots \qquad\quad \vdots \\ &a_{n1}\alpha_{nk}x_1 + \ldots a_{nk}\alpha_{nk}x_k + \ldots a_{nn}\alpha_{nk}x_n = \alpha_{nk}b_n \end{aligned}$$

Addition of all columns leads to

$$x_1\Big(\sum_{i=1}^{n} a_{i1}\alpha_{ik}\Big) + \ldots x_k\Big(\sum_{i=1}^{n} a_{ik}\alpha_{ik}\Big) + \ldots x_n\Big(\sum_{i=1}^{n} a_{in}\alpha_{ik}\Big) = \sum_{i=1}^{n} b_i\alpha_{ik}.$$

Comparison of these sums with those of generalized expansion theorem shows that on the left side only the sum at x_k differs from zero.

We get

$$x_k\left(\sum_{i=1}^{n} a_{ik}\alpha_{ik}\right) = \sum_{i=1}^{n} b_i\alpha_{ik}$$

The first sum is equal to $|A|$, the second one is $|A_k|$, hence it follows that $x_k = |A_k|/|A|$.

Problem: Solve the equations

$$\begin{aligned} 2x_1 + x_2 &= 4 \\ -x_1 + 3x_2 &= 0 \end{aligned}$$

Solution:

$$|A| = \begin{vmatrix} 2 & 1 \\ -1 & 3 \end{vmatrix} = 7; \quad |A_1| = \begin{vmatrix} 4 & 1 \\ 0 & 3 \end{vmatrix} = 12; \quad |A_2| = \begin{vmatrix} 2 & 4 \\ -1 & 0 \end{vmatrix} = 4;$$

hence

$$x_1 = 12/7$$
$$x_2 = 4/7$$

1.1.3 *Vector Algebra*

The properties of several physical quantities such as force, velocity, the electric or magnetic field, are not completely defined by their magnitude; their directions also have to be known. A mathematical theory of such quantities having a direction and a magnitude is known as vector algebra.

A vector is defined as a geometrical object which is given by its magnitude, its direction and its sense.

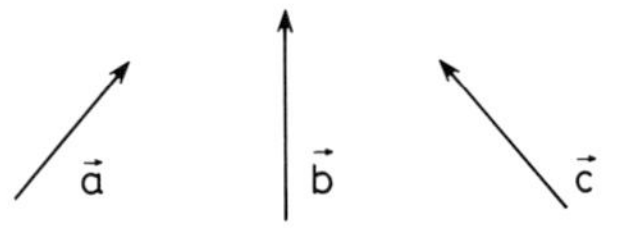

Fig. 1.2. Representation of vectors by arrows.

Quantities which are defined completely by a single number shall be denoted as scalars. They are sufficient to describe physical properties, such as temperature, density of a material, or energy. To distinguish clearly between vectors and scalars, we write **a**, **b**, **c** ... for vectors and a = |**a**|, b = |**b**|, c = |**c**|, ... for their magnitudes, which are scalar quantities. In a diagram a vector may be represented by an arrow (Fig. 1.2), the direction and the sense of the vector being given by the direction of the arrow, its magnitude given by the arrow's length.

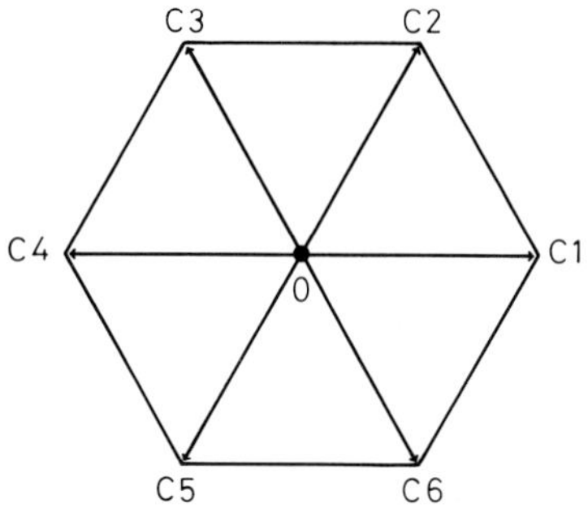

Fig. 1.3. Vector representation of the benzene ring atoms.

Example: Let us consider the hexagon of a benzene ring with the origin of a coordinate system in the center of the ring (the hydrogen atoms have been omitted for clarity). Each of the atoms C 1, . . . , C 6 can be represented by a vector from the origin to one corner of the hexagon, so that the atomic positions can be described by the vectors **C**1, ...,**C**6.

We can construct a large number of further vectors at the benzene hexagon, for instance those vectors which represent the bonds between the carbon atoms. Not quite trivial is the following

Problem:

(a) Find all vectors connecting two arbitrary atoms of the benzene molecule and draw them in a separate diagram as vectors fixed to the same origin. How many vectors can be found?

(b) Does the last diagram change if we transform the origin in Fig. 1.3 to one edge of the hexagon?

Solution:

(a) Fig. 1.4a shows all possible vectors between two atoms in the benzene ring and Fig. 1.4b shows these vectors drawn from the origin with the digits indicating their frequency. The sum of all frequency digitis is equal to the total number of possible vectors. If n atoms are present, every vector can have n connections (including to itself!), so that a total of n^2 vectors is possible.

(b) No. Since we have only difference vectors in Fig. 1.4b, this diagram does not depend on the choice of the origin.

Just as for matrices, we must now introduce algebraic operations for vectors. The sum vector **c** of two vectors **a** and **b**, $\mathbf{c} = \mathbf{a} + \mathbf{b}$, is defined geometrically in the following

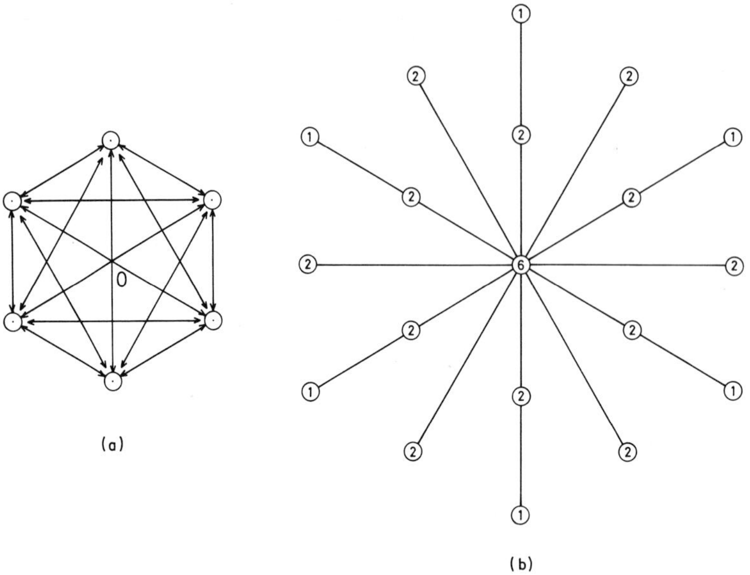

Fig. 1.4. (a) All vectors connecting a pair of benzene ring atoms, (b) all vectors of (a) drawn from a common origin. Digits indicate the frequency of appearance of each vector.

way (see Fig. 1.5). The origin of **b** is transferred to the top of **a**, then **c** is defined as the vector running from the origin of **a** to the top of **b**.

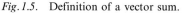

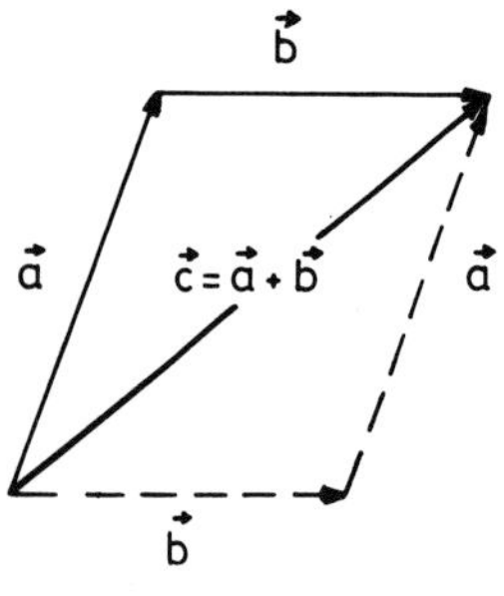

Fig. 1.5. Definition of a vector sum.

Completing the diagram of Fig. 1.5 to a parallelogram (dotted lines) it can immediately be seen that the vector sum is commutative, that is,

$$\mathbf{a} + \mathbf{b} = \mathbf{b} + \mathbf{a}$$

Before defining the difference of two vectors, we shall deal with the product of a scalar λ and a vector **a**. This product $\lambda\mathbf{a}$ defines a vector of magnitude $|\lambda||\mathbf{a}|$ having the direction of **a**. Its sense is that of **a**, if $\lambda > 0$, it is antiparallel to **a** if $\lambda < 0$ (see Fig. 1.6). If $\lambda = 0$, we get the null vector **0**, which is a vector of magnitude zero.

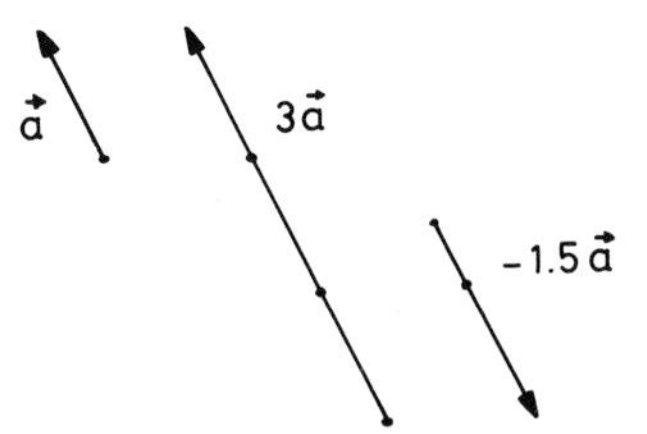

Fig. 1.6. Examples of various multiples of a vector **a**.

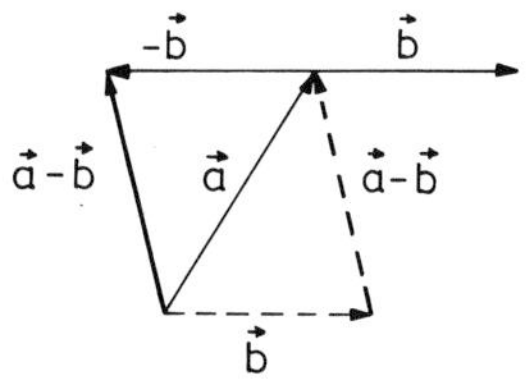

Fig. 1.7. Definition of the difference **a** − **b**.

The difference vector $\mathbf{c} = \mathbf{a} - \mathbf{b}$ can then be defined easily. It is given by the sum (see Fig. 1.7)

$$\mathbf{c} = \mathbf{a} - \mathbf{b} = \mathbf{a} + (-\mathbf{b})$$

Another definition of $\mathbf{a} - \mathbf{b}$ frequently used is shown in Fig. 1.7 by dotted lines: **a** and **b** are brought to the same origin. Then $\mathbf{a} - \mathbf{b}$ is obtained as the vector running from the top of **b** to the top of **a**.

Problem: Express the vectors between ortho-, meta- and parahydrogens of the benzene ring (Fig. 1.8) in terms of C-C and C-H vectors.

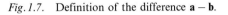

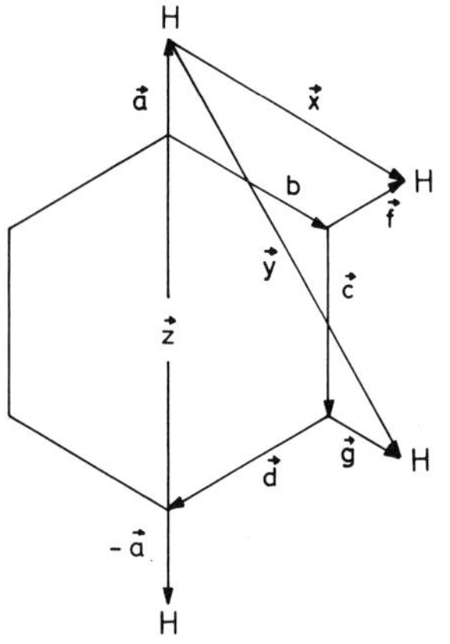

Fig. 1.8. Representation of H-H difference vectors by C-H and C-C vectors in the benzene molecule.

Solution: We get the ortho-vector **x** by the sum of $-\mathbf{a}$, **b** and **f**, hence $\mathbf{x} = -\mathbf{a} + \mathbf{b} + \mathbf{f}$. For **y** we get $\mathbf{y} = -\mathbf{a} + \mathbf{b} + \mathbf{c} + \mathbf{g}$, and for **z**, $\mathbf{z} = -\mathbf{a} + \mathbf{b} + \mathbf{c} + \mathbf{d} - \mathbf{a} = -2\mathbf{a} + \mathbf{b} + \mathbf{c} + \mathbf{d}$.

The product of vectors will be defined in two different ways. One will be the so-called *scalar product*, the other the *vector product*. The scalar product (or inner product) of two vectors **a** and **b** defines the scalar quantity

$$\mathbf{ab} = |\mathbf{a}||\mathbf{b}|\cos(\mathbf{a},\mathbf{b})$$

where $(\mathbf{a},\mathbf{b})$ is the angle between **a** and **b**. Note that the result of a scalar product is a *scalar*!

The properties of a scalar product are as follows:

(1) $\mathbf{ab} = \mathbf{ba}$
(2) $\lambda(\mathbf{ab}) = (\lambda\mathbf{ab}) = (\mathbf{a}\lambda\mathbf{b})$
(3) $(\mathbf{a} + \mathbf{b})\mathbf{c} = \mathbf{ac} + \mathbf{bc}$
(4) The scalar product is zero not only if **a** or **b** are null vectors but also if **a** is perpendicular to **b** (because $\cos 90^0 = 0$).
This property of the scalar product is frequently used to check whether two vectors are perpendicular to each other.
(5) The 'square' of a vector **a** is the scalar product **aa**. From the definition it follows $\mathbf{aa} = |\mathbf{a}||\mathbf{a}| = a^2$. On the other hand, the magnitude of a vector can be obtained from its square $|\mathbf{a}| = +\sqrt{\mathbf{aa}}$.

Problem: Calculate the magnitude of the ortho vector **x** in Fig. 1.8, if the C-C and C-H bonds have the values 1.4 Å and 1.0 Å.

Solution: We found $\mathbf{x} = -\mathbf{a} + \mathbf{b} + \mathbf{f}$. The scalar product is $\mathbf{xx} = (-\mathbf{a} + \mathbf{b} + \mathbf{f})(-\mathbf{a} + \mathbf{b} + \mathbf{f}) = \mathbf{a}^2 - \mathbf{ab} - \mathbf{af} - \mathbf{ba} + \mathbf{b}^2 + \mathbf{bf} - \mathbf{fa} + \mathbf{fb} + \mathbf{f}^2$.
With $a = f = 1.0$ and $b = 1.4$ we get

$$\begin{aligned}\mathbf{xx} &= 2a^2 - 2ab\cos 120^0 - 2a^2\cos 60^0 + 2ba\cos 60^0 + b^2\\ &= a^2 + ab + ab + b^2 = (a + b)^2,\end{aligned}$$

hence $x = 1.0 + 1.4 = 2.4$Å.

The vector product (or cross product) of two vectors **a** and **b** defines a vector **c**, $\mathbf{c} = \mathbf{a} \times \mathbf{b}$, with the following properties
(a) $|\mathbf{c}| = |\mathbf{a}||\mathbf{b}|\sin(\mathbf{a}, \mathbf{b})$
(b) **c** is perpendicular to **a** and **b**
(c) **a**, **b** and **c** taken in that order form a right-handed system.
Note that in contrast to the inner product which gives a scalar, the result of the vector product is a vector.

An illustrative way to demonstrate property (c) is given by the 'puppet rule' (see Fig. 1.9). Put the right foot of a puppet into the direction of **a**, the left into that of **b**, then **c** has the sense of the puppet's body.

Properties of the vector product:
(1) The vector product is not commutative, but

$$\mathbf{a} \times \mathbf{b} = -\mathbf{b} \times \mathbf{a}$$

(Remember the 'puppet rule'!)

Fig. 1.9. Illustration of a right-handed system by the "puppet rule".

(2) $\lambda(\mathbf{a} \times \mathbf{b}) = \lambda\mathbf{a} \times \mathbf{b} = \mathbf{a} \times \lambda\mathbf{b}$
(3) $(\mathbf{a} + \mathbf{b}) \times \mathbf{c} = \mathbf{a} \times \mathbf{c} + \mathbf{b} \times \mathbf{c}$
(4) $\mathbf{a} \times \lambda\mathbf{a} = \mathbf{0}$

The last property is frequently used to examine whether two vectors are parallel.

Problem: Let **a**, **b**, **c**, be three non-coplanar vectors with magnitudes different from zero, forming in that order a right-handed system. Calculate three unit vectors **x**, **y**, **z** of a right-handed system, perpendicular to each other with $\mathbf{x} = \lambda\ \mathbf{a}$ and **y** in the plane of **a** and **b** (Fig. 1.10).

Solution:

$$\mathbf{x} = \frac{\mathbf{a}}{|\mathbf{a}|};\ \mathbf{z} = \frac{\mathbf{a} \times \mathbf{b}}{|\mathbf{a} \times \mathbf{b}|};\ \mathbf{y} = \mathbf{z} \times \mathbf{x}$$

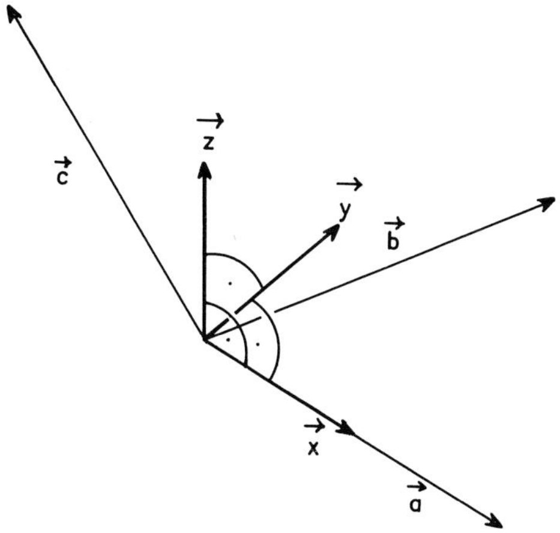

Fig. 1.10. Construction of a right-handed system by application of the vector product.

Since the result of the vector product is a vector, vector products of higher order can be obtained.

A *scalar triple product* is defined as a scalar product of a vector product $\mathbf{a} \times \mathbf{b}$ by another vector $\mathbf{c}$, $(\mathbf{a} \times \mathbf{b})\mathbf{c}$. It has an important geometrical interpretation: the absolute value of $(\mathbf{a} \times \mathbf{b})\mathbf{c}$ is the volume V of a parallelepiped bounded by the vectors $\mathbf{a}, \mathbf{b}, \mathbf{c}$. This is evident, because V is calculated as base plane G times altitude h (see Fig. 1.11). The magnitude of G is $G = ab\sin(\mathbf{a}, \mathbf{b}) = |\mathbf{a} \times \mathbf{b}|$; h is equal to $|\mathbf{c}||\cos\varphi|$. It follows that $V = Gh = |\mathbf{a} \times \mathbf{b}||\mathbf{c}||\cos\varphi| = |(\mathbf{a} \times \mathbf{b})\mathbf{c}|$. From this geometrical property it follows also that $|\mathbf{a}(\mathbf{b} \times \mathbf{c})|$ is equal to V and that this is true for an arbitrary order of $\mathbf{a}, \mathbf{b}$ and $\mathbf{c}$. A more detailed calculation shows that $\mathbf{a}(\mathbf{b} \times \mathbf{c}) = (\mathbf{a} \times \mathbf{b})\mathbf{c}$ always holds and that the scalar triple product is equal to V, if $\mathbf{a}$, $\mathbf{b}$ and $\mathbf{c}$, taken in that order, form a right-handed system, otherwise it is equal to -V.

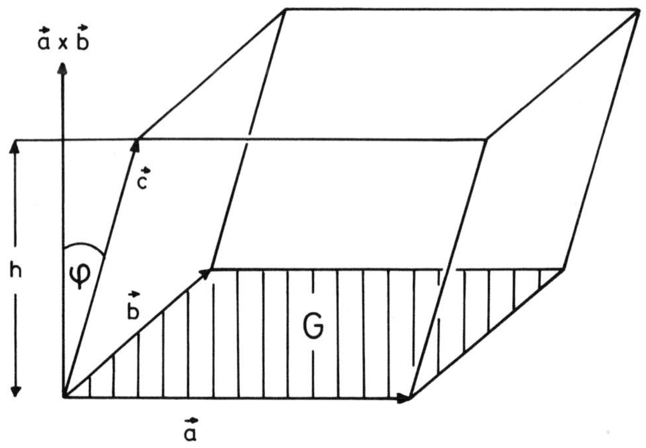

Fig. 1.11. Parallelepiped defined by three vectors **a**, **b**, **c**.

Since there is no need to mark the vector product and the scalar product separately, we can use the simple notation $(\mathbf{abc})$ for the scalar triple product.

A *vector triple product* is defined as $\mathbf{a} \times (\mathbf{b} \times \mathbf{c})$. If $\mathbf{b} \times \mathbf{c} \neq \mathbf{0}$, it follows from the definition of a vector product that $\mathbf{d} = \mathbf{a} \times (\mathbf{b} \times \mathbf{c})$ is a vector perpendicular to $\mathbf{b} \times \mathbf{c}$. Since $\mathbf{b} \times \mathbf{c}$ is perpendicular to $\mathbf{b}$ and $\mathbf{c}$, $\mathbf{d}$ must be coplanar to $\mathbf{b}$ and $\mathbf{c}$. It follows that $\mathbf{d} = \beta\mathbf{b} + \gamma\mathbf{c}$. A detailed calculation shows

$$\mathbf{d} = (\mathbf{ac})\mathbf{b} - (\mathbf{ab})\mathbf{c}$$

A *scalar four-fold product* is defined as $(\mathbf{a} \times \mathbf{b})(\mathbf{c} \times \mathbf{d})$. For this product the following equation holds

$$(\mathbf{a} \times \mathbf{b})(\mathbf{c} \times \mathbf{d}) = (\mathbf{ac})(\mathbf{bd}) - (\mathbf{ad})(\mathbf{bc})$$

Proof: Let $\mathbf{z} = \mathbf{c} \times \mathbf{d}$. Then we get $(\mathbf{a} \times \mathbf{b})\mathbf{z} = (\mathbf{abz}) = \mathbf{a}(\mathbf{b} \times \mathbf{z}) = \mathbf{a}(\mathbf{b} \times (\mathbf{c} \times \mathbf{d}))$ = (from the vector triple product) $\mathbf{a}((\mathbf{bd})\mathbf{c} - (\mathbf{bc})\mathbf{d}) = (\mathbf{ac})(\mathbf{bd}) - (\mathbf{ad})(\mathbf{bc})$.

A *vector four-fold product* is defined as $(\mathbf{a} \times \mathbf{b}) \times (\mathbf{c} \times \mathbf{d})$.
It holds that $(\mathbf{a} \times \mathbf{b}) \times (\mathbf{c} \times \mathbf{d}) = \mathbf{c}(\mathbf{abd}) - \mathbf{d}(\mathbf{abc}) = \mathbf{b}(\mathbf{acd}) - \mathbf{a}(\mathbf{bcd})$.
The proof follows immediately from the vector triple product. An immediate consequence of the vector four-fold product is the property of any four vectors, **a**, **b**, **c**, **d**:

$$\mathbf{a}(\mathbf{bcd}) - \mathbf{b}(\mathbf{acd}) + \mathbf{c}(\mathbf{abd}) - \mathbf{d}(\mathbf{abc}) = \mathbf{0} \tag{1.8}$$

1.1.4 *Linear Independency, Bases, Reciprocal Bases*

An important consequence of (1.8) must now be discussed, but first a new property of vectors is introduced. A set of n vectors $\mathbf{a}_1, \ldots, \mathbf{a}_n$ is said to be linearly dependent, if n scalars $\lambda_1, \ldots, \lambda_n$ can be found, not all being equal to zero, satisfying an equation of the form

$$\lambda_1 \mathbf{a}_1 + \lambda_2 \mathbf{a}_2 + \ldots \lambda_n \mathbf{a}_n = \mathbf{0}.$$

If an equation of that form does not exist, the vectors are called linearly independent.

It follows immediately from (1.8) that four vectors are always linearly dependent, if it can be shown that one of the scalar triple products is different from zero. The question of whether a scalar triple product is zero or not is easily decided. It holds that $(\mathbf{a}_1 \mathbf{a}_2 \mathbf{a}_3)$ is equal to zero if and only if $\mathbf{a}_1$, $\mathbf{a}_2$ and $\mathbf{a}_3$ are linearly dependent.

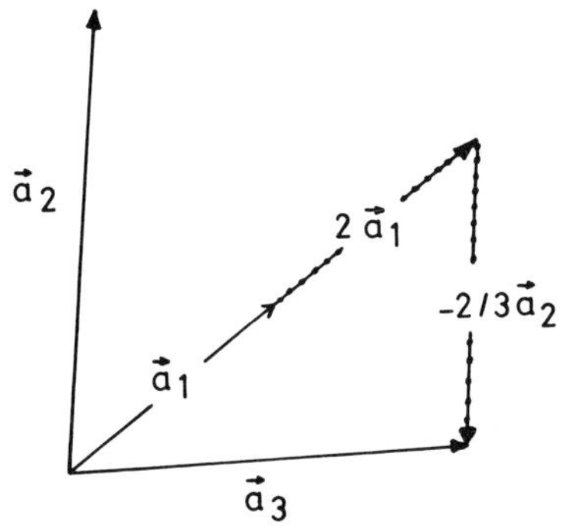

Fig. 1.12. Linear dependence of three coplanar vectors. $\mathbf{a}_3$ can be represented as $\mathbf{a}_3 = 2\mathbf{a}_1 - (2/3)\mathbf{a}_2$.

If $(\mathbf{a}_1 \mathbf{a}_2 \mathbf{a}_3) = 0$, the corresponding parallelepiped has no volume. This is possible only if $\mathbf{a}_1$, $\mathbf{a}_2$ and $\mathbf{a}_3$ are coplanar. In this case, however, one vector can always be represented as a linear combination of the two others (see the example in Fig. 1.12). Let $\mathbf{a}_3 = \lambda_1 \mathbf{a}_1 + \lambda_2 \mathbf{a}_2$. It follows that $\mathbf{a}_3 - \lambda_1 \mathbf{a}_1 - \lambda_2 \mathbf{a}_2 = \mathbf{0}$. Set $\lambda_3 = 1 \neq 0$ and we have the linear dependence of $\mathbf{a}_1$, $\mathbf{a}_2$ and $\mathbf{a}_3$. If, on the other hand, $\mathbf{a}_1$, $\mathbf{a}_2$, $\mathbf{a}_3$ are linearly dependent, we have an equation $\lambda_1 \mathbf{a}_1 + \lambda_2 \mathbf{a}_2 + \lambda_3 \mathbf{a}_3 = \mathbf{0}$ with, let us say $\lambda_1 \neq 0$. It follows that $\mathbf{a}_1 = -\lambda_2/\lambda_1 \mathbf{a}_2 - \lambda_3/\lambda_1 \mathbf{a}_3 = \alpha_1 \mathbf{a}_2 + \alpha_2 \mathbf{a}_3$. Then $(\mathbf{a}_1 \mathbf{a}_2 \mathbf{a}_3) = [(\alpha_1 \mathbf{a}_2$

$+\alpha_2 \mathbf{a}_3) \times \mathbf{a}_2]\mathbf{a}_3 = [\mathbf{0}$ (property 4 of the vector product) $+ \alpha_2(\mathbf{a}_3 \times \mathbf{a}_2)]\mathbf{a}_3 = 0$ (property 4 of the scalar product).

We can now say that in the set of vectors in three-dimensional space, a maximum of three vectors can be linearly independent, while four or more vectors are always linearly dependent. Three vectors $\mathbf{a}_1$, $\mathbf{a}_2$, $\mathbf{a}_3$ are linearly independent if they are not coplanar, since their scalar triple product is then different from zero. In this case they have the important property that an arbitrary vector **d** can be uniquely expressed as a linear combination of $\mathbf{a}_1$, $\mathbf{a}_2$, $\mathbf{a}_3$:

$$\mathbf{d} = \lambda_1 \mathbf{a}_1 + \lambda_2 \mathbf{a}_2 + \lambda_3 \mathbf{a}_3 \tag{1.9}$$

The existence of a linear combination for **d** follows from (1.8) (divide (1.8) by (**abc**), which is different from zero!). Assume that another linear combination $\mathbf{d} = \mu_1 \mathbf{a}_1 + \mu_2 \mathbf{a}_2 + \mu_3 \mathbf{a}_3$ exists. The difference between the two equations leads to $\mathbf{0} = (\lambda_1 - \mu_1)\mathbf{a}_1 + (\lambda_2 - \mu_2)\mathbf{a}_2 + (\lambda_3 - \mu_3)\mathbf{a}_3$. Since the $\mathbf{a}_i$ are linearly independent, their coefficients have to be zero, which means $\lambda_i - \mu_i = 0$ or $\lambda_i = \mu_i$.

A set of three vectors $[\mathbf{a}_1, \mathbf{a}_2, \mathbf{a}_3]$ which represents every vector **d** uniquely as in (1.9) is called a basis. Since every set of three linearly independent vectors is a basis of the vector set, a large number of bases can be found. However, it is convenient to use a special type of basis, the so-called *orthonormal bases*. An orthonormal basis is present if it consists of three vectors $[\mathbf{e}_1, \mathbf{e}_2, \mathbf{e}_3]$ having the following properties:

(a) $|\mathbf{e}_i| = 1$ for $i = 1, \ldots, 3$. (Vectors having unit length are called unit vectors.)

(b) The three vectors $\mathbf{e}_1$, $\mathbf{e}_2$ and $\mathbf{e}_3$ are pairwise perpendicular and form, in that order, a right-handed system.

In terms of an orthonormal basis, every vector **r** has a unique representation $\mathbf{r} = x\mathbf{e}_1 + y\mathbf{e}_2 + z\mathbf{e}_3$. The scalars x, y, z are equal to the coordinates of a point P

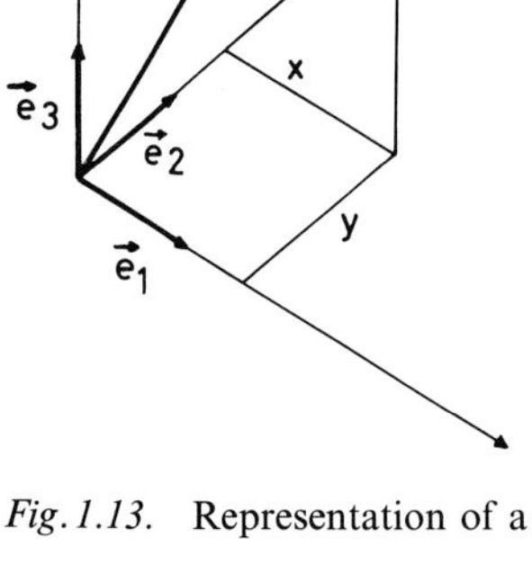

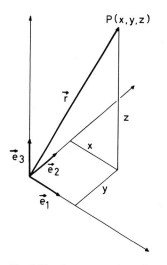

Fig. 1.13. Representation of a vector **r** in an orthonormal basis.

(see Fig. 1.13) in a cartesian system with its axes coinciding with the basis vectors $\mathbf{e}_i$. In this way the representation of an arbitrary point in three-dimensional space be done in two equivalent ways; either as a position vector **r** or in terms of the coordinates of a cartesian system. The representation as a vector has the clear advantage that all facilities of vector algebra can be applied in treating a problem.

The components x, y, z can be shown to have an important geometrical property in an orthonormal basis (see Fig. 1.14). If $\varphi_1, \varphi_2, \varphi_3$ are the angles between **r** and the basis vectors, we have

$$\begin{aligned} x &= r\cos\varphi_1 \\ y &= r\cos\varphi_2 \\ z &= r\cos\varphi_3 \end{aligned} \tag{1.10}$$

with $r = |\mathbf{r}|$. This follows immediately if we multiply **r** by $\mathbf{e}_1$:

$$\mathbf{r}\mathbf{e}_1 = x\mathbf{e}_1^2 + y\mathbf{e}_2\mathbf{e}_1 + z\mathbf{e}_3\mathbf{e}_1$$

$$r\,1\cos\varphi_1 = x\,1 + 0 + 0$$

or

$$r\cos\varphi_1 = x$$

We get the relations for y and z in the same way.

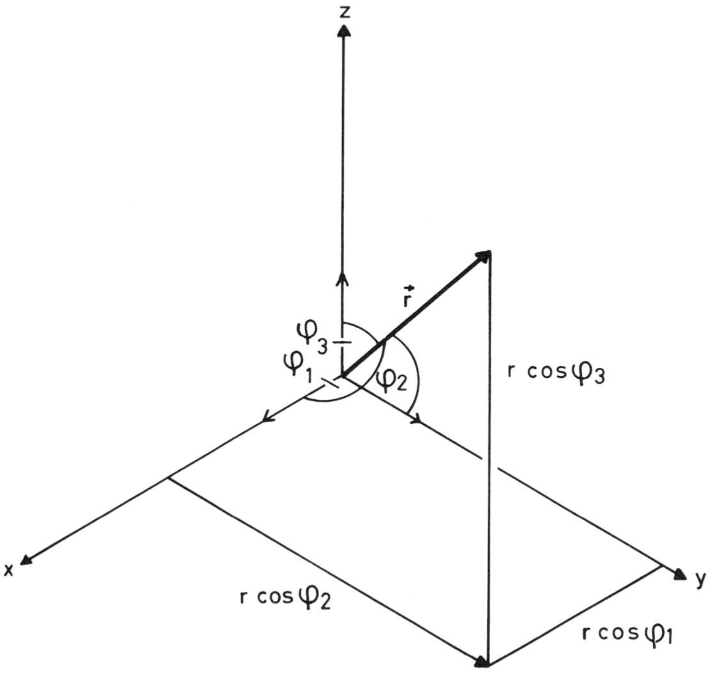

Fig. 1.14. Representation of vector components in terms of axial angles.

Since the representation of a vector **r** is unique relative to a fixed basis $\mathbf{a}_1, \mathbf{a}_2, \mathbf{a}_3$, the expression $\mathbf{r} = x\mathbf{a}_1 + y\mathbf{a}_2 + z\mathbf{a}_3$ can be replaced by the shorter form of a three-membered column,

$$\mathbf{r} = \begin{pmatrix} x \\ y \\ z \end{pmatrix}.$$

Then the vector arithmetic can be expressed in terms of columns and we get

(a) for the sum (or difference) of $\mathbf{r}_1$ and $\mathbf{r}_2$:

$$\mathbf{r}_1 \pm \mathbf{r}_2 = \begin{pmatrix} x_1 \pm x_2 \\ y_1 \pm y_2 \\ z_1 \pm z_2 \end{pmatrix}$$

(b) Let $g_{ij} = \mathbf{a}_i \mathbf{a}_j$, then for $\mathbf{r}_1 \mathbf{r}_2$ the following equation holds:

$$\begin{aligned} \mathbf{r}_1 \mathbf{r}_2 = & \, x_1 x_2 g_{11} + x_1 y_2 g_{12} + x_1 z_2 g_{13} \\ & + y_1 x_2 g_{12} + y_1 y_2 g_{22} + y_1 z_2 g_{23} \\ & + z_1 x_2 g_{13} + z_1 y_2 g_{23} + z_1 z_2 g_{33} \end{aligned}$$

(c) If the basis is orthonormal, we get

$$\mathbf{r}_1 \mathbf{r}_2 = x_1 x_2 + y_1 y_2 + z_1 z_2$$

and the following "formal" determinant for the cross product:

$$\mathbf{r}_1 \times \mathbf{r}_2 = \begin{vmatrix} \mathbf{a}_1 & \mathbf{a}_2 & \mathbf{a}_3 \\ x_1 & y_1 & z_1 \\ x_2 & y_2 & z_2 \end{vmatrix}$$

The verification of these formulae is left as an exercise to the reader. It is done very easily if all properties of the vector arithmetic are observed.

As shown in (c), the algebra of columns becomes easy if orthonormal bases are used, which is the reason why they are of such importance. For several problems, however, especially in crystallography, orthonormal bases are not suitable. Thus we have to keep in mind the algebra in non-orthonormal bases.

Problem (1): Calculate the components of (two-dimensional) position vectors of the benzene ring:

(a) in an orthonormal system,

(b) in a system corresponding to the symmetry of the benzene ring.

Answer:

(a) Let a be the bond distance between two carbon atoms. Then we have (see Fig. 1.15):

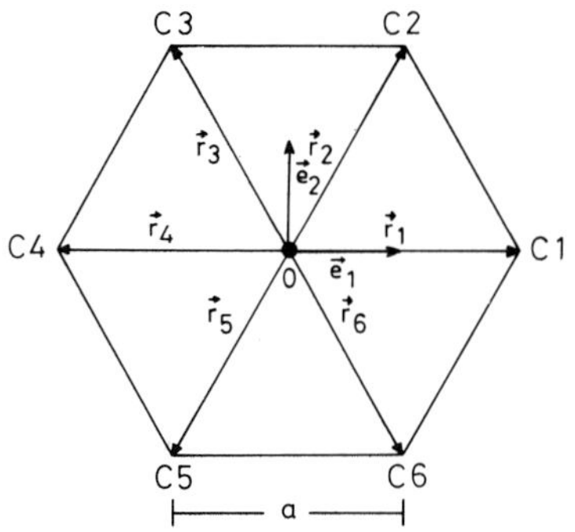

Fig. 1.15. Representation of the atomic vectors of benzene in different bases.

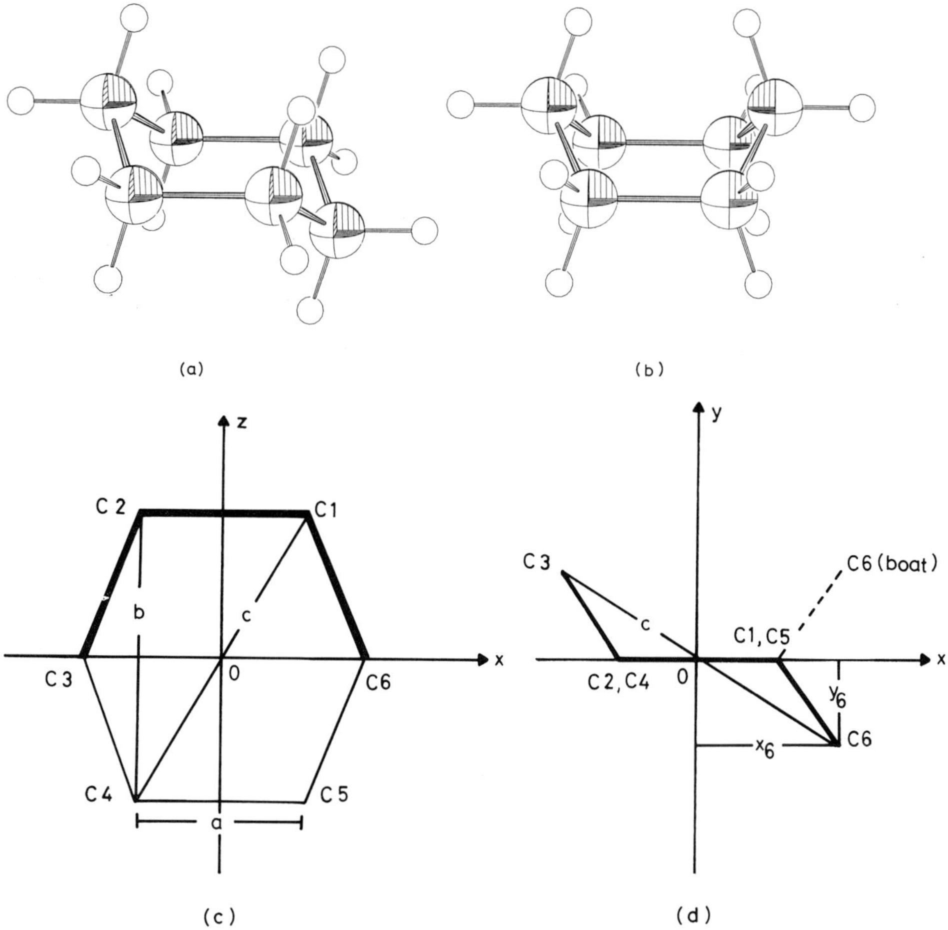

Fig. 1.16. The cyclohexane molecule:
(a) chair form; (b) boat form; (c) x-z plane, defined by C1, C2, C4 and C5;
(d) projection onto the x-y plane.

$$\mathbf{r}_1 = \begin{pmatrix} a \\ 0 \end{pmatrix} \quad \mathbf{r}_2 = \begin{pmatrix} a/2 \\ a\sqrt{2/2} \end{pmatrix} \quad \mathbf{r}_3 = \begin{pmatrix} -a/2 \\ a\sqrt{2/2} \end{pmatrix}$$

$$\mathbf{r}_4 = \begin{pmatrix} -a \\ 0 \end{pmatrix} \quad \mathbf{r}_5 = \begin{pmatrix} -a/2 \\ -a\sqrt{2/2} \end{pmatrix} \quad \mathbf{r}_6 = \begin{pmatrix} a/2 \\ -a\sqrt{2/2} \end{pmatrix}$$

(b) A basis closely related to the benzene ring symmetry is given by the vectors $\mathbf{a}_1 = \overline{OC1}$ and $\mathbf{a}_2 = \overline{OC3}$. In terms of this basis $\mathbf{r}_1$ to $\mathbf{r}_6$, read

$$\mathbf{r}_1 = \begin{pmatrix} 1 \\ 0 \end{pmatrix} \quad \mathbf{r}_2 = \begin{pmatrix} 1 \\ 1 \end{pmatrix} \quad \mathbf{r}_3 = \begin{pmatrix} 0 \\ 1 \end{pmatrix} \quad \mathbf{r}_4 = \begin{pmatrix} -1 \\ 0 \end{pmatrix} \quad \mathbf{r}_5 = \begin{pmatrix} -1 \\ -1 \end{pmatrix} \quad \mathbf{r}_6 = \begin{pmatrix} 0 \\ -1 \end{pmatrix}$$

Problem (2): Calculate the vectors of the cyclohexane carbon atoms in terms of a three-dimensional orthonormal basis.

Answer: Cyclohexane, C_6H_{12} can have a non-planar chair or boat-shaped six-membered ring (Fig. 1.16a and b). From these two possible conformations the chair form is that most frequently adopted. So let us calculate the carbon vectors of a cyclohexane chair. Four atoms are coplanar (those atoms in Fig. 1.16a being in a plane parallel to the plane of the table); we define this plane as the x – z plane of an orthonormal system (Fig. 1.16c). The center of the ring (intersection of the lines C1–C4 and C2–C5) is chosen as the origin. The components of the basis vectors are equal to the coordinates of the corresponding cartesian system and can be derived from the two projections on the x-z and the x-y plane (Figs. 1.16c and d). We can restrict ourselves to the determination of C1, C6 and C5; the three other vectors can then be derived from simple symmetry considerations.

Let a (a = 1.54 Å) be the bond distance between two carbon atoms. Without any calculation we get

$$C1 = \begin{pmatrix} a/2 \\ 0 \\ z_1 \end{pmatrix} \quad C6 = \begin{pmatrix} x_6 \\ y_6 \\ 0 \end{pmatrix} \quad C5 = \begin{pmatrix} a/2 \\ 0 \\ -z_1 \end{pmatrix}$$

The only quantities to be calculated are thus z_1, x_6 and y_6. z_1 is equal to b/2, with b the *meta* distance between two carbon atoms. b can be derived from the length of the difference vector between C2 and C4. Since all bond angles in the cyclohexane molecule are are tetrahedral and the cosine of a tetrahedral angle equals $-1/3$, we get

$$b^2 = [(C3 - C2) + (C4 - C3)]^2$$

Since

$|C3 - C2| = |C4 - C3| = a$ we get $b^2 = a^2 + a^2 + 2aa(1/3)$

and

$$b = a\sqrt{8/3}$$

hence $z_1 = a\sqrt{8/3}/2$. x_6 and y_6 can be obtained by using the *para* distance c ($c = a\sqrt{11/3}$ is obtained in the same way as b) and the following two equations (see Fig. 1.16d)

$$x_6^2 + y_6^2 = (c/2)^2$$

$$(C1\text{–}C6)\ (C5\text{–}C6) = aa(-1/3)$$

Using the column notation for the second equation, we get

$$\begin{pmatrix} a/2 - x_6 \\ 0 - y_6 \\ b/2 - 0 \end{pmatrix} \begin{pmatrix} a/2 - x_6 \\ 0 - y_6 \\ -b/2 - 0 \end{pmatrix} = -a^2/3$$

or $$(a/2 - x_6)^2 + y_6^2 - b^2/4 = -a^2/3.$$

Replacing y_6 by the first equation and b by $a\sqrt{8/3}$, we get, finally,

$$x_6 = (5a)/6$$

and then,

$$y_6 = \pm a\sqrt{8}/6.$$

The two signs for y_6 refer to the fact that C6 as well as C3 can have their y-coordinates opposite to that direction represented in Fig. 1.16d; however, this will only result in another orientation of the cyclohexane ring.

For our purpose, we choose the negative sign for y_6 and we get, finally,

$$C1 = \begin{pmatrix} a/2 \\ 0 \\ a\sqrt{8/3}/2 \end{pmatrix} \quad C6 = \begin{pmatrix} 5a/6 \\ -a\sqrt{8}/6 \\ 0 \end{pmatrix} \quad C5 = \begin{pmatrix} a/2 \\ 0 \\ -a\sqrt{8/3}/2 \end{pmatrix}$$

Then we can easily derive,

$$C2 = \begin{pmatrix} -a/2 \\ 0 \\ a\sqrt{8/3}/2 \end{pmatrix} \quad C3 = \begin{pmatrix} -5a/6 \\ a\sqrt{8}/6 \\ 0 \end{pmatrix} \quad C4 = \begin{pmatrix} -a/2 \\ 0 \\ -a\sqrt{8/3}/2 \end{pmatrix}$$

The bases we have introduced above are not sufficient to process all problems, especially in solid state physics. Therefore, it is necessary to deal with another type of basis, the so-called reciprocal basis.

Let $B = [\mathbf{a}_1, \mathbf{a}_2, \mathbf{a}_3]$ be a basis. The vectors defined by

$$\mathbf{a}_1^* = \frac{\mathbf{a}_2 \times \mathbf{a}_3}{(\mathbf{a}_1\,\mathbf{a}_2\,\mathbf{a}_3)}$$

$$\mathbf{a}_2^* = \frac{\mathbf{a}_3 \times \mathbf{a}_1}{(\mathbf{a}_1\,\mathbf{a}_2\,\mathbf{a}_3)} \tag{1.11}$$

$$\mathbf{a}_3^* = \frac{\mathbf{a}_1 \times \mathbf{a}_2}{(\mathbf{a}_1\,\mathbf{a}_2\,\mathbf{a}_3)}$$

are said to be reciprocal to the $\mathbf{a}_i$. They have the following properties:

(a) $\mathbf{a}_i\,\mathbf{a}_k^* = \Delta_{ik}$

(b) If $V = (\mathbf{a}_1\,\mathbf{a}_2\,\mathbf{a}_3)$, $V^* = (\mathbf{a}_1^*\,\mathbf{a}_2^*\,\mathbf{a}_3^*)$, then $V^* = 1/V$ (1.12)

(c) $B^* = [\mathbf{a}_1^*, \mathbf{a}_2^*, \mathbf{a}_3^*]$ is a basis, also, denoted as a reciprocal basis

(d) If $\alpha_1 = (\mathbf{a}_2, \mathbf{a}_3)$, $\alpha_2 = (\mathbf{a}_3, \mathbf{a}_1)$, $\alpha_3 = (\mathbf{a}_1, \mathbf{a}_2)$ are the angles between the basis vectors, then for the corresponding angles $\alpha_1^*, \alpha_2^*, \alpha_3^*$ of the reciprocal basis, the following equations hold:

$$\cos\alpha_1^* = \frac{\cos\alpha_2\cos\alpha_3 - \cos\alpha_1}{\sin\alpha_2\sin\alpha_3}$$

$$\cos\alpha_2^* = \frac{\cos\alpha_3\cos\alpha_1 - \cos\alpha_2}{\sin\alpha_3\sin\alpha_1} \tag{1.13}$$

$$\cos\alpha_3^* = \frac{\cos\alpha_1\cos\alpha_2 - \cos\alpha_3}{\sin\alpha_1\sin\alpha_2}$$

(a) follows immediately from the definition of the reciprocals.

(b)
$$\begin{aligned} V^* = (\mathbf{a}_1^*\,\mathbf{a}_2^*\,\mathbf{a}_3^*) &= (1/V^3)\,([\mathbf{a}_2 \times \mathbf{a}_3) \times (\mathbf{a}_3 \times \mathbf{a}_1)]\,(\mathbf{a}_1 \times \mathbf{a}_2)) \\ &= (1/V^3)\,([\mathbf{a}_3\,(\mathbf{a}_2\,\mathbf{a}_3\,\mathbf{a}_1) - \mathbf{a}_1\,(\mathbf{a}_2\,\mathbf{a}_3\,\mathbf{a}_3)]\,(\mathbf{a}_1 \times \mathbf{a}_2)) \\ &= (1/V^3)\,[\mathbf{a}_3\,(\mathbf{a}_2\,\mathbf{a}_3\,\mathbf{a}_1)\,(\mathbf{a}_1 \times \mathbf{a}_2)] = V^2/V^3 = 1/V \end{aligned}$$

(c) Since the $\mathbf{a}_i$ form a basis, V is different from zero.

Then, with (b), $V^* \neq 0$, hence B^* is a basis.

(d)
$$\begin{aligned} \mathbf{a}_1^*\,\mathbf{a}_2^* = a_1^*\,a_2^*\cos\alpha_3^* &= [(\mathbf{a}_2 \times \mathbf{a}_3)\,(\mathbf{a}_3 \times \mathbf{a}_1)]/V^2 \\ &= \text{(regarded as a scalar four-fold product)} \\ &= (a_2\,a_3\cos\alpha_1\,a_3\,a_1\cos\alpha_2 - a_2\,a_1\cos\alpha_3\,a_3^2)/V^2 \end{aligned}$$

Using (1.11) and the definition of the magnitude of a vector product, we get $a_1^*\,a_2^*\cos\alpha_3^* = (a_2\,a_3\sin\alpha_1\,a_3\,a_1\sin\alpha_2\cos\alpha_3^*)/V^2$.

Hence we get

$$a_2\,a_3\sin\alpha_1\,a_3\,a_1\sin\alpha_2\cos\alpha_3^* = a_2\,a_3\cos\alpha_1\,a_3\,a_1\cos\alpha_2 - a_2\,a_1\,a_3^2\cos\alpha_3$$

or

$$\cos\alpha_3^* = \frac{\cos\alpha_1\cos\alpha_2 - \cos\alpha_3}{\sin\alpha_1\sin\alpha_2}$$

The formulae for α_1^* and α_2^* are obtained in the same way.

It can be shown that the reciprocal of the reciprocal basis leads to the original basis. In dealing with the two basis types, the vector space referring to the basis B is frequently called "direct space", while the space of B* is called "reciprocal space". Since B and B* are reciprocal to each other, these notations are nonsense unless the vectors represent physical quantities. For instance, if the vectors of direct space have the dimension of a length, reciprocal vectors have the dimension of a reciprocal length. It is in this sense that the notations "direct" and "reciprocal" space must be understood. The scalar product of a vector **r** of direct and a vector **b** of reciprocal space leads to results frequently used in crystallography:

If $\mathbf{r} = x_1 \mathbf{a}_1 + x_2 \mathbf{a}_2 + x_3 \mathbf{a}_3$

and $\mathbf{b} = h_1 \mathbf{a}_1^* + h_2 \mathbf{a}_2^* + h_3 \mathbf{a}_3^*$,

we obtain

(a) $\mathbf{rb} = h_1 x_1 + h_2 x_2 + h_3 x_3$ (1.14)

(b) $x_i = \mathbf{r a}_i^* \qquad i = 1, \ldots, 3$ (1.15)

(c) $h_i = \mathbf{b a}_i \qquad i = 1, \ldots, 3$ (1.16)

1.1.5 *Basis Transformations*

As we have seen above, the choice of a basis can be made in several ways. It is therefore likely that two physicists will choose two different bases when dealing with the same problem. Can they still make sense to each other? They can, if one transforms his basis system so that it becomes identical to the other. It is necessary, therefore, to learn something about basis transformations. Assume that we have two bases, $[\mathbf{a}_1, \mathbf{a}_2, \mathbf{a}_3]$ and $[\mathbf{A}_1, \mathbf{A}_2, \mathbf{A}_3]$. Since every vector has a representation relative to a basis, we have on one hand,

$$\mathbf{A}_1 = s_{11} \mathbf{a}_1 + s_{12} \mathbf{a}_2 + s_{13} \mathbf{a}_3$$
$$\mathbf{A}_2 = s_{21} \mathbf{a}_1 + s_{22} \mathbf{a}_2 + s_{23} \mathbf{a}_3$$
$$\mathbf{A}_3 = s_{31} \mathbf{a}_1 + s_{32} \mathbf{a}_2 + s_{33} \mathbf{a}_3$$

On the other hand, we have a representation for the $\mathbf{a}_i$

$$\mathbf{a}_i = t_{i1} \mathbf{A}_1 + t_{i2} \mathbf{A}_2 + t_{i3} \mathbf{A}_3 \qquad i = 1, \ldots, 3$$

or by introducing a column

$$A = \begin{pmatrix} \mathbf{A}_1 \\ \mathbf{A}_2 \\ \mathbf{A}_3 \end{pmatrix} \quad \text{and} \quad a = \begin{pmatrix} \mathbf{a}_1 \\ \mathbf{a}_2 \\ \mathbf{a}_3 \end{pmatrix}$$

we get

$$A = Sa \tag{1.17a}$$

and

$$a = TA \tag{1.17b}$$

with

$$S = \begin{pmatrix} s_{11} & s_{12} & s_{13} \\ s_{21} & s_{22} & s_{23} \\ s_{31} & s_{32} & s_{33} \end{pmatrix} \quad \text{and} \quad T = \begin{pmatrix} t_{11} & t_{12} & t_{13} \\ t_{21} & t_{22} & t_{23} \\ t_{31} & t_{32} & t_{33} \end{pmatrix}$$

From what we have learned about matrix algebra, we find

$$S^{-1}A = S^{-1}Sa = a$$

Since the representation of a is unique (because A is a basis), we have $T = S^{-1}$.

Since we know how to transform the basis vectors, we have to study how to transform the components of an arbitrary vector in direct as well as in reciprocal space. Let $\mathbf{r}$ be a vector in direct space. It reads, in both bases,

$$\mathbf{r} = x_1 \mathbf{a}_1 + x_2 \mathbf{a}_2 + x_3 \mathbf{a}_3$$

and

$$\begin{aligned} \mathbf{r} &= X_1 \mathbf{A}_1 + X_2 \mathbf{A}_2 + X_3 \mathbf{A}_3 \\ &= X_1 (s_{11} \mathbf{a}_1 + s_{12} \mathbf{a}_2 + s_{13} \mathbf{a}_3) + X_2 (s_{21} \mathbf{a}_1 + s_{22} \mathbf{a}_2 + s_{23} \mathbf{a}_3) \\ &\quad + X_3 (s_{31} \mathbf{a}_1 + s_{32} \mathbf{a}_2 + s_{33} \mathbf{a}_3) \\ &= (s_{11} X_1 + s_{21} X_2 + s_{31} X_3) \mathbf{a}_1 + (s_{12} X_1 + s_{22} X_2 + s_{32} X_3) \mathbf{a}_2 \\ &\quad + (s_{13} X_1 + s_{23} X_2 + s_{33} X_3) \mathbf{a}_3 \end{aligned}$$

Since the representation in terms of the $\mathbf{a}_i$ is unique, we get

$$\begin{pmatrix} x_1 \\ x_2 \\ x_3 \end{pmatrix} = \begin{pmatrix} s_{11} & s_{21} & s_{31} \\ s_{12} & s_{22} & s_{32} \\ s_{13} & s_{23} & s_{33} \end{pmatrix} \begin{pmatrix} X_1 \\ X_2 \\ X_3 \end{pmatrix}$$

$$\begin{pmatrix} x_1 \\ x_2 \\ x_3 \end{pmatrix} = S' \begin{pmatrix} X_1 \\ X_2 \\ X_3 \end{pmatrix} \tag{1.18a}$$

Since each basis is equivalent to the other, we have at once,

$$\begin{pmatrix} X_1 \\ X_2 \\ X_3 \end{pmatrix} = T' \begin{pmatrix} x_1 \\ x_2 \\ x_3 \end{pmatrix} \tag{1.18b}$$

To get the transformation of reciprocal space, let us start with a vector $\mathbf{b} = h_1 \mathbf{a}_1^* + h_2 \mathbf{a}_2^* + h_3 \mathbf{a}_3^*$. We have to find out how to get the reciprocals $\mathbf{A}_i^*$ of $\mathbf{A}_i$ and the components H_i corresponding to a representation of $\mathbf{b}$ in terms of $\mathbf{A}_i^*$.

Let us start with the last problem. We write $\mathbf{b}$ in terms of the $\mathbf{A}_i^*$.

$$\mathbf{b} = H_1 \mathbf{A}_1^* + H_2 \mathbf{A}_2^* + H_3 \mathbf{A}_3^*.$$

Since $\mathbf{A}_i^* \mathbf{A}_k = \Delta_{ik}$, multiplication by $\mathbf{A}_i$ leads to

$$\mathbf{b}\mathbf{A}_1 = H_1$$

From the representation of **b** in terms of $\mathbf{a}_i^*$, however, we get with (1.14) and (1.16):

$$\mathbf{b}\mathbf{A}_1 = h_1 s_{11} + h_2 s_{12} + h_3 s_{13} = H_1$$

H_2 and H_3 are obtained in the same way, and we have

$$\begin{pmatrix} H_1 \\ H_2 \\ H_3 \end{pmatrix} = \begin{pmatrix} s_{11} & s_{12} & s_{13} \\ s_{21} & s_{22} & s_{23} \\ s_{31} & s_{32} & s_{33} \end{pmatrix} \begin{pmatrix} h_1 \\ h_2 \\ h_3 \end{pmatrix}$$

$$\begin{pmatrix} H_1 \\ H_2 \\ H_3 \end{pmatrix} = S \begin{pmatrix} h_1 \\ h_2 \\ h_3 \end{pmatrix} \tag{1.19a}$$

and

$$\begin{pmatrix} h_1 \\ h_2 \\ h_3 \end{pmatrix} = T \begin{pmatrix} H_1 \\ H_2 \\ H_3 \end{pmatrix} \tag{1.19b}$$

From this last equation, we get

$$\begin{aligned} \mathbf{b} &= h_1 \mathbf{a}_1^* + h_2 \mathbf{a}_2^* + h_3 \mathbf{a}_3^* \\ &= (t_{11} H_1 + t_{12} H_2 + t_{13} H_3) \mathbf{a}_1^* + \\ &\quad (t_{21} H_1 + t_{22} H_2 + t_{23} H_3) \mathbf{a}_2^* + \\ &\quad (t_{31} H_1 + t_{32} H_2 + t_{33} H_3) \mathbf{a}_3^* \\ &= (t_{11} \mathbf{a}_1^* + t_{21} \mathbf{a}_2^* + t_{31} \mathbf{a}_3^*) H_1 + \\ &\quad (t_{12} \mathbf{a}_1^* + t_{22} \mathbf{a}_2^* + t_{32} \mathbf{a}_3^*) H_2 + \\ &\quad (t_{13} \mathbf{a}_1^* + t_{23} \mathbf{a}_2^* + t_{33} \mathbf{a}_3^*) H_3 \end{aligned}$$

On the other hand, we have

$$\mathbf{b} = \mathbf{A}_1^* H_1 + \mathbf{A}_2^* H_2 + \mathbf{A}_3^* H_3$$

Since these equations hold for every H_i, it follows that

$$A^* = T' a^* \tag{1.20}$$

if A^* and a^* are defined like A and a.

Let us summarize the results: If S is a transformation matrix of a bais a to a basis A, T is the inverse of S, then the following transformations hold:

(a) $A = Sa; \quad a = TA$ (1.17)

(b) $A^* = T'a^*; \quad a^* = S'A^*$ (1.20)

(c) $$\begin{pmatrix} X_1 \\ X_2 \\ X_3 \end{pmatrix} = T' \begin{pmatrix} x_1 \\ x_2 \\ x_3 \end{pmatrix}; \quad \begin{pmatrix} x_1 \\ x_2 \\ x_3 \end{pmatrix} = S' \begin{pmatrix} X_1 \\ X_2 \\ X_3 \end{pmatrix} \qquad (1.18)$$

(d) $$\begin{pmatrix} H_1 \\ H_2 \\ H_3 \end{pmatrix} = S \begin{pmatrix} h_1 \\ h_2 \\ h_3 \end{pmatrix}; \quad \begin{pmatrix} h_1 \\ h_2 \\ h_3 \end{pmatrix} = T \begin{pmatrix} H_1 \\ H_2 \\ H_3 \end{pmatrix} \qquad (1.19)$$

In short, the direct basis vectors and the components of reciprocal vectors are transformed by the S-matrix. The reciprocal basis vectors and the components of direct space vectors are transformed by the transposed T-matrix.

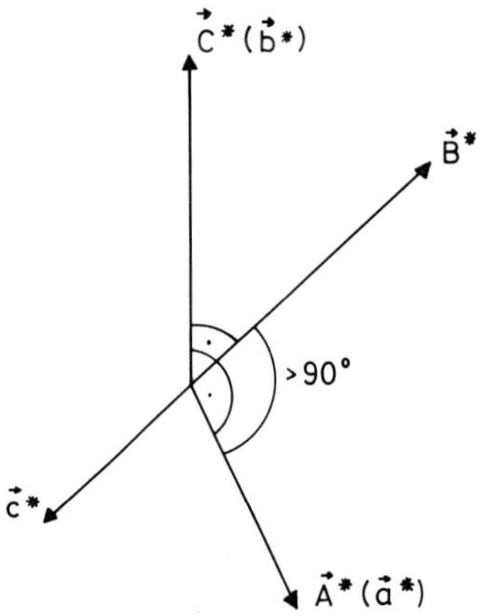

Fig. 1.17. Example of a basis transformation in reciprocal space.

Example: Let **A***, **B***, **C*** be given basis vectors in reciprocal space so that **C*** is perpendicular to **A*** and **B*** and the angle between **A*** and **B*** is greater than 90°. For special purposes, we need a reciprocal basis **a***, **b***, **c*** with **b*** having the direction of **C*** and a non-right angle less than 90°. These conditions are accomplished by the choice of **a***, **b***, and **c*** as shown in Fig. 1.17.

The transformation is

$$\mathbf{A}^* = \mathbf{a}^*$$
$$\mathbf{B}^* = -\mathbf{c}^*$$
$$\mathbf{C}^* = \mathbf{b}^*$$

In matrix notation,

$$\begin{pmatrix} \mathbf{A}^* \\ \mathbf{B}^* \\ \mathbf{C}^* \end{pmatrix} = \begin{pmatrix} 1 & 0 & 0 \\ 0 & 0 & -1 \\ 0 & 1 & 0 \end{pmatrix} \begin{pmatrix} \mathbf{a}^* \\ \mathbf{b}^* \\ \mathbf{c}^* \end{pmatrix}, \text{ hence } T' = \begin{pmatrix} 1 & 0 & 0 \\ 0 & 0 & -1 \\ 0 & 1 & 0 \end{pmatrix}$$

Notice that the determinant of T′ is > 0. If this does not hold, a right-handed system is transformed into a left-handed one. The S′-matrix is obtained by calculation of the inverse of T′:

$$S' = \begin{pmatrix} 1 & 0 & 0 \\ 0 & 0 & 1 \\ 0 & -1 & 0 \end{pmatrix}$$

Having the two matrices S′ and T′, all transformations in direct and reciprocal space can be performed using formulae (1.17) to (1.20).

1.1.6 *Lines and Planes*

Let us now discuss a geometrical problem which is very important in crystallography, the mathematical representation of planes. We shall see later that the X-ray diffraction of crystals can be expressed as a reflection from certain crystal planes and for this reason the study of that subject is very important.

In preparation, let us start with the vectorial representation of lines. A line is completely defined, if

(a) one point and its direction, or

(b) two different points

are given.

(a) If $\mathbf{r}_0$ is the vector of the given point and $\mathbf{n}$ the given direction, the line has the equation

$$\mathbf{r} = \mathbf{r}_0 + t\mathbf{n}$$

where the parameter t varies through all real numbers (see Fig. 1.18a).

(b) If $\mathbf{r}_1$ and $\mathbf{r}_2$ are the vectors of the given points, $\mathbf{r}_2 - \mathbf{r}_1$ is a vector defining the direction of the line. Then we have case (a), and we get

$$\mathbf{r} = \mathbf{r}_1 + t(\mathbf{r}_2 - \mathbf{r}_1)$$

A plane is completely defined if we have

(c) one point and two non-parallel directions, or

(d) three non-collinear points.

We shall treat (c) and (d) in the same way as (a) and (b) (see Fig. 1.19).

(c) Let $\mathbf{f}$ and $\mathbf{g}$ be the given directions and $\mathbf{r}_0$ the vector of the given point.

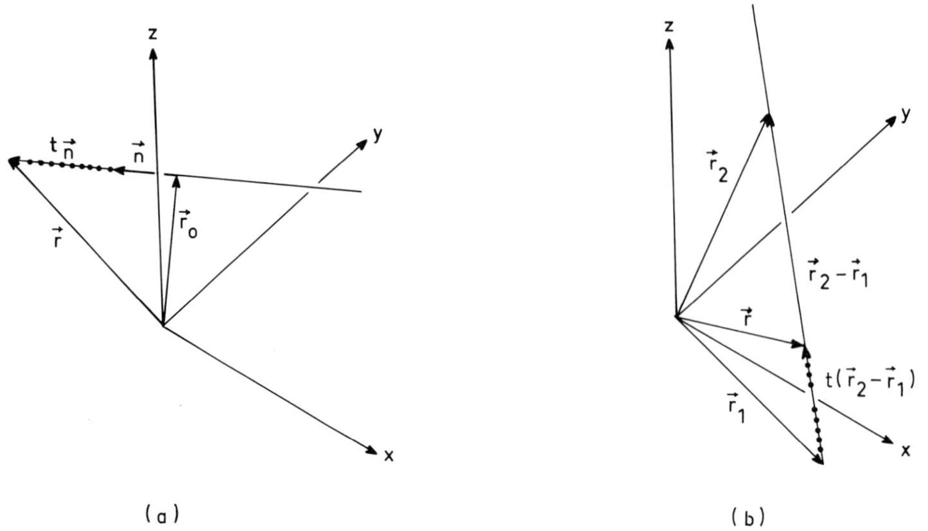

Fig. 1.18. Definition of a line:
(a) by one point and its direction; (b) by two points.

The vector $\mathbf{r}$ of an arbitrary point on the plane can then be expressed by

$$\mathbf{r} = \mathbf{r}_0 + t_1 \mathbf{f} + t_2 \mathbf{g}$$

So, if t_1 and t_2 vary through all real numbers, this equation describes all points of the plane.

(d) More frequently, the case of three given points, $\mathbf{r}_1, \mathbf{r}_2, \mathbf{r}_3$ must be considered. We can reduce this problem to case (c). Setting $\mathbf{f} = \mathbf{r}_2 - \mathbf{r}_1$ and $\mathbf{g} = \mathbf{r}_3 - \mathbf{r}_1$, we get

$$\mathbf{r} = \mathbf{r}_1 + t_1(\mathbf{r}_2 - \mathbf{r}_1) + t_2(\mathbf{r}_3 - \mathbf{r}_1)$$

By variation of t_1 and t_2, we get all points of the plane, especially with $(t_1, t_2) = (0, 0)$, $(1, 0)$ and $(0, 1)$ we get $\mathbf{r}_1$, $\mathbf{r}_2$ and $\mathbf{r}_3$.

Although we now have the equations of planes, they depend on the special choice of the given points. We shall now develop an equation containing more universal constants with important geometrical properties. Let us multiply the equation of (d) by the unit vector.

$$\mathbf{e} = \frac{(\mathbf{r}_2 - \mathbf{r}_1) \times (\mathbf{r}_3 - \mathbf{r}_1)}{|(\mathbf{r}_2 - \mathbf{r}_1) \times (\mathbf{r}_3 - \mathbf{r}_1)|} \tag{1.21}$$

in the sense of a scalar product. We get

$$\mathbf{r}\mathbf{e} = \mathbf{r}_1 \mathbf{e} + 0 + 0$$

or

$$(\mathbf{r} - \mathbf{r}_1)\mathbf{e} = 0$$

This equation is called the normal form of a plane equation. Let us write it in the form

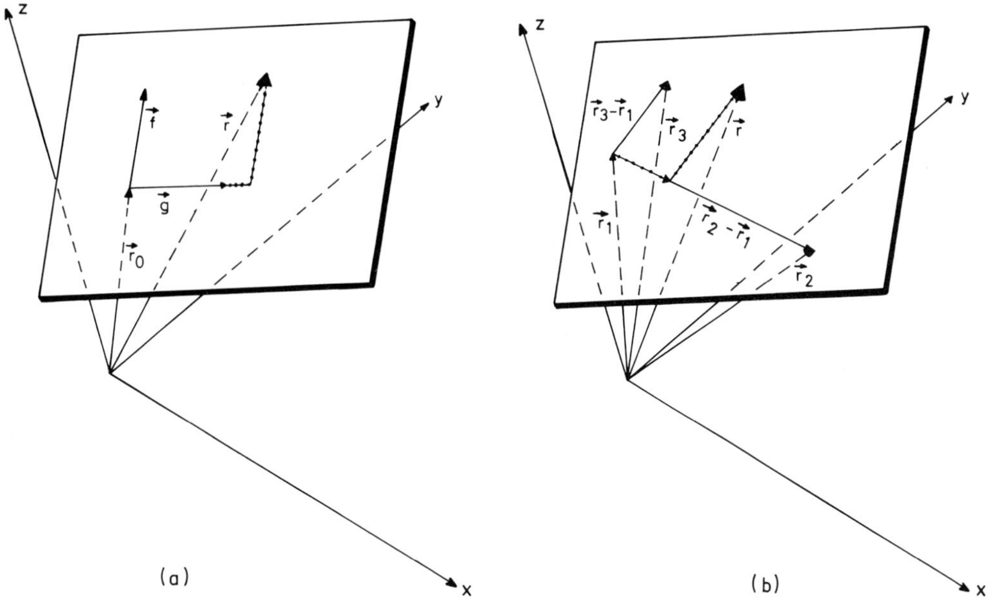

Fig. 1.19. Definition of a plane:
(a) by one point and two directions; (b) by three points.

$$\mathbf{r}\mathbf{e} = \mathrm{d} \tag{1.22}$$

with $\mathrm{d} = \mathbf{r}_1\mathbf{e}$ and study the properties of the quantities **e** and d. From the definition of **e** it is evident that the direction of **e** is normal to $\mathbf{r}_2 - \mathbf{r}_1$ and $\mathbf{r}_3 - \mathbf{r}_1$, hence **e** is normal to the plane. Since **e** is a unit vector and there is only one unit vector perpendicular to a plane (if we neglect its sense!), **e** is called "the normal vector" of the plane.

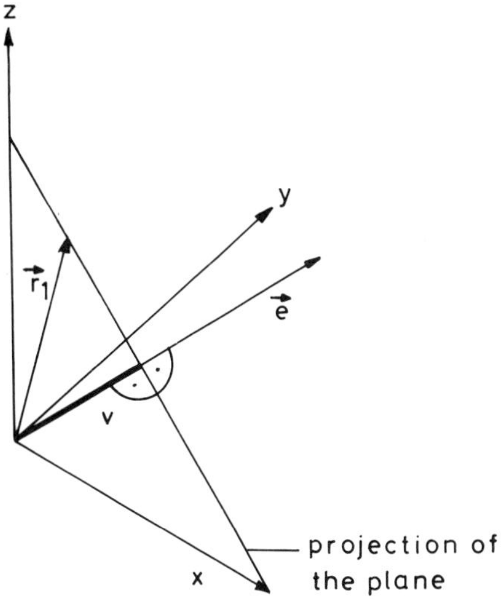

Fig. 1.20. Derivation of the property v = d.

In Fig. 1.20 we look into a direction parallel to the plane. Let v be the shortest distance form the origin to the plane.
Then $v = |\mathbf{r}_1| \cos(\mathbf{r}_1, \mathbf{e}) = |\mathbf{r}_1| |\mathbf{e}| \cos(\mathbf{r}_1, \mathbf{e}) = \mathbf{r}_1 \mathbf{e} = d$. We find that in the normal form of a plane equation, both constant quantities **e** and d have a special geometrical meaning.

As an exercise, let us now discuss two applications important in stereochemistry and crystallography.

(1) The torsional (or dihedral) angle σ of four atoms A, B, C, D with a chemical bond between AB, BC and CD, is defined as the angle between the two planes through A, B, C and B, C, D. For example, in Fig. 1.21a the torsional angle σ = H2–C1–C2–H5 is the angle between the planes through H2, C1, C2 and C1, C2, H5. A better demonstration of that quantity is shown in Fig. 1.21b. In this so-called "Newman projection" we look down the C–C bond. From the torsional angle, we know how much we have to turn the bond C1–H2 about the C–C bond until it is congruent with the C2–H5 bond. In the example of the ethane molecule, we see from Fig. 1.21b that σ must be 60°.

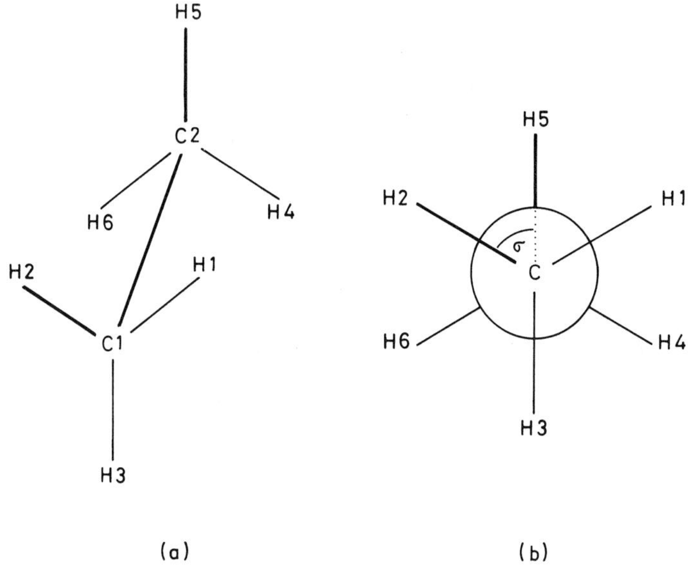

Fig. 1.21. (a) Definition of the torsional angle demonstrated in an example of the ethane molecule; (b) Representation of a torsional angle by a Newman projection.

To see how to calculate a torsional angle in a molecule with a more complicated geometry than ethane, let us calculate the torsional angle in the cyclohexane chair. Because of symmetry, all torsional angles are equal. For convenience, we choose the angle C6–C1–C2–C3 (see Fig. 1.16c) and take the atomic vectors of problem (2) in 1.1.4. We must calculate the normal vectors $\mathbf{e}_1$ and $\mathbf{e}_2$ of the planes C6–C1–C2 and

C1–C2–C3 and then the angle between the two normal vectors, because the angle between the planes is the same as that between these vectors. We already have the position vectors

$$\mathbf{r}_6 = \begin{pmatrix} 5a/6 \\ -a\sqrt{8}/6 \\ 0 \end{pmatrix} \quad \mathbf{r}_1 = \begin{pmatrix} a/2 \\ 0 \\ a\sqrt{8/3}/2 \end{pmatrix} \quad \mathbf{r}_2 = \begin{pmatrix} -a/2 \\ 0 \\ a\sqrt{8/3}/2 \end{pmatrix}$$

$$\mathbf{r}_3 = \begin{pmatrix} -5a/6 \\ a\sqrt{8}/6 \\ 0 \end{pmatrix}$$

$$\mathbf{e}_1 = \frac{(\mathbf{r}_2 - \mathbf{r}_1) \times (\mathbf{r}_6 - \mathbf{r}_1)}{|(\mathbf{r}_2 - \mathbf{r}_1) \times (\mathbf{r}_6 - \mathbf{r}_1)|}; \qquad \mathbf{e}_2 = \frac{(\mathbf{r}_3 - \mathbf{r}_2) \times (\mathbf{r}_1 - \mathbf{r}_2)}{|(\mathbf{r}_3 - \mathbf{r}_2) \times (\mathbf{r}_1 - \mathbf{r}_2)|}$$

If we denote the vector products in the numerators of $\mathbf{e}_1$ and $\mathbf{e}_2$ by $\mathbf{v}_1$ and $\mathbf{v}_2$, we get

$$\mathbf{v}_1 = \begin{pmatrix} -a \\ 0 \\ 0 \end{pmatrix} \times \begin{pmatrix} a/3 \\ -a\sqrt{8}/6 \\ -a\sqrt{8/3}/2 \end{pmatrix} = \begin{pmatrix} 0 \\ -a^2\sqrt{8/3}/2 \\ a^2\sqrt{8}/6 \end{pmatrix}$$

$$v_1^2 = \frac{a^4}{4}\frac{8}{3} + \frac{a^4}{36}8 = \frac{8}{9}a^4; \quad v_1 = \frac{a^2}{3}\sqrt{8}$$

$$\mathbf{e}_1 = \frac{\mathbf{v}_1}{v_1} = \begin{pmatrix} 0 \\ -3\sqrt{1/3}/2 \\ 1/2 \end{pmatrix}$$

$$\mathbf{v}_2 = \begin{pmatrix} -a/3 \\ a\sqrt{8}/6 \\ -a\sqrt{8/3}/2 \end{pmatrix} \times \begin{pmatrix} a \\ 0 \\ 0 \end{pmatrix} = \begin{pmatrix} 0 \\ -a^2\sqrt{8/3}/2 \\ -a^2\sqrt{8}/6 \end{pmatrix}$$

$$v_2 = v_1 = a^2\sqrt{8}/3$$

$$\mathbf{e}_2 = \begin{pmatrix} 0 \\ -3\sqrt{1/3}/2 \\ -1/2 \end{pmatrix}$$

Since $\mathbf{e}_1$ and $\mathbf{e}_2$ are unit vectors, we get

$$\cos\sigma = \mathbf{e}_1\mathbf{e}_2 = 3/4 - 1/4 = 1/2$$

and from this, $\sigma = \pm 60°$.

The sign of a torsional angle is a matter of convention. We give here the convention of Klyne and Prelog [Klyne, W. & Prelog, V., Exper. *16*, 521 (1960)], since that is most frequently used (Fig. 1.22):

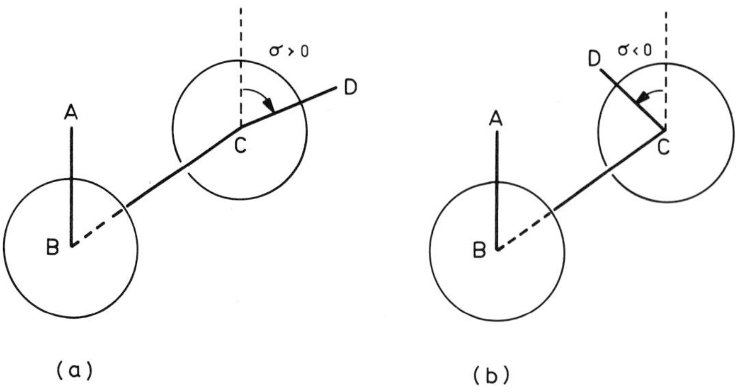

Fig. 1.22. KLYNE & PRELOG sign definition of torsional angles; (a) σ is positive; (b) σ is negative.

"The torsional angle is considered positive when it is measured clockwise from the front substituent A to the rear substituent D (Fig. 1.22a) and negative when it is measured anti-clockwise (Fig. 1.22b)."

Let us now proceed to the boat form of cyclohexane and look at what changes of torsional angles can be observed (see dotted lines in Fig. 1.16d). The only change in atomic vectors is the y-component of $\mathbf{r}_6$, which changes its sign. We have

$$\mathbf{r}_{6,\,boat} = \begin{pmatrix} 5a/6 \\ a\sqrt{8}/6 \\ 0 \end{pmatrix}$$

Only a new $\mathbf{e}_{1,boat}$ has to be calculated:

$$\mathbf{v}_{1,boat} = \begin{pmatrix} 0 \\ -a^2\sqrt{8/3}/2 \\ -a^2\sqrt{8}/6 \end{pmatrix} \quad \text{and} \quad \mathbf{e}_{1,boat} = \begin{pmatrix} 0 \\ -3\sqrt{1/3}/2 \\ -1/2 \end{pmatrix}$$

From $\mathbf{e}_{1,boat} = \mathbf{e}_2$ and from $\mathbf{e}_2\mathbf{e}_2 = 1$, it follows that $\sigma = 0$. We see that the different conformations of the cyclohexane ring are immediately recognizable if the torsional angles are known. That is why the calculation of these angles is frequently used in the stereochemistry of carbohydrates and other complicated non-planar molecules.

Another example shows that even the sign of a torsional angle may be a helpful information. In organic chemistry the pyranosyl ring is an important structural element. Its form is similar to that of cyclohexane, so the chair form is adopted in most cases. However, the existence of the hetero atom allows more than one chair form for the pyranosyl fragment. The most important forms are shown in Fig. 1.23a–d. All four molecules have equal bond lengths and valence angles and even the magnitudes of ring torsional angles are equal. But the signs of ring torsional angles of the $^4C_1$(D) and the $^1C_4$(L) form are opposite to those of the $^1C_4$(D) and the $^4C_1$(L) form.

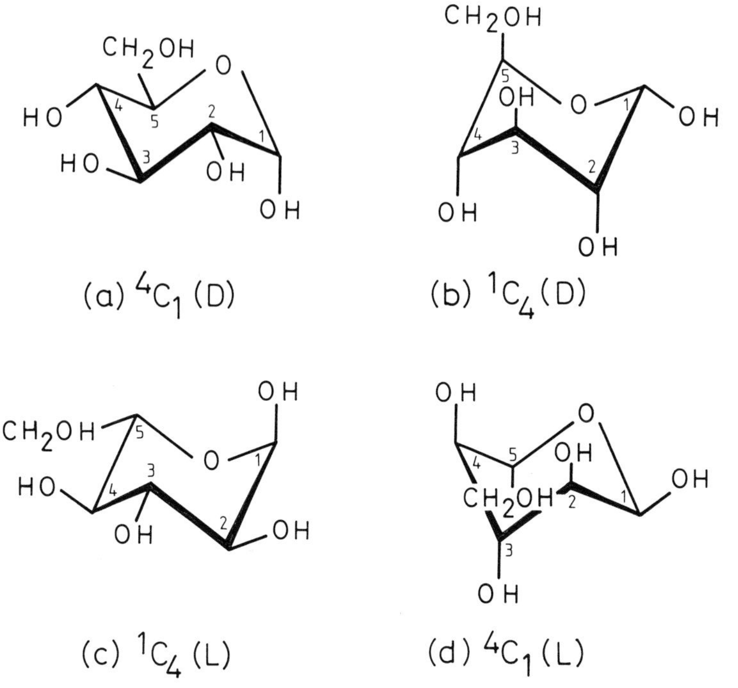

Fig. 1.23. Four different forms of the glucopyranosyl ring;
(a) $^4C_1$(D), the torsional angle $\sigma = $ C1–C2–C3–C4 is negative;
(b) $^1C_4$(D), σ is positive; (c) $^1C_4$(L), σ is positive; (d) $^4C_1$(L), σ is negative.

If a chemist has the atomic coordinates of a pyranosyl ring at hand and if he wants to known which of the four forms his coordinates describe he may calculate for instance the torsional angle $\sigma = $ C1–C2–C3–C4. He must then have an additional information from chemical considerations. If he knows, for example, whether he has a D- or L-

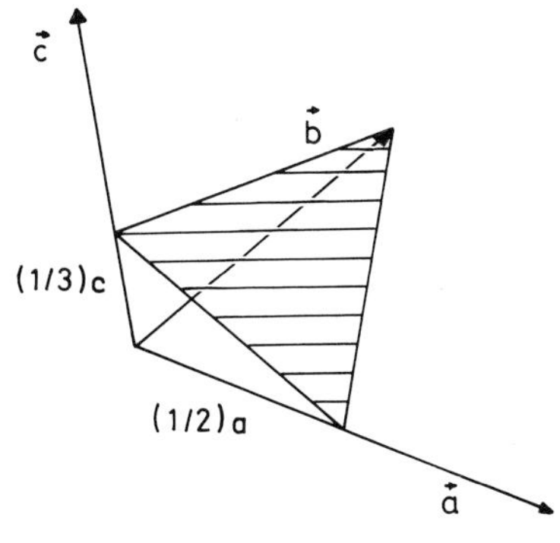

Fig. 1.24. Example of a lattice plane. Actual values of hkl are h = 2, k = 1, l = 3.

compound, the sign of σ tells him whether the pyranosyl ring has the $^4C_1$ or the $^1C_4$ conformation. If he knows, however, that the molecule must have $^4C_1$ or $^1C_4$ conformation, then he learns from the sign of σ whether the atomic coordinates describe a molecule belonging to the D- or the L-series.

(2) Three arbitrary (not necessarily orthonormal!) base vectors **a**, **b**, **c** may be given. Find the normal vector **e** and the origin distance d for a plane intersecting the base vectors at 1/h **a**, 1/k **b**, 1/l **c**, with h, k, l being arbitrary non-zero scalars. In crystallography, h, k, l will always be integers (see Fig. 1.24).

A vector in the direction of **e** is given by the vector product of two vectors on the planes, eg. 1/l **c** – 1/h **a** and 1/k **b** – 1/h **a**. Setting $\mathbf{v} = (1/l\,\mathbf{c} - 1/h\,\mathbf{a}) \times (1/k\,\mathbf{b} - 1/h\,\mathbf{a})$, we get

$$\begin{aligned}\mathbf{v} &= 1/(kl)\,\mathbf{c} \times \mathbf{b} - 1/(hl)\,\mathbf{c} \times \mathbf{a} - 1/(hk)\,\mathbf{a} \times \mathbf{b} \\ &= \text{(with the definition of reciprocal basis)} \\ &= -\,(\mathbf{abc})\,[1/(kl)\,\mathbf{a}^* + 1/(hl)\,\mathbf{b}^* + 1/(hk)\,\mathbf{c}^*] \\ &= -\frac{(\mathbf{abc})}{hkl}\,(h\mathbf{a}^* + k\mathbf{b}^* + l\mathbf{c}^*)\end{aligned}$$

Now let us consider the vector

$$\mathbf{h} = h\mathbf{a}^* + k\mathbf{b}^* + l\mathbf{c}^*$$

(This is an exception to our convention, in that h is not the magnitude of the vector **h**. The reason is that crystallographers denote vectors of the above kind by the symbol **h** and the defining integers by h, k, l with h being not equal to $|\mathbf{h}|$.)

Since **h** is a multiple of **v**, **h** has the direction of **e**, and we get

$$\mathbf{e} = \mathbf{h}/|\mathbf{h}|$$

The normal equation of the plane is then

$$\mathbf{re} = (\mathbf{rh})/|\mathbf{h}| = d$$

or

$$\mathbf{rh} = |\mathbf{h}|\,d$$

If we express **r** in terms of the direct basis

$$\mathbf{r} = x\mathbf{a} + y\mathbf{b} + z\mathbf{c}$$

and suppose **r** to be the vector of a point in the plane, the vectors

$$\mathbf{r}_1 = 1/h\,\mathbf{a} - \mathbf{r}$$

$$\mathbf{r}_2 = 1/k\,\mathbf{b} - \mathbf{r}$$

and $\quad \mathbf{r}_3 = 1/l\,\mathbf{c} - \mathbf{r}$

are coplanar and therefore their scalar triple product has to be zero:

$$
\begin{aligned}
(\mathbf{r}_1\mathbf{r}_2\mathbf{r}_3) = 0 &= \big((1/h - x)\mathbf{a} - y\mathbf{b} - z\mathbf{c}\big)[(-x\mathbf{a} + (1/k - y)\mathbf{b} - z\mathbf{c}) \\
&\quad \times (-x\mathbf{a} - y\mathbf{b} + (1/l - z)\mathbf{c})] \\
&= \big((1/h - x)\mathbf{a} - y\mathbf{b} - z\mathbf{c}\big)(xy\mathbf{a}\times\mathbf{b} - x(1/l - z)\mathbf{a}\times\mathbf{c} \\
&\quad - x(1/k - y)\mathbf{b}\times\mathbf{a} + (1/k - y)(1/l - z)\mathbf{b}\times\mathbf{c} + zx\mathbf{c}\times\mathbf{a} + zy\mathbf{c}\times\mathbf{b})
\end{aligned}
$$

Since all scalar triple products containing the same vector twice vanish, we get

$$
\begin{aligned}
(\mathbf{r}_1\mathbf{r}_2\mathbf{r}_3) = (\mathbf{abc})&[(1/h - x)(1/k - y)(1/l - z) - (1/h - x)zy - 2xyz \\
&- xy(1/l - z) - xz(1/k - y)] = 0
\end{aligned}
$$

Since $(\mathbf{abc}) \neq 0$, because they are basis vectors, the expression in square brackets vanishes. Elementary calculation leads to

$$\frac{1}{hkl} - \frac{z}{hk} - \frac{y}{hl} - \frac{x}{kl} = 0$$

or

$$hx + ky + lz = 1 \qquad (1.23)$$

On the other hand, it follows from (1.14) for the scalar product

$$\mathbf{rh} = hx + ky + lz$$

So we get $\mathbf{rh} = |\mathbf{h}|d = 1$

$$|\mathbf{h}| = 1/d \qquad (1.24)$$

Let us summarize this result since it is of major importance in crystallography:

For a plane given by the vectors $1/h\,\mathbf{a}$, $1/k\,\mathbf{b}$, $1/l\,\mathbf{c}$ with $\mathbf{a}$, $\mathbf{b}$, $\mathbf{c}$ being arbitrary basis vectors and h, k, l scalars, the vector $\mathbf{h} = h\mathbf{a}^* + k\mathbf{b}^* + l\mathbf{c}^*$ has the following properties:

(a) $\mathbf{h}$ has the direction of the normal vector of the plane,

(b) the magnitude of $\mathbf{h}$ is the reciprocal of the origin distance d.

With these important geometrical properties, every plane in three-dimensional space can be expressed by its vector $\mathbf{h}$ in reciprocal space. Since it is more advantageous to proceed with vectors than with equations of planes in arithmetical calculations, we have a good reason for the introduction of reciprocal bases.

In the special case of planes being parallel to one or more basis vectors, the intersection on the basis vectors is infinite. Since the inverse is zero, we have to use a zero for the corresponding index of $\mathbf{h}$. For instance, the plane parallel to the $\mathbf{a} - \mathbf{b}$ plane has the vector

$$\mathbf{h} = 0\mathbf{a}^* + 0\mathbf{b}^* + l\mathbf{c}^* = l\mathbf{c}^*.$$

Closely related is the following problem: For a given set of basis vectors $\mathbf{a}$, $\mathbf{b}$, $\mathbf{c}$, consider all points in space having the position vectors

$\mathbf{n} = m\mathbf{a} + n\mathbf{b} + p\mathbf{c}$, where m, n, p are integers.

These vectors define a lattice, and the vectors **n** are called lattice vectors. Similarly, we have a lattice vector $\mathbf{h} = h\mathbf{a}^* + k\mathbf{b}^* + l\mathbf{c}^*$ of reciprocal space if h, k, l are integers. So we have always two lattices, that of direct space, usually called the direct lattice, and that of reciprocal space, called the reciprocal lattice.

A given reciprocal lattice vector **h** defines a plane in direct space intersecting the basis vectors at $1/h\mathbf{a}$, $1/k\mathbf{b}$, $1/l\mathbf{c}$. It is an interesting problem in crystallography to determine a set of planes, parallel to this given plane, with the additional property that each plane contains at least one lattice point of direct lattice and that each lattice point lies on one plane.

It is immediately clear that each of the desired planes has a normal vector $c\mathbf{h}$ where c is a scalar factor. Now let us consider instead of $\mathbf{h} = h\mathbf{a}^* + k\mathbf{b}^* + l\mathbf{c}^*$ the vector $\mathbf{h}' = h'\mathbf{a}^* + k'\mathbf{b}^* + l'\mathbf{c}^*$, with h′, k′, l′ having the property to divide h, k, l by a common factor and the greatest common divisor (g. c. d.) of h′, k′, l′ equal to 1. Then it is evident that each normal vector of the desired planes is a multiple of $\mathbf{h}'$, since **h** and $\mathbf{h}'$ are parallel. One example of the desired planes can easily be obtained. Let f be the least common multiple (l. c. m.) of h′, k′ and l′. If we multiply $1/h'$, $1/k'$, $1/l'$ by f we get three integers,

$$m = f/h', \quad n = f/k', \quad p = f/l'.$$

The plane having the axial components $m\mathbf{a}$, $n\mathbf{b}$, $p\mathbf{c}$ is then one example of the desired lattice planes, since it contains the lattice points (m, 0, 0), (0, n, 0), (0, 0, p) and its normal vector is parallel to

$$\mathbf{v} = \begin{pmatrix} 1/m \\ 1/n \\ 1/p \end{pmatrix} = \begin{pmatrix} h'/f \\ k'/f \\ l'/f \end{pmatrix} = 1/f \begin{pmatrix} h' \\ k' \\ l' \end{pmatrix} = 1/f\mathbf{h}'$$

We can now solve the problem by two propositions.

(1) The set of planes desired can be expressed by the equation

$$\mathbf{h}'\mathbf{r} = n \qquad n = 0, \pm 1, \pm 2, \ldots \tag{1.25}$$

with $\mathbf{h}' = h'\mathbf{a}^* + k'\mathbf{b}^* + l'\mathbf{c}^*$.

(2) The distance d between two neighboured lattice planes is given by

$$d = 1/|\mathbf{h}| \tag{1.26}$$

Since the proof of this is not well-known in the crystallographic literature it is given below, however, if the reader is afraid of to many mathematical details he is recommended to pass over immediately to section 1.2. (I am grateful to Prof. Wendling, Free University, Berlin, for his formulation of this proof.)

To prove (1) we state at first that all planes parallel to the given one can be espressed by the equation

$$\mathbf{h}'\mathbf{r} = C, \qquad \text{with C being an arbitrary constant.}$$

It follows from (1.24) that every plane having the desired properties must satisfy an equation

$$\mathbf{d}\mathbf{r} = 1$$

with $\mathbf{d}$ parallel to $\mathbf{h}$. That means $\mathbf{d} = C'\mathbf{h}$. Since $\mathbf{h}'$ is parallel to $\mathbf{h}$, we get

$$\mathbf{d} = C'\mathbf{h} = C''\mathbf{h}'$$

and

$$1 = \mathbf{d}\mathbf{r} = C''\mathbf{h}'\mathbf{r}$$

or

$$\mathbf{h}'\mathbf{r} = 1/C'' = C$$

The desired set of lattice planes is therefore a subset of the set of planes defined by the equation above.

If a lattice point given by the integers (m', n', p') satisfies this equation, we get

$$\mathbf{h}'\mathbf{r} = h'm' + k'n' + l'p' = C$$

It follows immediately that C must be an integer. To prove (1.25), we have to show that the set of all integers C appearing on the right side of the last equation is equal to the set of all integers. Let us denote that set by Z. Then we have,

$$Z = [h'm' + k'n' + l'p'/m', n', p' \text{ integers}]$$

The plane having the equation $\mathbf{h}'\mathbf{r} = 0$ contains the origin, which is a lattice point with $m' = n' = p' = 0$, so zero is an element of Z. Since Z is a set of integers, there exists an integer $u \neq 0$ with $|u|$ being the smallest positive integer. Then we shall show that u is the greatest common divisor (g.c.d.) of h', k', l', and $Z = [nu/n = 0, \pm 1, \pm 2, \ldots]$.

First we need the property of the sum (difference) of two elements C, C′ of Z to be an element of Z, also. This is trivial, since C and C′ are of the form

$$\begin{aligned} C &= h'm' + k'n' + l'p' \\ C' &= h'm'' + k'n'' + l'p'' \end{aligned}$$

hence

$$\begin{aligned} C \pm C' &= h'(m' \pm m'') + k'(n' \pm n'') + l'(p' \pm p'') \\ &= h'm''' + k'n''' + l'p''' \end{aligned}$$

which is again an element of Z.

Let C be an arbitrary element of Z. Suppose

$$C = nu + r, \quad \text{with} \quad |r| < |u|$$

Since u is an element of Z, nu is an element of Z and the same holds for $C - nu$. That means, $r = C - nu$ is an element of Z. Since $|r| < |u|$ and u was that element of Z with the smallest magnitude, it follows that $r = 0$, hence $C = nu$. Then it follows that $Z = [nu/n = 0, \pm 1, \pm 2, \ldots]$. Now it remains to show that u = g.c.d. of h', k', l'. Suppose u' to be the greatest common divisor of h', k', l'. Since u is an element of Z, it can be written

$$u = h'm' + k'n' + l'p' = u'm^*m' + u'n^*n' + u'p^*p'$$
$$\text{(since } u' \text{ is a common divisor)}$$
$$= u'(m^*m' + n^*n' + p^*p')$$

The last expression in brackets is an integer, say r, so we get

$$u = u'r, \quad \text{hence } |u'| \leqq |u|$$

Since $h' = h'1 + k'0 + l'0$ is an element of Z and $Z = [nu]$ was already shown, an integer m' exists with

$$h' = m'u$$

Similarly we get, that u is a divisor of k' and l'. Since u' is the greatest common divisor, it follows that $|u| \leqq |u'|$. Finally, we get

$$u' = \pm u$$

Since in our case, the g.c.d. of (h', k', l') was equal to 1, we get

$$Z = [n/n = 0, \pm 1, \pm 2, \ldots]$$

and thus (1.25) holds.

If we write (1.25) in the form

$$\frac{\mathbf{h}'}{n}\mathbf{r} = 1 \quad \text{for } n \neq 0$$

we get from (1.24), for the origin distance d' of all planes except that passing the origin,

$$d' = n/|\mathbf{h}'|$$

and the distance between two neighboured planes is then

$$d = 1/|\mathbf{h}'|$$

1.2 Fundamental Results of Diffraction Theory

The aim of this section is to survey the results of diffraction theory. The proofs of several propositions will not be given. More detailed descriptions of this topic are given in: Hosemann, R. & Bagchi, S. N., "Direct Analysis of Diffraction by Matter", (1962), Amsterdam: North Holland Publishing Company; or von Laue, M., "Röntgenstrahlinterferenzen", (1960), 3rd ed., Frankfurt/M.: Akademische Verlagsgesellschaft.

1.2.1 *Fourier Transforms and Convolution Operations*

It is well known, from differential and integral calculus, that a function f of the real variable x, which is defined in an interval (−l, l) can be evaluated, under certain conditions, into a Fourier series of the form

$$f(x) = a_0/2 + \Sigma_k a_k \cos((\pi k x)/l) + b_k \sin((\pi k x)/l)$$

The coefficients a_k and b_k are called Fourier coefficients. They are obtained by

$$a_k = 1/l \int_{-l}^{+l} f(x) \cos((\pi k x)/l)\, dx$$

$$b_k = 1/l \int_{-l}^{+l} f(x) \sin((\pi k x)/l)\, dx$$

Under certain conditions (which shall not concern us at the moment), the Fourier series can be expanded into functions of three-dimensional arguments **r** and to infinite domains of definition. As we shall see, the knowledge of reciprocal space is essential for the development of Fourier theory in three variables. In the one-dimensional Fourier series, we have the argument x/l. Since the variable x will be replaced by a vector **r**, the quantity 1/l will be replaced by a vector **b** of reciprocal space. This is physically reasonable, because the argument of a trigonometric function cannot contain a physical unit.

In three-dimensional space, a Fourier transform is then defined as follows:

(a) Let f be a function of **r**. The function G is called the Fourier transform of f, if

$$G(\mathbf{b}) = \int_V f(\mathbf{x})\, e^{2\pi i(\mathbf{bx})}\, dV \qquad (1.27)$$

(b) Let H be a function of **b**. The function g is called the Fourier inverse transform of H, if

$$g(\mathbf{r}) = \int_{V^*} H(\mathbf{b})\, e^{-2(\mathbf{bx})}\, dV^* \qquad (1.28)$$

The integration with respect to V or V* has to be calculated over the whole space, which is direct space in (a) and reciprocal space in (b).

An important relation between the Fourier transforms is given by the *Fourier theorem*: If G is the Fourier transform of f, then f is the inverse transform of G and vice versa. If we denote the Fourier and its inverse transform by the symbols **F** and $\mathbf{F}^{-1}$, Fourier theorem reads

$$f = \mathbf{F}^{-1}(\mathbf{F}(f))$$

and

$$G = \mathbf{F}(\mathbf{F}^{-1}(G))$$

Example: Fourier transform of a one-dimensional normalized Gaussian function. A normalized Gaussian function is of the form

$$g(x) = \frac{a}{\sqrt{\pi}}\, e^{-a^2x^2} \qquad a > 0$$

The Fourier transform is

$$G(b) = \mathbf{F}(g(x)) = \int_{-\infty}^{-\infty} \frac{a}{\sqrt{\pi}}\, e^{-a^2x^2}\, e^{2\pi i b x}\, dx$$

$$= \int_{-\infty}^{+\infty} \frac{a}{\sqrt{\pi}}\, e^{-a^2\left(x^2 - \frac{2\pi i b x}{a^2} + \frac{\pi^2 i^2 b^2}{a^4}\right) + \frac{\pi^2 i^2 b^2}{a^2}}\, dx$$

$$= \frac{a}{\sqrt{\pi}}\, e^{-\frac{\pi^2 b^2}{a^2}} \int_{-\infty}^{+\infty} e^{-a^2\left(x - \frac{i\pi b}{a^2}\right)^2} dx$$

Substituting

$$u = x - \frac{i\pi b}{a^2}, \quad du = dx,$$

we get

$$G(b) = \frac{a}{\sqrt{\pi}}\, e^{-\frac{\pi^2 b^2}{a^2}} \int_{-\infty}^{+\infty} e^{-a^2u^2}\, du = \frac{a}{\sqrt{\pi}}\, e^{-\frac{b^2\pi^2}{a^2}} \cdot \frac{\sqrt{\pi}}{a}$$

Hence

$$G(b) = \mathbf{F}(g(x)) = e^{\frac{-\pi^2}{a^2} b^2}$$

is a Gaussian function, too, but it is no longer a normalized one, because

$$\int_{-\infty}^{+\infty} G(b)\, db = \int_{-\infty}^{+\infty} e^{\frac{-\pi^2}{a^2} b^2}\, db = \sqrt{\pi}\, \frac{a}{\pi} = \frac{a}{\sqrt{\pi}}$$

Shift-Theorem: If $G(\mathbf{b}) = \mathbf{F}(f(\mathbf{r}))$ then

and
$$\mathbf{F}(f(\mathbf{r} - \mathbf{r}_0)) = G(\mathbf{b})\, e^{2\pi i (\mathbf{b}, \mathbf{r}_0)}$$
$$\mathbf{F}^{-1}(G(\mathbf{b} - \mathbf{b}_0)) = f(\mathbf{r})\, e^{-2\pi i (\mathbf{b}_0, \mathbf{r})} \qquad (1.29)$$

As an example of the typical method of calculation in Fourier theory, let us present the proof:

$$\mathbf{F}(f(\mathbf{r} - \mathbf{r}_0)) = \int_V f(\mathbf{r} - \mathbf{r}_0)\, e^{2\pi i (\mathbf{b}, \mathbf{r})}\, dV$$

$$= \int_V f(\mathbf{r} - \mathbf{r}_0)\, e^{2\pi i (\mathbf{b}, (\mathbf{r} - \mathbf{r}_0))}\, dV\, e^{2\pi i (\mathbf{b}, \mathbf{r}_0)}$$

$$= e^{2\pi i (\mathbf{b}, \mathbf{r}_0)} \int_V f(\mathbf{t})\, e^{2\pi i (\mathbf{b}, \mathbf{t})}\, dV \qquad (\text{setting } \mathbf{t} = \mathbf{r} - \mathbf{r}_0)$$

$$= G(\mathbf{b})\, e^{2\pi i (\mathbf{b}, \mathbf{r}_0)}$$

The formula for the inverse transform is obtained in the same way.

By splitting up a function $f(\mathbf{r})$ into its odd and even part, some symmetry properties of Fourier transforms can be obtained. Denoting by f^- the function defined by $f^-(\mathbf{r}) = f(-\mathbf{r})$ and f^* the conjugate complex of f, and if $G(\mathbf{b}) = \mathsf{F}(f(\mathbf{r}))$, then

(1) $\mathsf{F}(f^-(\mathbf{r})) = G^-(\mathbf{b})$

(2) $\mathsf{F}(f^*(\mathbf{r})) = G^*(-\mathbf{b})$

(3) $\mathsf{F}(f^*(-\mathbf{r})) = G^*(\mathbf{b})$

If f is a real function, it follows immediately from (1) and (3),

$$G^-(\mathbf{b}) = G^*(\mathbf{b}) \tag{1.30}$$

This important property of Fourier transforms is the well-known "Friedel's law" in crystallography.

A further important integral transformation frequently used in crystallography is the convolution product. If f and g are two functions of $\mathbf{r}$ then the function $h = f \frown g$ is called the convolution product of f and g, if

$$h(\mathbf{r}) = (f \frown g)(\mathbf{r}) = \int_V f(\mathbf{t})\, g(\mathbf{r}-\mathbf{t})\, dV(\mathbf{t}) \tag{1.31}$$

If $f = g$, $f^{\frown 2} = f \frown f$ is called the second convolution power of f, and $f^{\frown n} = f \frown f \frown \ldots f$ (n times) the nth convolution power.

In contrast to the convolution power, the convolution square is defined by

$$f^{\sim 2}(\mathbf{r}) = \int_V f(\mathbf{t})\, f(\mathbf{t}-\mathbf{r})\, dV(\mathbf{t}) \tag{1.32}$$

The connection between $f \frown f$ and $f^{\sim 2}$ is given by

$$f^{\sim 2} = f \frown f^-$$

If f is a centrosymmetric function, which means $f = f^-$, then

$$f^{\sim 2} = f^{\frown 2} \tag{1.33}$$

An example of a function which is always centrosymmetric is $f^{\sim 2}$ itself. If $Q = f^{\sim 2}$ of an arbitrary function f, $Q = Q^-$ always holds.

Example: Convolution square of a normalized Gaussian function. Let

$$f(x) = \frac{a}{\sqrt{\pi}}\, e^{-a^2x^2}.$$

Then

$$h(u) = f^{\sim 2}(u) = \int_{-\infty}^{+\infty} \frac{a}{\sqrt{\pi}}\, e^{-a^2x^2}\, \frac{a}{\sqrt{\pi}}\, e^{-a^2(u-x)^2}\, dx$$

$$= \frac{a^2}{\pi} \int_{-\infty}^{+\infty} e^{-a^2x^2} e^{-a^2u^2 - a^2x^2 + 2a^2ux} dx$$

$$= \frac{a^2}{\pi} \int_{-\infty}^{+\infty} e^{-2a^2(x^2 - ux) - a^2u^2} dx$$

$$= \frac{a^2}{\pi} \int_{-\infty}^{+\infty} e^{-2a^2(x^2 - ux + \frac{u^2}{4}) - a^2u^2 + \frac{2a^2u^2}{4}} dx$$

$$= \frac{a^2}{\pi} \int_{-\infty}^{+\infty} e^{-2a^2\left(x - \frac{u}{2}\right)^2 - \frac{a^2}{2}u^2} dx$$

Transforming $(x - u/2) = v$ and $dx = dv$, we get

$$h(u) = \frac{a^2}{\pi} e^{-\frac{a^2}{2}u^2} \int_{-\infty}^{+\infty} e^{-2a^2v^2} dv = \frac{a^2}{\pi} \frac{\sqrt{\pi}}{a\sqrt{\pi}} e^{-\frac{a^2}{2}u^2} = \frac{a}{\sqrt{2\pi}} e^{-\frac{a^2}{2}u^2}$$

With $b = \frac{a}{\sqrt{2}}$,

we have

$$h(u) = f^{\sim 2}(u) = \frac{b}{\sqrt{\pi}} e^{-b^2u^2}$$

Hence the convolution square of a normalized Gaussian function is also normalized.

Convolution Theorem: Let f_1 and f_2 be arbitrary functions of $\mathbf{r}$, $G_1(\mathbf{b})$ and $G_2(\mathbf{b})$ their Fourier transforms. Then the following equations hold:

(a) $\mathsf{F}(f_1 \frown f_2) = G_1\ G_2$ (1.34a)

(b) $\mathsf{F}^{-1}(G_1 \frown G_2) = f_1\ f_2$ (1.34b)

(c) $\mathsf{F}(f_1\ f_2) = G_1 \frown G_2$ (1.34c)

(d) $\mathsf{F}^{-1}(G_1\ G_2) = f_1 \frown f_2$ (1.34d)

A consequence of convolution theorem can be derived if $f(\mathbf{r})$ is a real function, $G(\mathbf{b})$ its Fourier transform. Then we get

$$\mathsf{F}(f(\mathbf{r})^{\sim 2}) = G(\mathbf{b})\, G(\mathbf{b})^* = |G(\mathbf{b})|^2 \tag{1.35a}$$

and

$$\mathsf{F}^{-1}(G(b)\, G^*(b)) = f(r)^{\sim 2} \tag{1.35b}$$

Finally, let us calculate some limiting conditions which lead to interesting aspects of crystallographic structures. Supposing f is a function of $\mathbf{r}$, and $G = \mathsf{F}(f)$, then

(1) $G(\mathbf{0}) = \lim_{\mathbf{b}\to 0} G(\mathbf{b}) = \lim_{\mathbf{b}\to 0} \int_V f(\mathbf{r})\, e^{2\pi i(\mathbf{b},\mathbf{r})} dV = \lim_{\mathbf{b}\to 0} \int_V f(\mathbf{r})\, dV,$

hence

$$G(\mathbf{0}) = \int_V f(\mathbf{r})\, dV \tag{1.36a}$$

Similarly, we get,

(1a) $$f(\mathbf{0}) = \int_{V^*} G(\mathbf{b})\, dV^* \tag{1.36b}$$

(2) Let

$$\mathbf{b} = \begin{pmatrix} b_1 \\ b_2 \\ b_3 \end{pmatrix}, \quad \mathbf{r} = \begin{pmatrix} x_1 \\ x_2 \\ x_3 \end{pmatrix}$$

be vectors of reciprocal and direct space, then

$$G(0, b_2, b_3) = \lim_{b_1 \to 0} G(\mathbf{b}) = \lim_{b_1 \to 0} \int_V f(\mathbf{r})\, e^{2\pi i(b_1 x_1 + b_2 x_2 + b_3 x_3)}\, dV$$

$$= \int_{-\infty}^{+\infty} \int_{-\infty}^{+\infty} [\int_{-\infty}^{+\infty} f(\mathbf{r})\, dx_1]\, e^{2\pi i(b_2 x_2 + b_3 x_3)}\, dx_2\, dx_3$$

Setting $\int_{-\infty}^{+\infty} f(\mathbf{r})\, dx_1 = f_1(x_2, x_3)$, we have f_1 as a function of x_2 and x_3 only, and f_1 is the projection of f onto the $\mathbf{a}_2$, $\mathbf{a}_3$ plane. It follows that

$$G(0, b_2, b_3) = \mathsf{F}(f_1),$$

which means $G(0, b_2, b_3)$ is the Fourier transform of the projection of f onto the $\mathbf{a}_2$, $\mathbf{a}_3$ plane. In the same way,

(2a) $f(0, x_2, x_3) = \mathsf{F}^{-1}(G_1)$,

where G_1 is defined analogous to f_1.

1.2.2 *Electron Density and Related Functions*

From these properties of Fourier theory we can understand some important results of diffraction theory. Diffraction occurs when light interacts with matter. From experiments with diffraction from a slit or a lattice, we learn that diffraction is observable, if the geometrical dimensions of diffracting material are of the same order of magnitude as the wavelength of the light. When X-rays are the source of radiation with wavelengths of about one Angstrom ($1\ \text{Å} = 10^{-8}$ cm), diffraction can be observed when the geometry of the diffracting matter is of atomic dimensions. The best materials fulfilling these conditions are crystals, and that is why diffraction on crystals is of such importance.

Since X-rays are electromagnetic waves which interact with the electrons of matter, all the results of X-ray diffraction experiments must be due to the electron distribution of the diffracting material. So we have to be concerned with the electron distribution relative to the material volume, hence with the electron density function denoted by $\varrho(\mathbf{r})$.

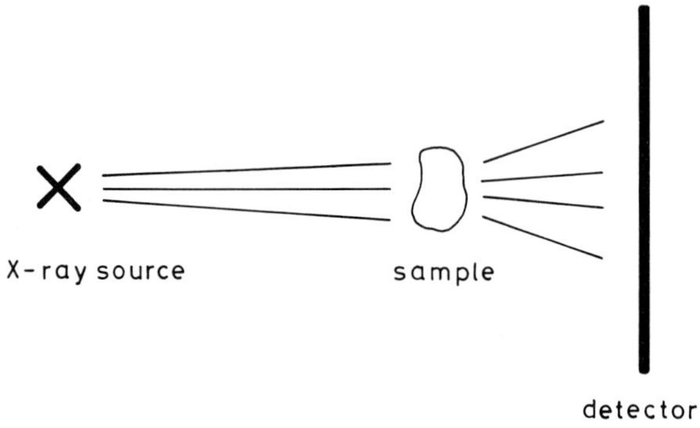

Fig. 1.25. General X-ray diffraction experiment.

The most general diffraction experiment is that in which X-rays from an X-ray source interact with a sample of arbitrary material and the diffracted radiation is recorded on a detector as shown in Fig. 1.25. The aim of such an experiment is to get information about the electron density ϱ of the sample from the intensity, I, of the diffracted radiation.

The relationship between I and ϱ is complex and we shall not deal with its theoretical derivation, but the result is simple to understand, using the Fourier and convolution formalism described previously.

In the experiment shown in Fig. 1.25, an arbitrary basis **a**, **b**, **c**, of the direct space can describe the electron density ϱ in terms or **r**. Since all physics occurs in this space, we call this space, relative to **a**, **b**, **c**, not only direct space, as in the section on vector algebra, but also the "physical space". Its vectors **r** have the dimension of a length, e.g. Å. Together with $\varrho(\mathbf{r})$, we have to consider the Fourier transform $\mathsf{F}(\varrho)$. The function $\mathrm{F} = \mathsf{F}(\varrho)$, a function of reciprocal space with

$$\mathrm{F}(\mathbf{b}) = \int_{V} \varrho(\mathbf{r})\, e^{2\pi i(\mathbf{b},\mathbf{r})}\, dV \tag{1.37}$$

is called the "structure factor". The reciprocal space with the basis vectors $\mathbf{a}^*$, $\mathbf{b}^*$, $\mathbf{c}^*$ (see 1.1.4) is called "Fourier space", and its vectors have the dimension of a reciprocal length (e.g. $Å^{-1}$).

Now we can present the main result of diffraction theory. The diffracted intensity I is proportional to the square of the magnitude of F, i.e. |F|, the structure amplitude:

$$\mathrm{I}(\mathbf{b}) \sim \mathrm{F}(\mathbf{b})\, \mathrm{F}^*(\mathbf{b}) = |\mathrm{F}(\mathbf{b})|^2 \tag{1.38}$$

This is the main result of diffraction theory, and at the same time, its main problem. The aim of the experiment is to get the electron density function ϱ. If we had its Fourier transform F, we could calculate ϱ by inverse transformation, but unfortunately experiments provide us with the magnitude $|F|$ from I, and not with the phase of this complex function. This "phase problem" is the central problem of every diffraction experiment and no direct experimental solution has yet been found (except from the very special case of a "three-beam diffraction experiment" which was reported recently by B. Post [Post, B., Acta Cryst. *A35*, 17 (1979)]. This experiment has, at least in principle, the possibility of direct experimental phase measurement. However its general application to practical problems of single crystal analysis seems scarcely possible for the present).

Since we cannot obtain the inverse transformation of F, let us examine the inverse transformation of what we get from experiment, that is, $\mathbf{F}^{-1}(FF^*)$. The function $P = \mathbf{F}^{-1}(FF^*)$ given by

$$P(\mathbf{u}) = \int_{V^*} F(\mathbf{b})\, F^*(\mathbf{b})\, e^{-2\pi i(\mathbf{b},\mathbf{u})}\, dV^* \tag{1.39}$$

with its argument **u** being a vector in direct space, is of great importance in crystallography. It is called the "Patterson function", since it was Patterson (1935) who introduced this function for the first time.

Note that the argument of P is usually denoted by a "**u**" and not by an "**r**", although it is a vector of physical space as is the argument **r** of $\varrho(\mathbf{r})$. When we discuss the properties of P in another section (5.2.1), we shall see that it is useful to distinguish between **u** and **r** because there is a geometric difference between these two quantities.

For the moment, let us only point to a mathematical property of P, which can easily be derived from (1.35b). The Patterson function is the convolution square of the electron density, hence

$$P(\mathbf{u}) = \int_{V} \varrho(\mathbf{r})\, \varrho(\mathbf{r}-\mathbf{u})\, dV \tag{1.40}$$

This last equation will be discussed in detail since it provides an important method for the solution of the phase problem. All procedures, called "heavy atom" or "Patterson" methods (see 5.2.2) are based on (1.40).

Now we have introduced the three important functions of diffraction theory which are the two functions of direct space,

(a) the electron density $\varrho(\mathbf{r})$

(b) the Patterson function $P(\mathbf{u})$

and the function of reciprocal space,

(c) the structure factor $F(\mathbf{b})$.

An elegant summary of the results of diffraction theory has been given by Hosemann & Bagchi in their famous book (Direct Analysis ..., see above) in the form of a diagram:

$$\begin{array}{ccccc} I(\mathbf{b}) & \longleftrightarrow & FF^*(\mathbf{b}) & \leftrightarrows & F(\mathbf{b}) \\ & & \updownarrow & & \updownarrow \\ & & P(\mathbf{u}) & \leftrightarrows & \varrho(\mathbf{r}) \end{array}$$

The arrows drawn as full lines indicate the possible operations. The arrows drawn as dotted lines show the phase problem. We see that there is no path from the result of experiment, I, to the right side of the diagram, to the desired electron density ϱ.

1.2.3 *Diffraction Conditions for Single Crystals*

A single crystal is a sample of material with an electron density function which is periodic in three dimensions; that is, (Fig. 1.26) there are three non-coplanar vectors **a**, **b**, **c** with

$$\varrho(\mathbf{r}) = \varrho(\mathbf{r} + n\mathbf{a}) = \varrho(\mathbf{r} + m\mathbf{b}) = \varrho(\mathbf{r} + p\mathbf{c}) \quad n, m, p \text{ are integers} \tag{1.41}$$

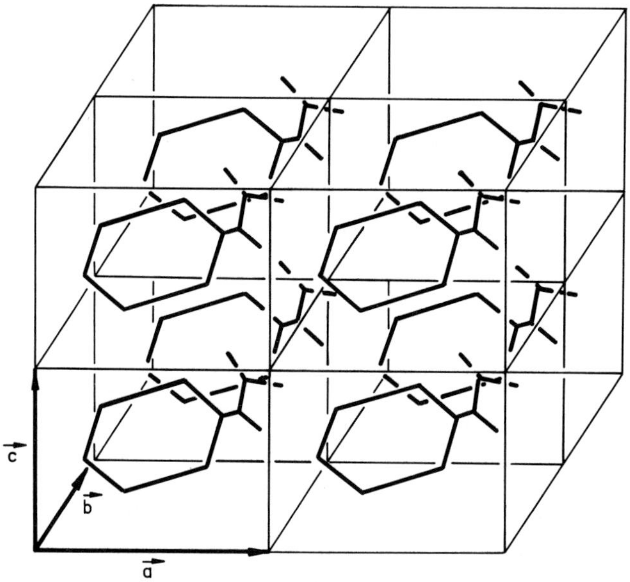

Fig. 1.26. Model of a single crystal.

The volume element defined by **a**, **b**, **c** is called the unit cell. It represents the non-periodic unit. The whole crystal lattice is obtained by periodic sequences of unit cells in all three dimensions. The vectors **a**, **b**, **c**, forming a basis of direct space in a mathematical sense, are called unit cell vectors. Their magnitudes a, b, c, together with the angles between them, $\alpha = (\mathbf{b}, \mathbf{c})$, $\beta = (\mathbf{c}, \mathbf{a})$, $\gamma = (\mathbf{a}, \mathbf{b})$ are called lattice constants. The reciprocal basis, being present together with every basis, is called the basis of reciprocal lattice. The quantities a*, b*, c*, α^*, β^*, γ^*, defined by 1.11, are called reciprocal lattice constants.

Because of the periodicity of the crystal lattice, it is evident that the electron density has to be determined only for the unit cell. A further simplification is that the diffracted intensity is no longer spread continuously in space but becomes discrete. This is due to the famous *Laue diffraction condition:* A single crystal may have the unit cell vectors **a**, **b**, **c** and the corresponding reciprocal cell vectors **a***, **b***, **c***. Then

$$I(\mathbf{b}) \neq 0 \quad \text{only if } \mathbf{b} = \mathbf{h} = h\mathbf{a}^* + k\mathbf{b}^* + l\mathbf{c}^* \quad \text{with h, k, l integers} \tag{1.42}$$

otherwise $I(\mathbf{b}) = 0$. Two important properties derive from the Laue condition. The first is that the intensity distribution is discontinuous. Of course, this is an important advantage for the experimental situation. If diffraction intensities from a single crystal are to be taken they appear only at discrete points in space and no continuous intensity diagram has to be taken. The second is that each diffraction spectrum is related uniquely to a lattice plane, which intersects the unit cell vectors on 1/h**a**, 1/k**b**, 1/l**c**.

We know from the Laue condition that only the lattice planes are responsible for diffraction; the geometrical conditions under which this will occur are most simply understood through the *Ewald diffraction condition* (Fig. 1.27).

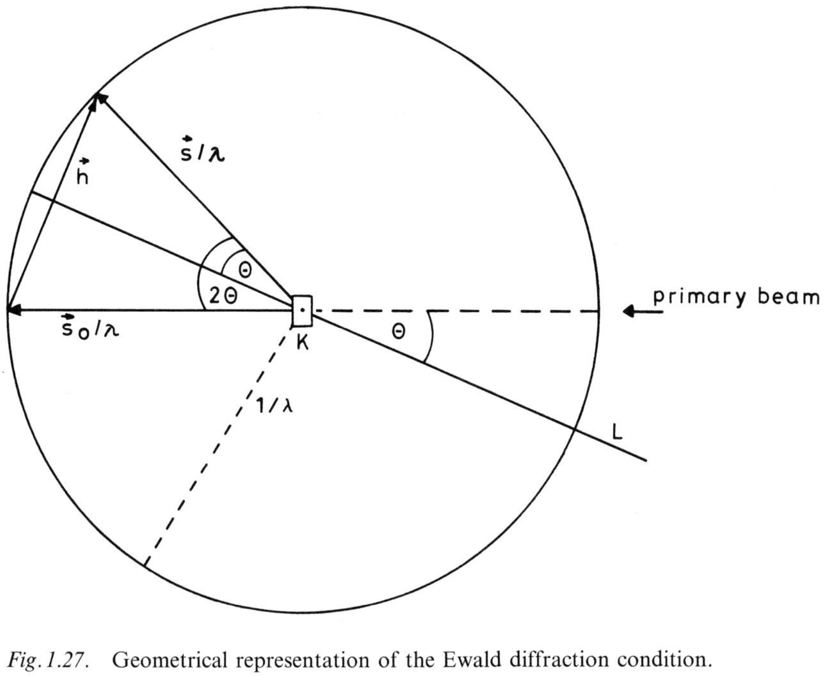

Fig. 1.27. Geometrical representation of the Ewald diffraction condition.

For a given crystal K and an X-ray beam of wavelength λ, let $\mathbf{s}_0$ be the unit vector in the direction of the primary beam. If $\mathbf{s}$ is the unit vector of the diffracted beam concerning a lattice plane with the normal vector $\mathbf{h}$, the three vectors, $\mathbf{h}$, $\mathbf{s}_0$, and $\mathbf{s}$ satisfy the equation

$$\mathbf{h} = \frac{\mathbf{s} - \mathbf{s}_0}{\lambda} \qquad (1.43)$$

Before interpreting the Ewald condition, let us explain the role of the angle θ in Fig. 1.27. We find that

$$\frac{|\mathbf{h}|/2}{1/\lambda} = \sin\theta$$

or with $d = 1/|\mathbf{h}|$,

$$\lambda = 2d\sin\theta \qquad (1.44)$$

This famous equation is called Bragg's law, and the angle θ is said to be the "Bragg angle".

Now we can discuss both the Ewald condition and Bragg's law. From the Ewald condition, it follows that a lattice plane L must be in a special position to cause diffraction (the diffraction position). That position is realized if its normal vector $\mathbf{h}$ is on the surface of a sphere of radius $1/\lambda$ around K. This sphere is the "Ewald sphere".

Diffraction of a lattice plane L happens only if d, the reciprocal of the normal vector's magnitude, satisfies Bragg's equation. In this case, the plane L behaves like a mirror with respect to the X-ray beam, since the diffracted beam behaves like a reflection from a mirror, with the incident and reflected beams having the same angle to the plane. Now we can understand why crystallographers refer to the discrete diffracted intensities as "reflections", and we can now express in a few words the diffraction of X-rays on single crystals: X-rays are reflected by the lattice planes of a single crystal if, and only if, the angle of the incident beam to the plane satisfies Bragg's law. In that case, the incident beam, the diffracted beam, and the vector normal to the plane satisfy Ewald's condition.

Another important consequence derives from Bragg's law. Since $|\sin\theta| \leqq 1$, we get

$$\lambda/(2d) \leqq 1$$

or

$$|\mathbf{h}| \leqq 2/\lambda \qquad (1.45)$$

It follows that, for a given radiation with fixed wavelength λ, the number of possible reflections is limited. Only those reflections in reciprocal space inside the sphere of radius $2/\lambda$ can be observed. This sphere is called the "limiting sphere". The limiting sphere has twice the radius of the Ewald sphere (see Fig. 1.28). Since we have one reflection per one reciprocal unit cell, the number of reflections inside the limiting sphere is M if its volume is equal to M times V*. It follows that

$$MV^* = \frac{4}{3}\pi\left(\frac{2}{\lambda}\right)^3$$

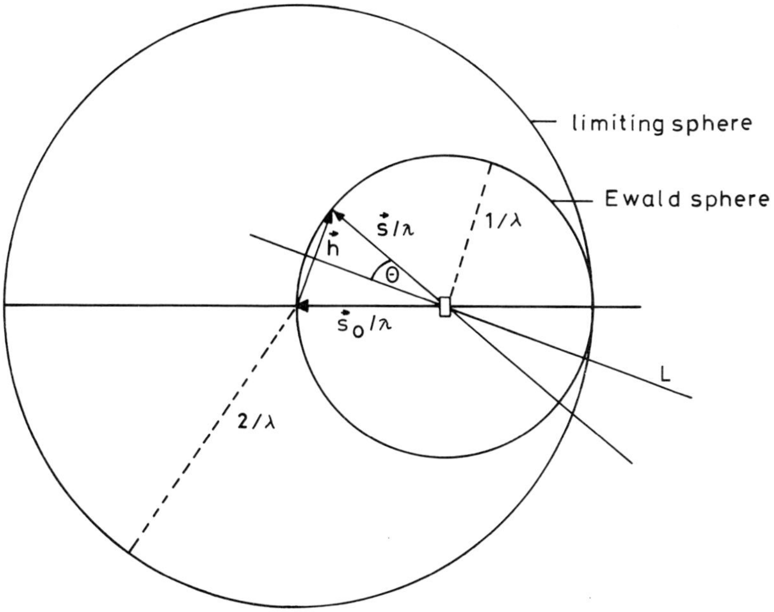

Fig. 1.28. Ewald sphere (radius $1/\lambda$) and limiting sphere (radius $2/\lambda$).

$$M = \frac{32\pi}{3} \frac{1}{\lambda^3} \frac{1}{V^*}$$

or

$$M = \frac{33.5}{\lambda^3} V \tag{1.46}$$

with $V = 1/V^*$ = cell volume. For the wavelengths of CuKα and MoKα radiation, which are most frequently used, we have $\lambda(\text{Cu}) \approx 1.542$ Å and $\lambda(\text{Mo}) \approx 0.711$ Å. Then we get $M \approx 9$ V for the copper sphere and $M \approx 93$ V for Mo radiation, i.e. the number of reflections to be taken with Mo radiation is about a factor ten larger than for Cu radiation.

Usually the experimental conditions only allow a reflection measurement below a given θ_{max}. Then (1.45) has to be replaced by

$$|\mathbf{h}| < (2/\lambda) \sin \theta_{max} \tag{1.45a}$$

and (1.46) becomes

$$M = \frac{33.5}{\lambda^3} V \sin^3 \theta_{max} \tag{1.46a}$$

If, for instance, $\theta_{max} = 70°$ must be chosen for CuKα radiation, we get $M \approx 7.6$ V. The same magnitude of M for the same cell volume V is obtained, if $\theta_{max} \approx 26°$ is taken if MoKα radiation is used.

2 Preliminary Experiments

We can now proceed to the first experiments in a single crystal structure analysis. To complete our theoretical knowledge by practical experience we shall describe in the course of this book the structure determination of three compounds in detail. These three compounds are

(1) potassium hydrogen tartrate, $C_4H_5O_6K$, (KAMTRA)

(2) ammonium tetrasulfurpentanitride oxide, $NH_4[S_4N_5O]$, (NITROS)

(3) sucrose, $C_{12}H_{22}O_{11}$, (SUCROS)

We have chosen these compounds for several reasons. Good single crystals of all three compounds can be obtained and the molecules are relatively small. The data measurement procedures and computer calculations will provide typical examples of the problems which arise with an organic or inorganic structure analysis. We can use KAMTRA, for instance, as an example of an organic structure in which the phase problem can be solved by application of the "heavy atom method" (see 5.2.2). NITROS which was first synthesized and crystallized by Steudel [Steudel, R., Z. f. Naturforsch. B 24, 934 (1969)] is a structure to be solved by "direct methods" (see. 5.3) in a centro-symmetrical space group and an example of an inorganic structure. SUCROS is an optically active organic structure to which we can apply "direct methods" to a structure with an acentric space group.

If the reader becomes experienced with all the problems arising from these three structure analyses, or if he even redetermines one or two of them himself, he should be able to solve most of the more straight-forward structural problems arising in his laboratory.

When starting a structure analysis, we have to assume that single crystals of the compound are available. The first information we can get is that of the crystal symmetry and lattice dimensions. For this purpose we must describe the film techniques which are the first experimental applications of single crystal diffraction.

2.1 Film Methods

2.1.1 *The Rotation Method*

From the Ewald condition, shown in Figs. 1.27 and 1.28, we see that we get diffraction of a lattice plane if, and only if, its vector **h** intersects Ewald's sphere. If a crystal is oriented in an arbitrary orientation towards the X-ray beam, this condition will only be fulfilled for a few planes.

One way to obtain diffraction by every reflection in the limiting sphere is to rotate the crystal about an axis normal to the direction of the incident beam. This is the fundamental concept of most film methods used in single crystal diffractometry which uses monochromatic radiation. (Another way to obtain a large number of reflections is to vary the wavelength to get a variety of Ewald spheres. This method was used in the first diffraction experiments of von Laue and coworkers and is called the Laue Method.)

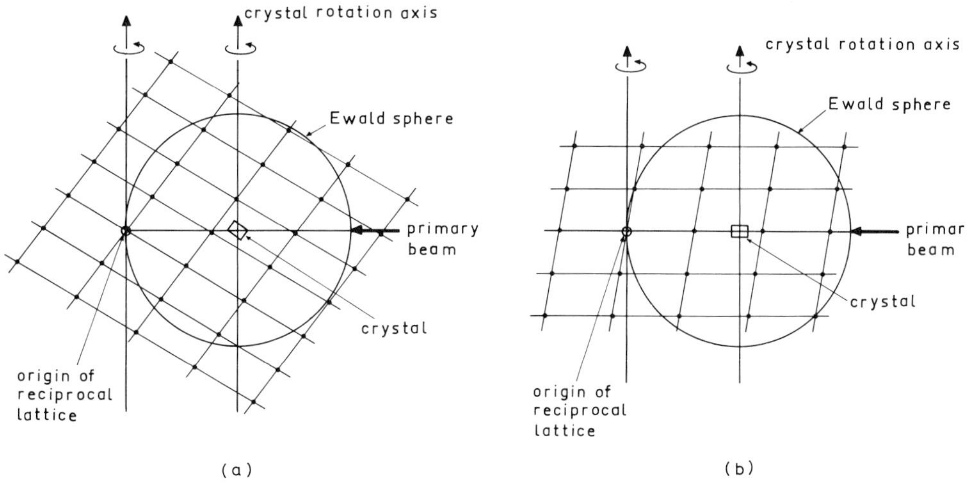

Fig. 2.1. (a) Random orientation of a crystal with respect to the X-ray beam; (b) Orientation of reciprocal lattice planes in the direction of the primary beam.

Since all diffraction spots can be assigned to a lattice vector **h** in reciprocal space, all single crystal diffraction experiments can be descibed in terms of the reciprocal lattice. That is why we have plotted only the reciprocal lattice in Fig. 2.1. Let us now examine two possible orientations of this lattice with respect to the incident beam. (We have placed the origin of the reciprocal lattice on the intersection of the incident beam direction and the Ewald sphere.) In Fig. 2.1a, the reciprocal lattice is in an arbitrary orientation to the incident beam. Rotation of the crystal (and the reciprocal lattice) gives rise to several intersections of lattice points with the Ewald sphere. However, the reflected beams will have no special directions and will therefore show a distribution on a film which bears no obvious relationship to the crystal symmetry or the crystal axes. The result of such a rotation photograph is shown in Fig. 2.2a; it is like the "stars in the sky".

We get a more interpretable diffraction pattern if the crystal is adjusted, as shown in Fig. 2.1b, so that it rotates about an axis which is normal to a set of reciprocal

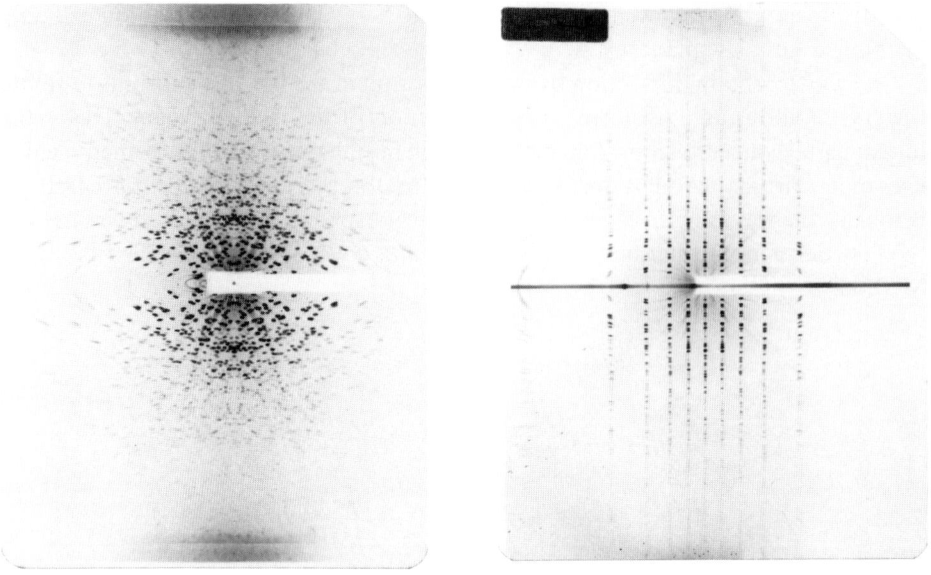

(a) (b)

Fig. 2.2. (a) Rotation photograph of a randomly oriented crystal; (b) Rotation photograph of a properly aligned crystal.

lattice planes and normal to the X-ray beam. The reciprocal lattice planes then intersect the Ewald sphere on parallel circles. The reflected beams are then all gathered on different cones, each cone being associated with one reciprocal lattice plane. These cones, named *Laue cones*, have the crystal rotation axis as the common cone-axis (Fig. 2.3).

If the film is positioned cylindrically around the crystal with the cylinder axis coinciding with the rotation axis of the crystal, these cones appear as parallel circles on the cylindrical film, becoming straight lines after the film is unrolled.

An example of such a rotation photograph is shown in Fig. 2.2b. Since all reflections of one reciprocal layer are present on one line of the film, the lines are called "layer lines". That in the plane of the incident beam is the zero layer line, the next the first layer line, and so forth. From the rotation photograph of an aligned crystal, we can measure the distance D* between the lattice layers.

From Fig. 2.3, we find

$$\tan \mu_1 = l_1/R_F \qquad R_F = \text{radius of film camera}$$

and

$$\sin \mu_1 = \frac{D^*}{1/\lambda}$$

or

$$D^* = \frac{\sin \mu_1}{\lambda}$$

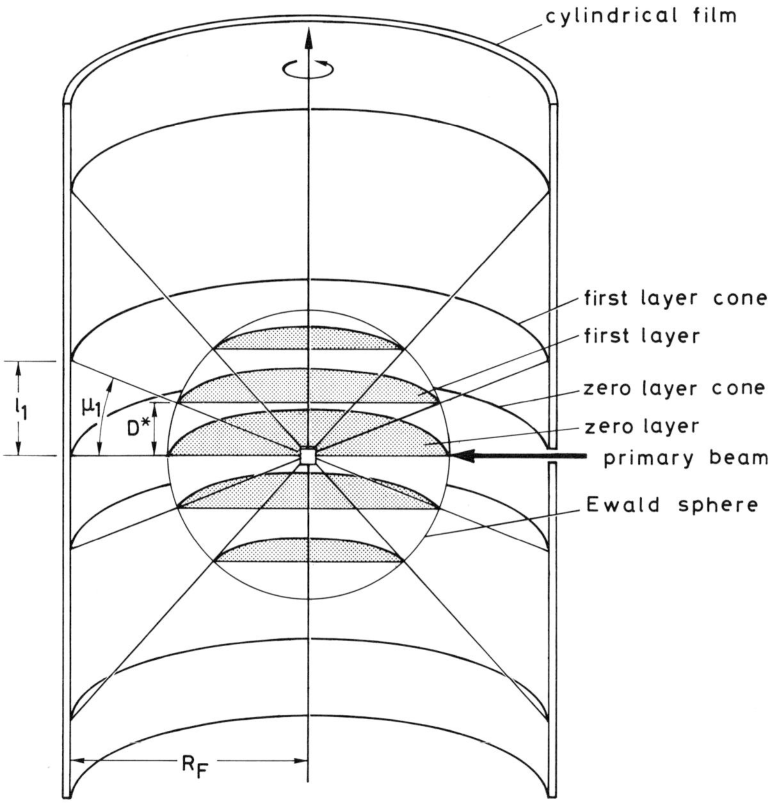

Fig. 2.3. Laue cones and their registration on a rotation photograph.

Since R_F is known, and l_1 can be measured from the photograph, μ_1 and then D* can be calculated.

In general for the nth layer line we get

$$\tan \mu_n = l_n / R_F \tag{2.1}$$

and from $\sin \mu_n = \dfrac{nD^*}{1/\lambda}$ we find

$$D^* = \frac{\sin \mu_n}{n\lambda} \tag{2.2}$$

Equation (2.2) applies to every rotation axis and for every kind of lattice layer. Consider a rotation axis of the form $\mathbf{r} = u\mathbf{a} + v\mathbf{b} + w\mathbf{c}$ with u, v, w being integers with no common factor. Now, since direct and reciprocal space are completely equivalent, we can use formula (1.25) of 1.1.6 for planes in reciprocal space and their normal vectors in direct space. From the definition of **r** it follows that **r** is normal to a set of planes satisfying the equation

$$\mathbf{h}_n\mathbf{r} = n \qquad n = 0, 1, 2, \ldots$$

The distance between these planes is $D^* = 1/|\mathbf{r}|$ (see 1.26). Since $\mathbf{r}$ is the rotation axis, these planes are identical to the layer lines. That means for a reflection $\mathbf{h}$ lying in the nth layer line, we get the so-called "layer line condition":

$$\mathbf{hr} = n \qquad n = 0, 1, 2, \ldots \tag{2.3}$$

From (2.2) we get

$$D^* = \frac{1}{r} = \frac{\sin \mu_n}{n\lambda}$$

or with $D = r$

$$D = \frac{n\lambda}{\sin \mu_n} \tag{2.4}$$

If the rotation axis is a crystal axis, say one of the base vectors, for instance $\mathbf{a}$, then we have $\mathbf{r} = 1\mathbf{a} + 0\mathbf{b} + 0\mathbf{c}$ and (2.3) reduces to $h1 + k0 + l0 = n$.

It follows that the zero layer line includes the reflections of the $\mathbf{b}^* - \mathbf{c}^*$ plane. They all have the indices 0kl. The first layer line contains reflections of type 1kl, in general reflections on the nth layer line are of the type nkl.

From (2.4) we then get, with $D = |\mathbf{a}| = a$

$$a = \frac{n\lambda}{\sin \mu_n} \tag{2.5}$$

Thus we can measure the magnitude of the lattice constant, a, from an aligned rotation photograph, if the rotation axis is parallel to the lattice direction. The lattice constant derived from a rotation photograph is, in fact, the only information of direct space that we get from film techniques. All further information will be of reciprocal space.

Although the rotation photograph is the first technique to provide information about the crystal lattice, it has one disadvantage. It provides no information relating to the reflections within the layers. To overcome this difficulty, we use other film methods. Those most frequently used in single crystal analysis are the Weissenberg, the de Jong-Bouman, and the precession methods.

2.1.2 *Zero Level Weissenberg and Normal Beam Method*

The experimental conditions for this technique are similar to those for the rotation method. As before, it is necessary to rotate a well-aligned crystal about an axis perpendicular to the direction of the primary beam. There are two main differences from the rotation method. A slotted screen is used which permits the recording of only one layer line, and the film holder makes a translational movement in the direction of the crystal's rotation axis (Fig. 2.4), as the crystal rotates.

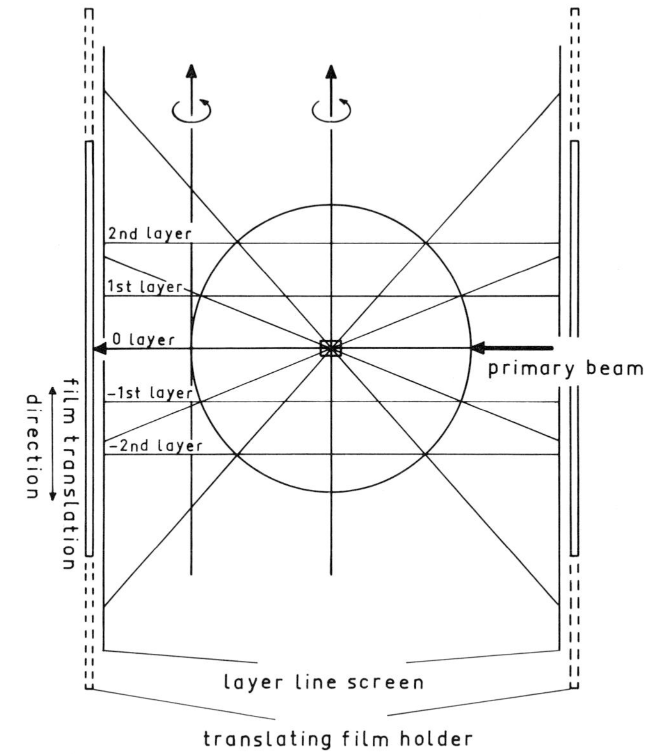

Fig. 2.4. Principle of a Weissenberg camera, normal beam technique.

With this experimental arrangement, reflections intersecting the Ewald sphere at different times will be registered at different positions on the film. Thus a complete layer of reciprocal space which appears as one single line (layer line) on a rotation film will be distributed over the whole Weissenberg film. In contrast to the rotation technique, all reflections are separated and can be identified. These experimental conditions give a distorted representation of a layer of the reciprocal lattice, in contrast to the other two moving film methods which will be described later.

An example of a Weissenberg film is shown in Fig. 2.5a. The typical arrangement of diffraction spots on "festoon"-like curves can clearly be seen (see also Fig. 2.5b). To measure detailed information about the geometry of the reciprocal lattice from the reflection positions, the experimental conditions and their geometrical consequences have to be known.

Fig. 2.6 shows the experimental conditions for a zero-level Weissenberg exposure

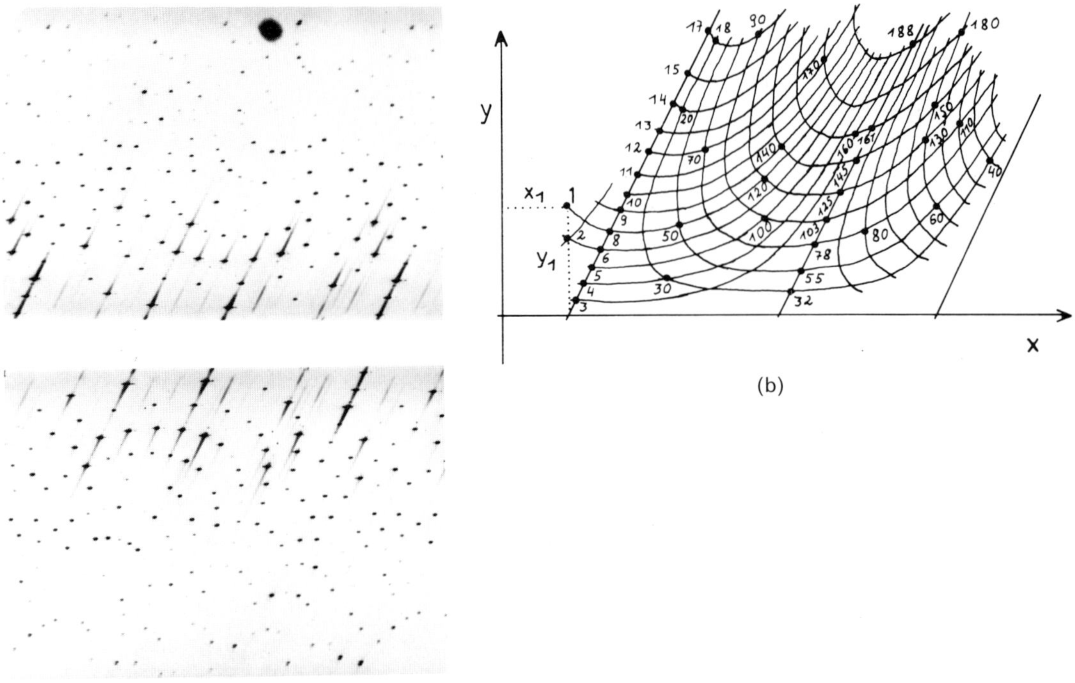

Fig. 2.5. (a) Example of a Weissenberg exposure (sucrose, zero layer);
(b) Schematic representation of festoon-like arrangement of reflections. The numbers correspond to the enumeration of reflections in Fig. 2.7.

in the plane of the zero layer; this is perpendicular to the projection drawn in Fig. 2.4. The crystal rotation axis is directed perpendicular to the plane of the paper.

Consider a two-dimensional coordinate system in the plane having its origin, N, at the intersection of the primary beam and the Ewald sphere and its x-axis perpendicular to the direction of the primary beam. Relative to this coordinate system we choose polar coordinates so that to every lattice point P having the position vector **h**, a polar radius r and a polar angle ω can be assigned.

If the crystal is rotated about its axis, P turns by an angle $\omega + \beta$ to intersect the Ewald sphere at P′, resulting in a reflection. This reflection will appear on the extension of the vector $\mathbf{s}/\lambda$ on the film at B. Let t be the time necessary to move P to P′ and let us then denote the position vector of P′ by $\mathbf{h}(t)$. Let y be the circular arc AB which will be a straight line if the film is unrolled. If R_F is the radius of the cylindrical film holder, we get,

$$y/(2\pi R_F) = (2\theta)/360$$

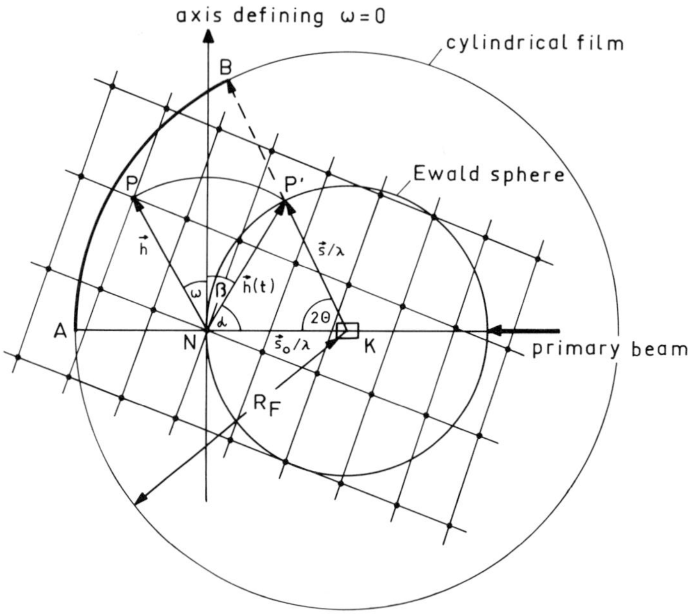

Fig. 2.6. Experimental conditions of the Weissenberg technique in the zero layer plane, projected in the direction of the crystal rotation axis.

or

$$\theta = (y\,180)/(2\pi R_F) \tag{2.6}$$

Most Weissenberg cameras have a film cassette radius of

$$R_F = 180/(2\pi) \approx 28.6 \text{ mm}$$

Then we get

$$\theta \text{ (in degrees)} = y \text{ (in mm)}$$

The crystal rotation is coupled via a fixed gearing, g (to be given in mm/degrees) with the translation x of the film holder. Then we get for x,

$$x = g(\omega + \beta)$$

From Fig. 2.6, it follows from simple geometric considerations,

$$\beta = 90^\circ - \alpha$$

$$2\alpha = 180^\circ - 2\theta$$

hence $\beta = 90^\circ - (90^\circ - \theta) = \theta$, or

$$x = g(\omega + \theta)$$

The quantities x and y can be measured directly on a Weissenberg film (see Fig. 2.5b).

x has to be measured relative to an origin which can be chosen arbitrarily, in principle. However, for convenience, the point $x = 0$ should be positioned at the left margin. Using Bragg's equation, $r = |\mathbf{h}|$ can be calculated and the polar coordinates (r, ω) of a lattice point can be obtained from x and y by

$$r = |\mathbf{h}| = (2 \sin \theta)/\lambda \tag{2.7}$$

with $\theta = (180\,y)/(2\pi R_F)$ ($= y$, if $R_F = 28.6$ mm), and

$$\omega = x/g - \theta \tag{2.8}$$

By application of the transform $(x, y) \longrightarrow (r, \omega)$ an undistorted reciprocal lattice layer is obtained, from which the distance between reciprocal lattice points can be calculated. In our laboratory we have developed a computer program which calculates the corresponding polar coordinates from a given set of (x, y) values. Fig. 2.7 shows a graphical representation of the undistorted reciprocal layer obtained from the exposure of Fig. 2.5a.

A special feature of a Weissenberg film is that reflections having the same ω-values (for instance the reflections 3, 4, 5, ... in Fig. 2.7) are situated on a straight line passing through the origin of the reciprocal lattice. For these reflections, it follows from (2.7) and (2.8) that

$$x = g(\omega + \theta) = g\omega + (yg\;180)/(2\pi R_F)$$

Setting $c = g\omega$, and $d = (g\;180)/(2\pi R_F)$, then c and d are constants and we get

$$y = (1/d)x - c/d \tag{2.9}$$

Thus y and x are related by the equation of a straight line, which implies that these reflections are positioned on a straight line on the Weissenberg exposure (see Fig. 2.5b for comparison). Thus we have shown that reflections situated on a straight line on a Weissenberg film belong to a reciprocal lattice line passing through the origin. This fact is frequently utilized to determine reciprocal lattice constants directly from the Weissenberg film. If the rotation axis is identical to one of the unit cell vectors, say **c**, the zero layer Weissenberg exposure contains the hk0 reflections. Since the reciprocal lattice lines h00 and 0k0 pass through the origin of the reciprocal lattice, they appear as lines on the film. They can be identified by symmetry considerations as we shall discuss in Section 3. These reflections, which are multiples of the reciprocal basis vectors, are said to be "axial reflections". For a h00 reflection we have

$$\mathbf{h} = h\mathbf{a}^* + 0\mathbf{b}^* + 0\mathbf{c}^* = h\mathbf{a}^*$$

Then Bragg's equation is

$$\lambda = 2 \sin \theta/|\mathbf{h}| = 2 \sin \theta/(|h|a^*)$$

We get

$$a^* = 2 \sin \theta/|h|\lambda \tag{2.7a}$$

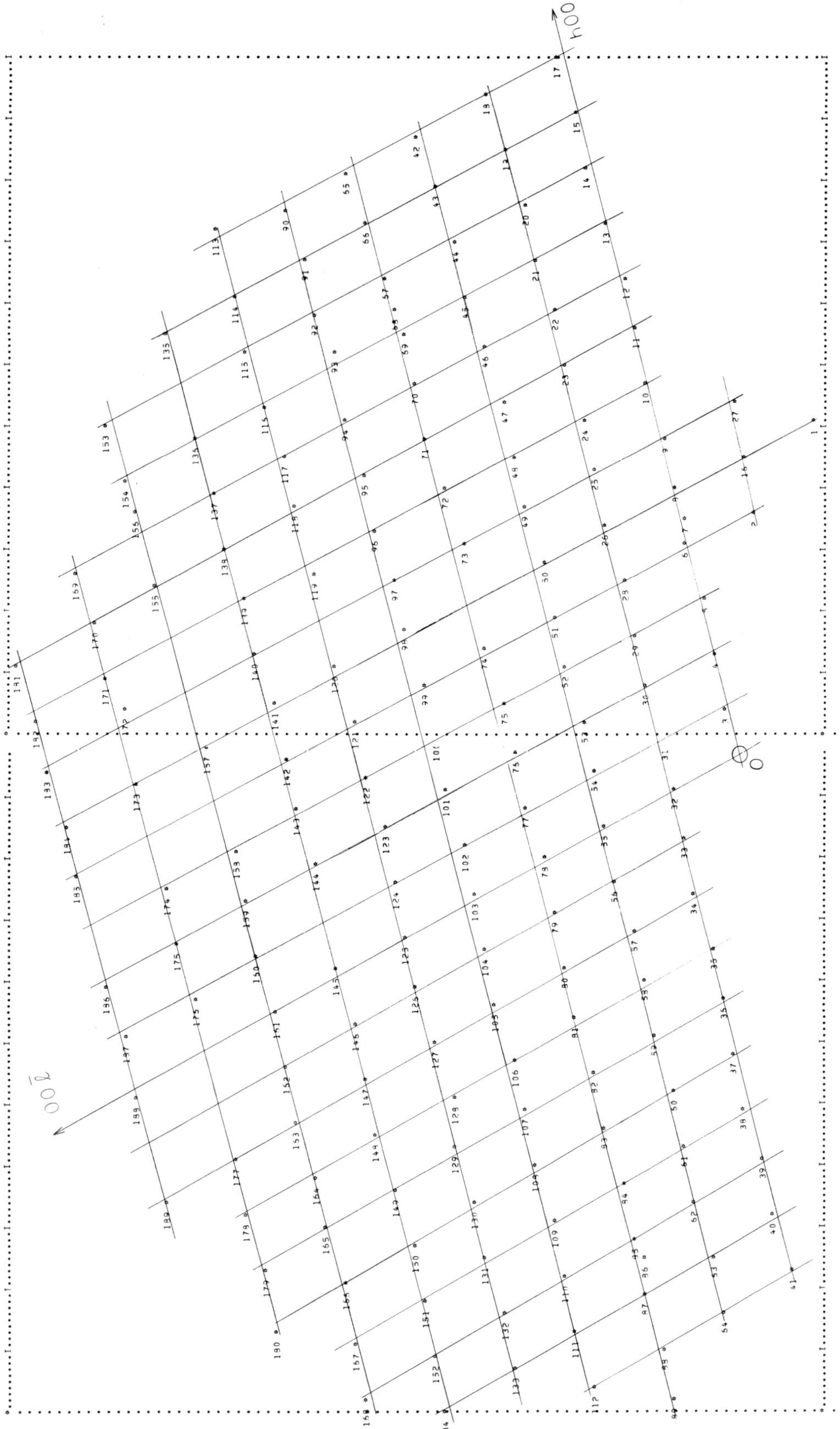

Fig. 2.7. Computer output of an undistorted lattice layer obtained from the Weissenberg exposure of Fig. 2.5.

Since θ is given by its y-value the reciprocal lattice constant a* is obtained from a Weissenberg exposure.

Let us now proceed to the upper level Weissenberg technique. The experimental conditions are similar to those for the zero layer, except that the layer line screen has to be shifted to prevent all but the desired nth layer to be registered on the film. This technique is denoted "normal beam method", since the incident beam is still normal to the crystal rotation axis. The shift t_n of the layer line screen (see Fig. 2.3 and 2.4) can be calculated from

$$\tan\mu_n = \frac{t_n}{R_s}$$

or

$$t_n = R_s \tan\mu_n$$

where R_s is the radius of the layer-line screen.

A disadvantage of the normal-beam method is that it allows the registering of only those layer which would appear on a rotation photograph. As can be seen from (2.4), that means (since $\sin\mu_n \leqq 1$) that

$$n \leqq D/\lambda \tag{2.10}$$

2.1.3 *Upper Level Weissenberg – Equi-inclination Method*

With minor changes in the geometrical conditions, the number of layers being registered can be enlarged by a factor of two. Therefore, in practice it is customary to prefer the so-called "equi-inclination method" rather than the normal-beam method. When using this technique the primary beam is no longer chosen to be perpendicular to the crystal rotation axis, but it is turned by an angle φ_n from that direction (Fig. 2.8a). As before, the origin N of the reciprocal lattice is at the intersection of the primary beam and the Ewald sphere. The central plane c is perpendicular to the crystal rotation axis and passes through the crystal center. Then the distance z between zero layer and c-plane is given by (see Fig. 2.8a)

$$\frac{z}{1/\lambda} = \sin\varphi_n$$

φ_n, the "equi-inclination angle", is now chosen so that the nth layer has the same distance to the c-plane as the zero layer, and since $(1/2)nD^* = z$ (see Fig. 2.3),

$$\frac{(1/2)nD^*}{1/\lambda} = \sin\varphi_n$$

$$\frac{n\lambda}{2D} = \sin\varphi_n$$

or, using (2.4),

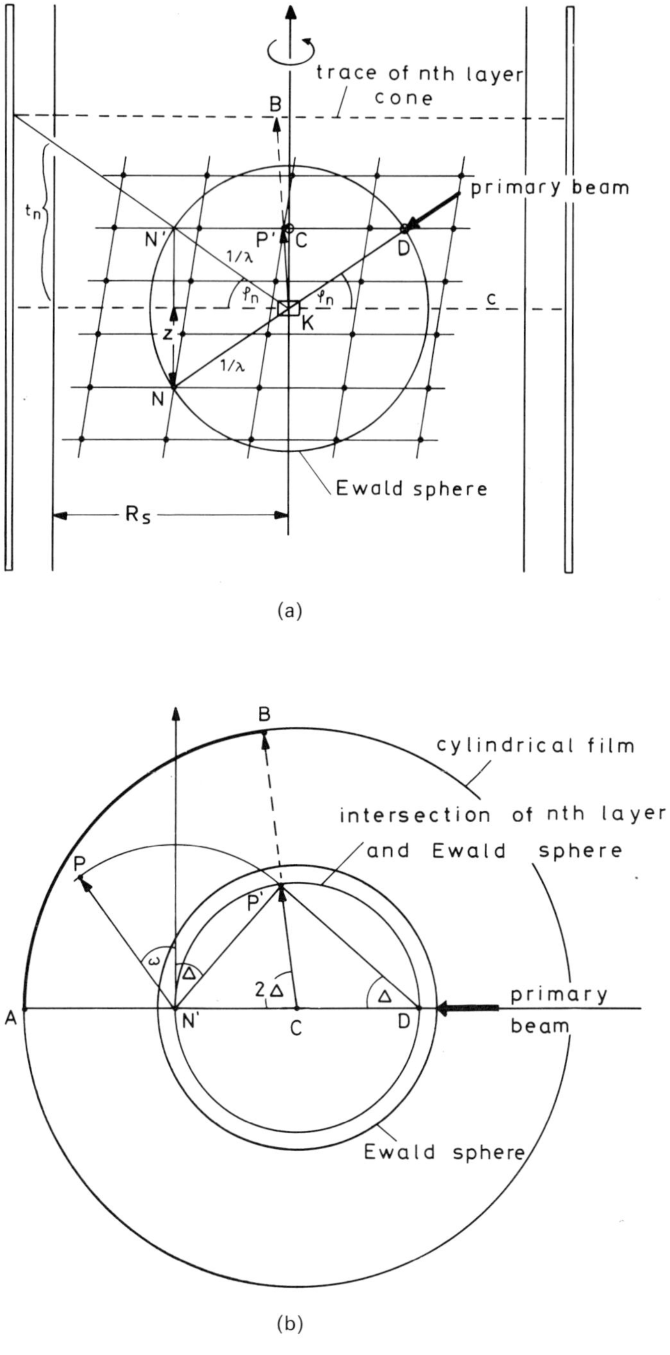

Fig. 2.8. Experimental conditions for the equi-inclination method; (a) view perpendicular to the crystal rotation axis (as in Fig. 2.4); (b) view in the direction of the crystal rotation axis (as in Fig. 2.6).

$$\sin \varphi_n = (\sin \mu_n)/2 \tag{2.11}$$

The reflections of the nth layer having their reflection position at P′ are registered on the film at B (Fig. 2.8a and b). To suppress all but the reflections of the nth layer, the layer line screen has to be shifted by an amount t_n (Fig. 2.8a), with

$$t_n = R_s \tan \varphi_n \tag{2.12}$$

To obtain the polar coordinates r and ω using the upper level Weissenberg technique, let us look again at Fig. 2.8b, which shows the experimental conditions in the plane of the nth layer. It is perpendicular to Fig. 2.8a. There is one significant difference from the situation represented in Fig. 2.6. The reflection angle 2θ is replaced by an angle 2Δ, with 2Δ = angle N′ C P′, which is the projection of 2θ = angle N K P′ onto the plane of the nth layer. All further results derived from Fig. 2.6 are still valid, so we have

$$\omega = x/g - \Delta$$

r can again be derived from the distance $y = \overline{AB}$ on the film. However, y has to be transformed into Δ after (2.6) and $r = \overline{N'P'}$ is no longer equal to $|\mathbf{h}|$, but r is the projection of $|\mathbf{h}|$ onto the plane of the nth layer. We get

$$r = \overline{N'D} \sin \Delta = 2\,\overline{N'C} \sin \Delta$$

From Fig. 2.8a, it can be seen that

$$\overline{N'C} = (1/\lambda) \cos \varphi_n$$

holds. Hence,

$$r = (2/\lambda) \sin \Delta \cos \varphi_n$$

Let us summarize. For examination of n layers (n = 0, 1, 2, ...) of reciprocal lattice by the "equi-inclination" Weissenberg technique, the following procedure must be followed:

(1) After setting the crystal about a special rotation axis, a rotation photograph is taken. The distance D of reciprocal layers perpendicular to the rotation axis is given by

$$D = \frac{n\lambda}{\sin \mu_n} \quad (2.4)$$

with

$$\tan \mu_n = (2l_n)/(2R_F) \quad (2.1)$$

If the rotation axis has the direction of one unit cell vector, (2.4) gives the length of this unit cell vector.

(2) To obtain a Weissenberg photograph of the nth layer of reciprocal lattice, the inclination angle φ_n and the displacement t_n of the layer line screen have to be calculated from

$$\sin \varphi_n = (\sin \mu_n)/2 \quad (2.11)$$

$$t_n = R_s \tan \varphi_n \quad (2.12)$$

For $n = 0$, $\varphi_0 = t_0 = 0$ holds.

(3) The transformation to an undistorted reciprocal lattice can be performed by taking the coordinates (x, y) of each reflection on the film, followed by the calculation of the corresponding polar coordinates (r, ω):

$$\Delta = (y\,180)/(2\pi R_F)\ (= y, \text{ if } R_F = 28.6 \text{ mm})$$
$$\omega = x/g - \Delta$$
$$r = (2 \sin \Delta \cos \varphi_n)/\lambda \qquad (2.13)$$

If $n = 0$, $\Delta = \theta$.

To complete this description, two advantages of the equi-inclination technique against the normal-beam method should be emphasized. From (2.4) and (2.11), it follows that φ_n has to satisfy

$$(n\lambda)/(2D) = \sin \varphi_n \leqq 1$$

It follows that

$$n \leqq (2D)/\lambda \qquad (2.14)$$

A comparison with (2.10) shows that in the case of equi-inclination techniques, the maximum number of layers which can be recorded is twice the number of those in the case of the normal-beam method. A further advantage is illustrated in Fig. 2.9. In the normal-beam method there is a torus-shaped volume element concentric to the rotation axis, in which reflections can never intersect the Ewald sphere. This so-called

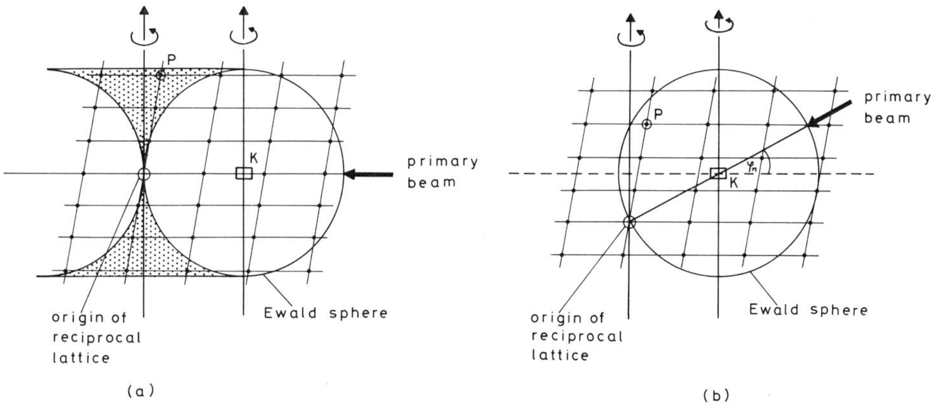

Fig. 2.9. Illustration of the shadowed area:
(a) normal beam method; the lattice point P cannot penetrate the Ewald sphere;
(b) equi-inclination method; P does penetrate the Ewald sphere.

"shadowed area" can be avoided with the equi-inclination technique, as shown in Fig. 2.9b. In this case, the projection of the rotation axis on the nth layer is situated on the surface of the Ewald sphere and therefore every reflection within the limit of Bragg's equation can pass through the Ewald sphere.

2.1.4 *Precession and De Jong-Bouman Technique*

The great disadvantage of the Weissenberg method is that it records a distorted representation of the reciprocal lattice. This can be avoided by the two film methods now to be described. However, these methods also have disadvantages, so that no film method is ideal and the actual problem to be solved often decides which of the various methods is the best. The distortion of the reciprocal lattice representation on a Weissenberg film results from the fact that the movement of the crystal, and therefore the movement of the reciprocal lattice, is different from that of the film. It follows that an undistorted representation can only be obtained if the crystal and the film

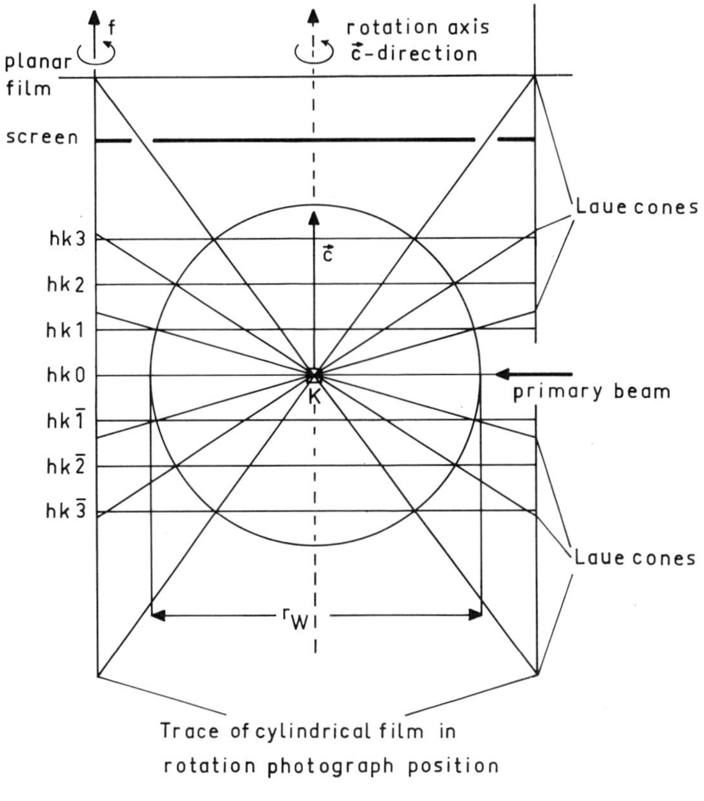

Fig. 2.10. Crystal alignment by the rotation method.

movement are synchronous. This condition is realized in different ways in the de Jong-Bouman and the precession methods.

Let us discuss the de Jong-Bouman method by recalling the experimental conditions for the rotation method which are represented schematically in Fig. 2.10. The crystal is assumed to rotate about a basis vector, say **c**, with **c** aligned perpendicular to the primary beam direction. Then the layer lines hk0, hk1, ... are perpendicular to **c** and, in accordance with Ewald's condition, their intersections with the Ewald sphere result in the reflections being located on a special Laue cone. The registration of the Laue cones on a cylindrical film when unrolled appear as straight lines.

With the same experimental arrangement for the crystal and the primary beam, we can also get an undistorted diffraction pattern with a planar film (see Fig. 2.10) if it is placed parallel to the reciprocal lattice layers and rotated about an axis f (which must be parallel to the crystal rotation axis) synchronous with the crystal. Then with a screen which allows only the Laue cone of one lattice layer to pass, the film registers an undistorted image of one layer (which, in the example in Fig. 2.10, is the third layer line).

This is the concept of the de Jong-Bouman method. However, in practice, this method is realized in a slightly different way, because with the experimental conditions of Fig. 2.10, it would be impossible to take the zero layer without placing the film in the zero layer plane, which cannot be done.

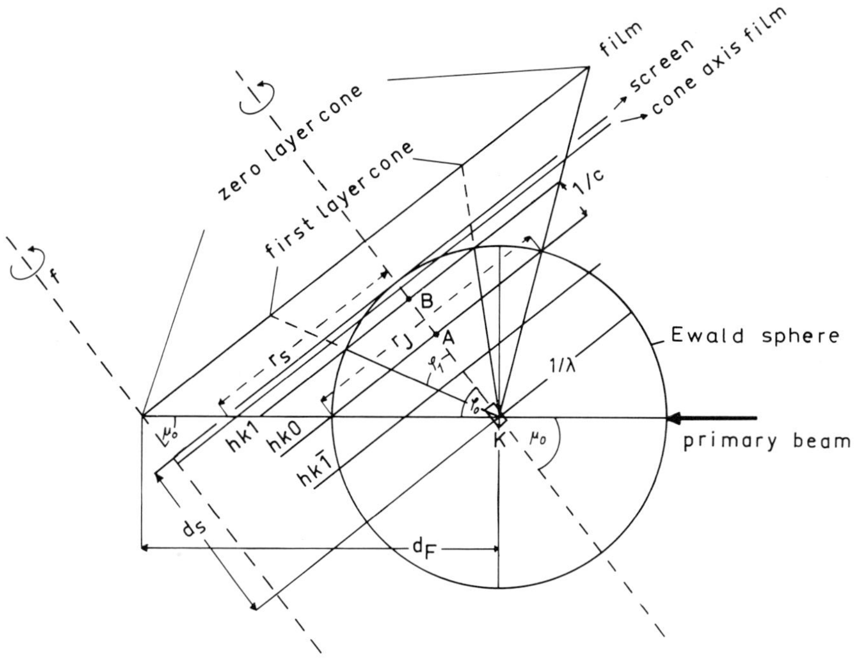

Fig. 2.11. De Jong-Bouman geometry, zero layer.

Therefore, the de Jong-Bouman technique uses the experimental situation shown in Fig. 2.11. It differs from that in Fig. 2.10 only by inclination of the crystal rotation axis against the primary beam direction by an angle μ_0, which is not 90°. If we denote the angles subtended by the Laue cones at the crystal by $2\varphi_0$, $2\varphi_1$, etc., we get a cone opening angle of $2\varphi_0 = 2\mu_0$ for the zero layer. The film is then rotated about an axis f which is inclined by an angle φ_0 against the primary beam direction. To prevent all but the zero layer cone from reaching the film, a screen having an annular opening of radius r_s is placed at a distance d_s from the crystal with

$$d_s = r_s/\tan\varphi_0 \tag{2.15}$$

If crystal and film are rotated synchronously, then an undistorted image of the zero layer appears on the film. The magnification follows from simple geometric proportionality. The ratio of a distance x taken on the film (in mm) to a corresponding distance d^* on the zero layer (in $Å^{-1}$) is equal to the ratio of d_F and $1/\lambda$ (d_F in mm, λ in Å):

$$\frac{x}{d^*} = \frac{d_F}{1/\lambda}$$

or

$$\frac{1}{d^*} = \frac{d_F \lambda}{x} \ [Å] \tag{2.16}$$

On a de Jong-Bouman exposure of zero layer, a spherical area of radius r_J is recorded. It holds that

$$\frac{r_J/2}{1/\lambda} = \sin\varphi_0$$

or

$$r_J = (2/\lambda)\sin\varphi_0$$

For a Weissenberg exposure (see Fig. 2.10), the corresponding area has a radius r_W, equal to the diameter of the Ewald sphere, hence

$$r_W = 2/\lambda$$

Since $|\sin\varphi_0| < 1$ for $\varphi_0 = \mu_0 \neq 90°$, the area covered by a de Jong-Bouman film is always smaller than for a Weissenberg film. In addition, μ_0 cannot be chosen too close to 90°, since this would require a very large film (see Fig. 2.11). Usually, $\mu_0 = \varphi_0 = 45°$ is chosen, for which $\sin\varphi_0 = \sqrt{2}/2$ and

$$\frac{r_J}{r_W} = \frac{\sqrt{2}}{2}$$

Since the corresponding areas F_J and F_W are related by the square of r, it follows that

$$\frac{F_J}{F_W} = \frac{1}{2} \tag{2.17}$$

Under normal experimental conditions, the area recorded on a film of a de Jong-Bouman type has only half the size of the corresponding Weissenberg film. Hence the advantage of getting an undistorted lattice is off-set by a loss of information. For a crystal with small lattice constants, it is therefore recommended that a radiation of short wavelength (eg. Mo$K\alpha$) be used for the de Jong-Bouman technique.

As Fig. 2.11 shows, not only the zero layer but also higher layer Laue cones are present (as long as the corresponding layers intersect the Ewald sphere). It is therefore usual to place a second film between the screen and the crystal to record all Laue cones on one film. An exposure of that type is called a "cone-axis exposure" and contains the same information as a rotation photograph in the Weissenberg arrangement. The only difference is that one layer line appears as a line on a rotation photograph but as a circle on a cone-axis exposure.

The determination of the lattice constant in the rotation direction can then be made from a cone-axis exposure. Suppose that the rotation axis is **c**. Then the distance d* between the zero and first layer is 1/c. From Fig. 2.11 we get

$$\frac{\overline{AK}}{1/\lambda} = \cos\varphi_0$$

$$\frac{\overline{BK}}{1/\lambda} = \cos\varphi_1$$

$$1/c = \overline{BK} - \overline{AK}$$

$$1/c = \frac{\cos\varphi_1}{\lambda} - \frac{\cos\varphi_0}{\lambda} \tag{2.18}$$

If the radii of the zero and first layer circle on the cone-axis photograph are r_0 and r_1 we get

$$\tan\varphi_i = r_i/\Delta \qquad i = 0, 1$$

where Δ is the film to crystal distance. Since Δ is not usually very precisely known, whereas φ_0 is known from the experimental conditions, we get

$$\Delta = r_0/\tan\varphi_0$$

and

$$\tan\varphi_1 = r_1/\Delta = \frac{r_1}{r_0}\tan\varphi_0 \tag{2.19}$$

So we get φ_1 from (2.19) and then c from (2. 18).

The first or a higher layer could be obtained in principle only by adaption of the screen radius to the opening angle of the desired layer. However, as can be seen from Fig. 2.11, the opening angle $2\varphi_1$ of the first layer differs from $2\varphi_0$. So the size of the area of the first layer appearing on the film is quite different from that of the zero layer. Since it is desirable to have areas of similar size for each layer on the film, the

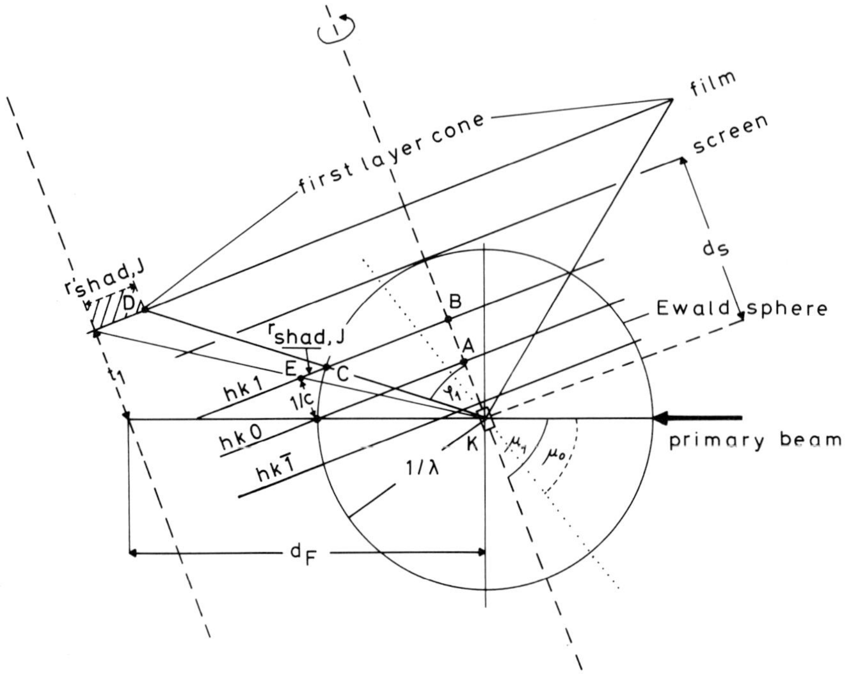

Fig. 2.12. De Jong-Bouman technique for taking first layer photographs.

experimental conditions are altered with each exposure to get equal opening angles for all cones. This can be achieved by enlarging the inclination angle (Fig. 2.12). It follows from (2.18) that φ_1 depends on φ_0, but φ_0 is equal to the inclinaton angle μ_0. If we denote the inclination angle for the first layer by μ_1, then from (2.18), when $\varphi_1 = \varphi_0$

$$\cos\mu_1 = \cos\varphi_0 - \lambda/c \tag{2.20a}$$

or for an arbitrary layer,

$$\cos\mu_n = \cos\varphi_0 - (n\lambda)/c \tag{2.20b}$$

For $\varphi_0 = \mu_0 = 45°$ we get all opening angles $2\varphi_n = 90°$ if

$$\cos\mu_n = \sqrt{2}/2 - (n\lambda)/c \tag{2.20c}$$

Since the layer to crystal distance is different for the first layer from that for the zero layer, a different magnification factor would result if the film were left at the same distance as for the zero layer. Therefore we displace the film by a shift t_1 (see Fig. 2.12), which is given by

$$\frac{t_1}{1/c} = \frac{d_F}{1/\lambda}$$

$$t_1 = \frac{\lambda d_F}{c} \tag{2.21a}$$

or for the nth layer

$$t_n = \frac{n\lambda d_F}{c} \tag{2.21b}$$

Then the ratio of a distance x taken on the film to the corresponding distance d* is given by

$$\frac{x}{d^*} = \frac{\overline{DK}}{\overline{CK}} = \frac{d_F}{1/\lambda}$$

or

$$\frac{1}{d^*} = \frac{d_F \lambda}{x}$$

which is equal to (2.16), and hence we have the same magnification factor.

A further advantage follows from always using the same opening angle. The position of the annular screen need not be changed once it has been calculated from (2.15).

Let us summarize for the de Jong-Bouman technique:

(1) Adjust the crystal with a rotation axis perpendicular to the primary beam as in the Weissenberg technique.

(2) Choose an inclination angle μ_0 ($\mu_0 = 45°$ is a good choice) for the zero layer, place the film at a distance d_F, calculate the screen position from (2.15), and place a second film on the crystal side of the screen to obtain a cone-axis exposure simultaneously.

(3) To get the instrumental setting constants for the first (and higher layers), determine the circle radii from the cone-axis exposure, calculate c [(2.19) and (2.18)]. Then calculate μ_n and t_n from (2.20) and (2.21) and use them when taking the upper layers.

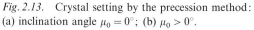

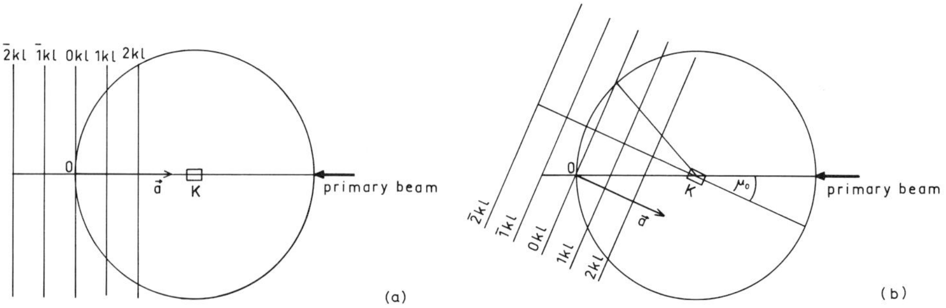

Fig. 2.13. Crystal setting by the precession method: (a) inclination angle $\mu_0 = 0°$; (b) $\mu_0 > 0°$.

The other film method which allows the recording of the undistorted lattice is the "precession method" developed by Buerger in 1942 (a detailed description is given in: M. J. Buerger, "The Precession Method", (1964), New York: John Wiley & Sons). The basic experimental situation is illustrated in Fig. 2.13. In contrast to all methods discussed so far, the crystal is assumed to be aligned with one axis, say **a**, in the direction of the primary beam (Fig. 2.13a). The reciprocal lattice layers 0kl, 1kl, etc., are then oriented perpendicular to the primary beam.

Consider the zero layer. To make this layer intersect the Ewald sphere, it is necessary to incline the crystal with its a-axis by an angle μ_0 against the X-ray beam direction (Fig. 2.13b). Since reflections of zero layer lattice points are obtained only if these points cut the surface of the Ewald sphere, movement of the crystal is necessary.

In the Buerger precession method, the crystal executes a precession movement about the primary beam axis keeping a constant opening angle μ_0 between the crystal a-axis and the primary beam direction. One cycle of crystal movement on a precession apparatus is illustrated in Fig. 2.14 a–e.

Suppose that the a-axis is situated in the plane of the paper, so that the vector **k** from the crystal to the surface of the Ewald sphere has no component outside the plane of the paper. The top of **k** is then identical with the point B_1 (Fig. 2.14a). The position of this vector is given in a separate diagram on the right showing the situation when looking from O to the crystal. The zero layer is perpendicular to the line through A_3, O and A_1, and is shown in the plane by rotation about a right angle along this line. It can now be seen that the area of the reciprocal lattice layer inside the Ewald sphere is given by the small circle of diameter $\overline{OA_1}$.

A quarter cycle of motion changes the situation to that represented in Fig. 2.14b. The vector **k** now points to B_2 situated above the plane of the paper. The portion of the reciprocal lattice plane inside the Ewald sphere is now given by the small circle drawn as a full line having the diameter $\overline{OA_2}$. After a further quarter cycle, the situation is opposite to that of the start of the experiment (Fig. 2.14c). The vector **k** pointing to B_3 is again situated totally in the plane of the paper. The circle of diameter $\overline{OA_3}$ indicates the area inside the Ewald sphere.

The situation after the execution of three quarters of one cycle is illustrated in Fig. 2.14d. The vector **k** now points to B_4 being below the plane of the paper. The circle having the diameter $\overline{OA_4}$ indicates that section of the reciprocal lattice plane inside the Ewald sphere.

Finally, we find the situation after one complete cycle has been executed (Fig. 2.14e). It is identical to that of the starting situation. Note that in all representations of Fig. 2.14, we always observe the same crystal face due to the absence of any crystal rotation.

As can be seen from the reciprocal lattice sections drawn on the left, every lattice point inside the *large* circle of diameter $\overline{A_2A_4}$ (or $\overline{A_1A_3}$) penetrates the Ewald sphere once and leaves it once if one cycle of precession movement is executed. So we get reflections of all lattice points inside this circle. During one cycle, each lattice point is in reflection position twice. Denoting the radius $\overline{OA_i}$ (i = 1, ..., 4) of the large circle

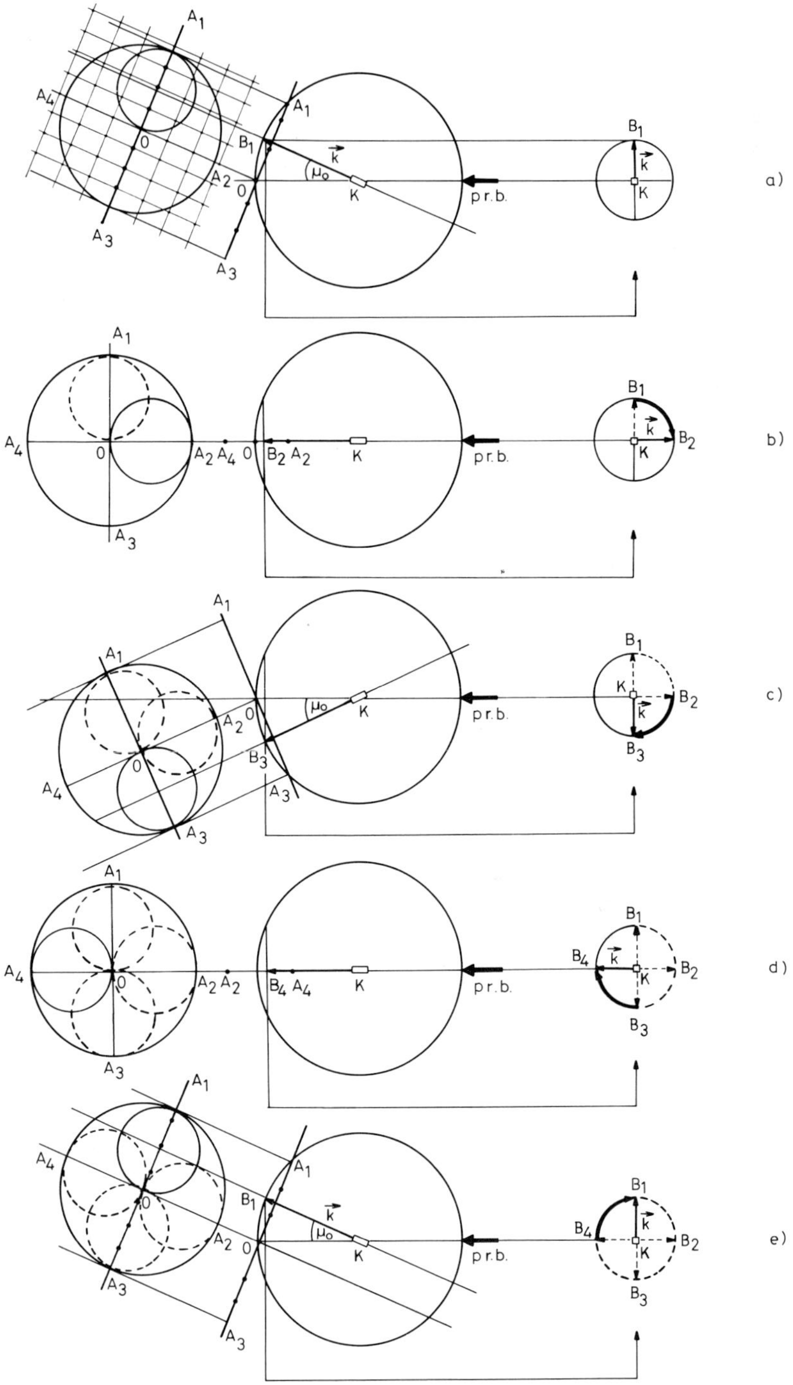

Fig. 2.14. One cycle of crystal and corresponding reciprocal lattice plane movement on a precession device.

by r_B, we get $r_B = (2\sin\mu_0)/\lambda$, which is identical to the formula for the de-Jong-Bouman radius r_J.

We can now explain the function of a precession camera. As in the de Jong-Bouman method, an undistorted exposure of one reciprocal plane is obtained if a screen is positioned so that only reflections of one lattice plane are able to pass and if a planar film simultaneously performs the same motion as the crystal.

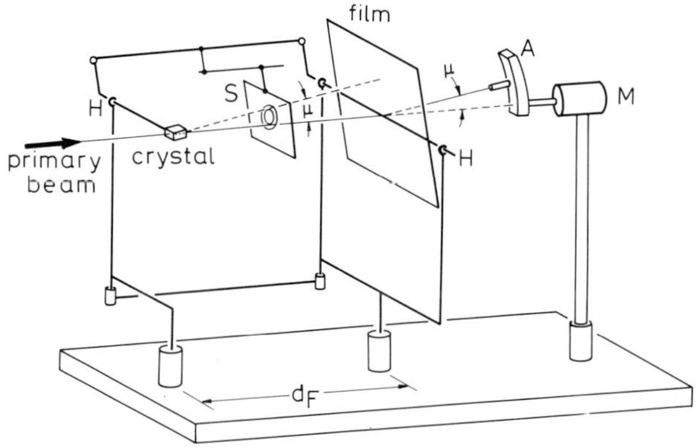

Fig. 2.15. Schematic representation of a Buerger precession camera.

A schematic representation of a Buerger precession camera is given in Fig. 2.15. The principal parts of this camera are the crystal and the film holders H, which are coupled to ensure a simultaneous motion. An annular screen S, which is attached rigidly to the crystal holder, obstructs all but the reflections of one layer. The inclination angle μ is adjusted via an arc A, which is connected with a motor M. The rotation of the motor shaft then causes the precession motion of the complete system of film, crystal and screen. To choose the geometric dimensions, let us look at Fig. 2.16, which shows the geometry in the case of a zero layer precession exposure. The arrangement is similar to that shown in Fig. 2.11 and we get nearly the same geometrical properties. The screen distance is again

$$d_s = \frac{r_s}{\tan\mu_0} \quad (2.22a)$$

and the magnification factor is $(d_F\lambda)$, so that for the relation of a distance d* on the zero layer to the corresponding distance x on the film (2.16) is valid:

$$\frac{1}{d^*} = \frac{d_F\lambda}{x}\ [\text{Å}] \quad (2.16)$$

or

$$d^* = \frac{x}{d_F\lambda}\ [\text{Å}^{-1}] \quad (2.16a)$$

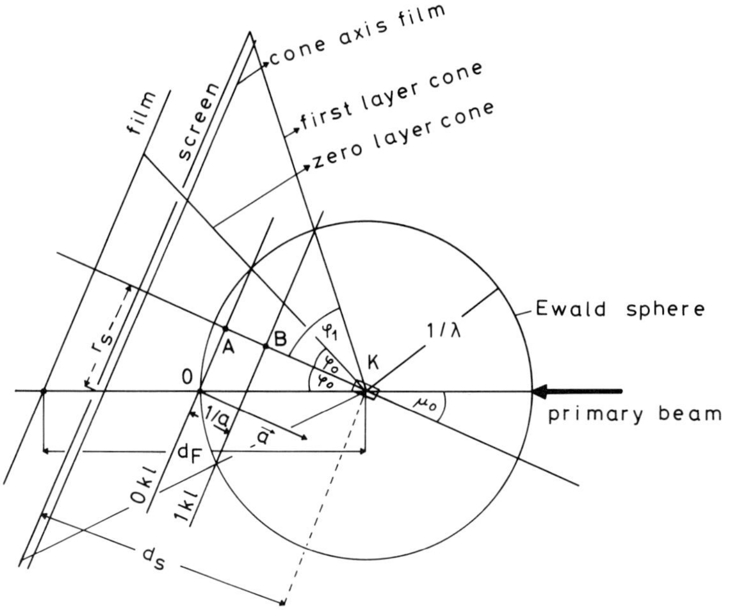

Fig. 2.16. Geometrical situation for taking a zero-layer precession exposure.

As in the case of the de Jong-Bouman technique, a cone-axis exposure can be taken together with the zero-layer precession exposure, if a second film is placed before the screen as illustrated in Fig. 2.16. Since the first layer is now supposed to be closer to the crystal, and its cone opening angle $2\varphi_1$ is larger than that of the zero layer, we get

$$1/a = \overline{AK} - \overline{BK} = \frac{\cos\mu_0}{\lambda} - \frac{\cos\varphi_1}{\lambda} \tag{2.23}$$

As for the de Jong-Bouman technique, we have $\varphi_0 = \mu_0$. Getting φ_1 from (2.19) we then get a from (2.23).

For an upper level precession photograph (see Fig. 2.17), the precession angle μ_1 can be chosen arbitrarily. However, it is usually chosen to be smaller than μ_0, since otherwise the large cone opening angle would require too large a film holder, resulting in a collision during the precession motion. The screen distance d_s for a screen of radius r_s is now given by

$$d_{s,1} = \frac{r_s}{\tan\varphi_1} \tag{2.22b}$$

where φ_1 is the opening angle of the first layer cone obtained as in (2.20) from

$$\cos\varphi_1 = \cos\mu_1 - \lambda/a \tag{2.24a}$$

So that the scale of the zero- and first-level film is identical, the film must be displaced

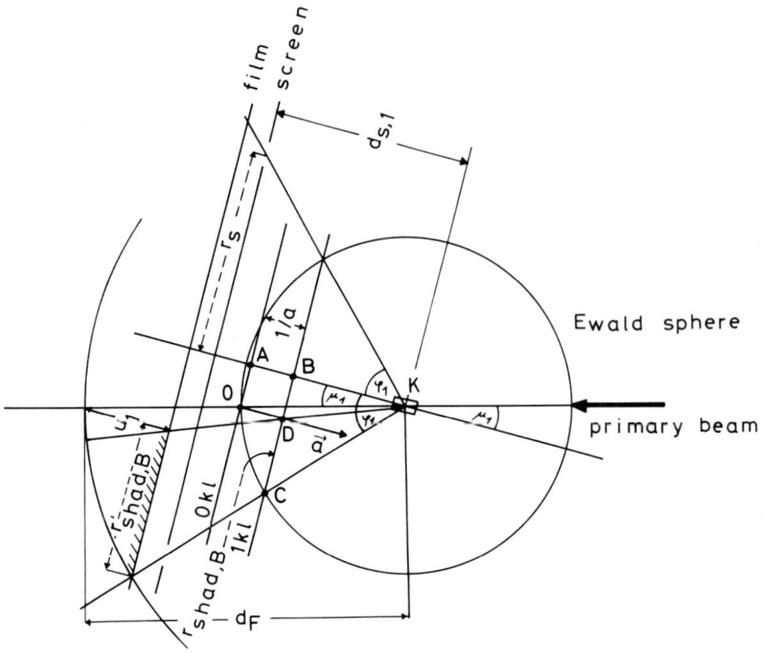

Fig. 2.17. Geometrical conditions for a first-layer precession exposure.

towards the crystal by an increment u_1, which is identical in magnitude to t_1 given by (2.21); hence

$$u_1 = \frac{\lambda d_F}{a} \tag{2.25a}$$

Note that the displacements t_1 and u_1 have *opposite* directions. Generalizing the results for the first layer to the nth layer, we get

$$d_{s,n} = \frac{r_s}{\tan \varphi_n} \tag{2.22c}$$

with

$$\cos \varphi_n = \cos \mu_n - (n\lambda)/a \tag{2.24b}$$

$$u_n = \frac{n \lambda d_F}{a} \tag{2.25b}$$

Let us now summarize the procedure for obtaining a series of precession phographs:

(1) Adjust the crystal with one axis, say **a**, *parallel* to the primary beam.

(2) Choose an inclination angle μ_0 ($\mu_0 = 30°$ is a good choice), place a film in the film holder at a distance d_F (d_F is an instrument constant, usually $d_F = 60$ mm). Choose an annular screen of radius r_s (screens with radii from 20 to 60 mm are generally

available) and calculate d_s from (2.22a). Place a second film on the crystal side of the screen to obtain a cone-axis exposure simultaneously.

(3) To get the instrumental setting constants for the first and higher layers, determine the circle radii from the cone-axis exposure and calculate a from (2.23).

(4) Choose an inclination angle μ_n (smaller than μ_0) for the upper layers. Calculate $d_{s,n}$ from (2.22c) and u_n from (2.25b) and use them when taking the upper layers. Note that it may be necessary to vary the screen radius r_s when recording the upper layers.

It is of interest to compare the areas of the reciprocal lattice recorded when using these different techniques. The area covered by a de Jong-Bouman exposure with a reasonable inclination angle of 45° is only half that obtained when using the Weissenberg technique. The radius r_B of the circle limiting that portion of the zero layer to be recorded by the precession technique is given by the same formula $[r_B = (2\sin\mu_0)/\lambda]$ as in the case of the de Jong-Bouman technique. However, the precession angle μ_0 can be smaller and is generally not larger than 30°, to avoid collisions of the moving parts of the instrument. With $\sin 30° = 0.5$, we get $r_B = 1/\lambda$ in comparsion with $r_W = 2/\lambda$ and $r_J = \sqrt{2}/\lambda\,(\mu_0 = 45°)$. So we get the relations

$$r_B : r_J : r_W = 1 : \sqrt{2} : 2 \qquad (2.26a)$$

and for the corresponding areas,

$$F_B : F_J : F_W = 1 : 2 : 4 \qquad (2.26b)$$

if the usual inclination angles $\mu_0 = 30°$ for the precession and $\mu_0 = 45°$ for the de Jong-Bouman technique are assumed and if the wavelength is kept constant. For both techniques, which provide undistorted representations of the reciprocal lattice layers, there is significant loss of information. This disadvantage can be compensated for by using a shorter wavelength. If for instance a Weissenberg exposure is taken with CuKα radiation ($\lambda = 1.5418$ Å), a precession photograph of the same reciprocal lattice layer would record the same number of reflections if MoKα radiation ($\lambda = 0.7107$ Å) were used, since $\lambda_{Mo} : \lambda_{Cu} \approx 1 : 2$.

A further drawback of the de Jong-Bouman and precession methods should be mentioned. For both techniques, central circular areas exist for all the upper layers which do not intersect the Ewald sphere (see an example in Fig. 2.18). It is therefore impossible to record these "shadowed areas" by either of these methods. In the case of the de Jong-Bouman method, this shadowed area, marked by its radius $r_{shad,J} = \overline{EC}$ on Fig. 2.12, is situated outside the Ewald sphere and can never penetrate when the crystal and its reciprocal lattice are rotated. $r_{shad,J}$ is given by

$$r_{shad,J} = \overline{EC} = \overline{EB} - \overline{CB} = \overline{OA} - \overline{CB} = \frac{\sin\mu_1}{\lambda} - \frac{\sin\varphi_1}{\lambda}$$

The image of the shadowed circle then has the radius $r'_{shad,J}$ with

$$r'_{shad,J} : r_{shad,J} = d_F : 1/\lambda$$

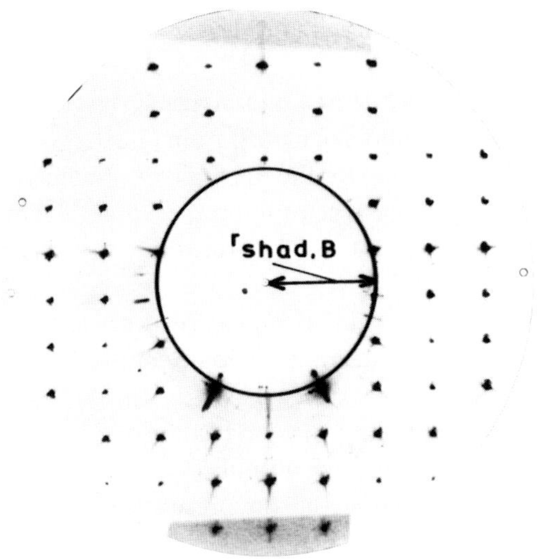

Fig. 2.18. Example of an upper level precession exposure, showing a large central shadowed area.

So we get, for the de Jong-Bouman method,

$$r_{\text{shad, J}} = 1/\lambda(\sin\mu_1 - \sin\varphi_1) \qquad (2.27a)$$

$$r'_{\text{shad, J}} = d_F\,(\sin\mu_1 - \sin\varphi_1) \qquad (2.27b)$$

In the case of the precession technique, a circular area of radius $r_{\text{shad, B}} = \overline{CD}$ (see Fig. 2.17) is always situated *inside* the Ewald sphere. We get,

$$r_{\text{shad, B}} = \overline{CB} - \overline{DB} = \overline{CB} - \overline{OA} = \frac{\sin\varphi_1}{\lambda} - \frac{\sin\mu_1}{\lambda}$$

and

$$r'_{\text{shad, B}} : r_{\text{shad, B}} = d_F : 1/\lambda$$

Hence, for the precession method

$$r_{\text{shad, B}} = 1/\lambda\,(\sin\varphi_1 - \sin\mu_1) \qquad (2.28a)$$

$$r'_{\text{shad, B}} = d_F\,(\sin\varphi_1 - \sin\mu_1) \qquad (2.28b)$$

The question can arise: which method should be used for an actual problem? A general answer cannot be given. With modern techniques, the film methods are used only for space group determination and for examination of crystal quality. Therefore, the information obtained from precession or de Jong-Bouman exposures is sufficient in most cases, even if CuKα radiation is used. On the other hand, for an experienced crystallographer, the distorted representation of a Weissenberg exposure causes no great problem, and this technique might be preferred since a Weissenberg instrument is less troublesome to operate. Because the mechanical parts of a Weissenberg camera

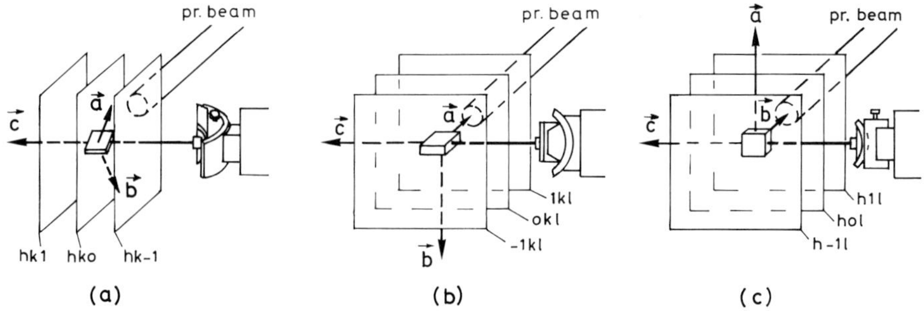

Fig. 2.19. Combination of different film methods for a crystal with orthogonal lattice constants. (a) Weisssenberg and de Jong-Bouman technique: $\vec{c}$ perpendicular to the primary beam, hk0, hk1, ... layers are recorded;
(b) precesssion technique: $\vec{a}$ parallel to the primary beam, 0kl, 1kl, ... layers are recorded;
(c) precession technique: $\vec{b}$ parallel to the primary beam, h0l, h1l, ... layers are recorded.

are less complicated than the other two instruments, the problems of misalignment, due to mechanical breakdown or mechanical wear, are less frequent.

The most effective film technique, however, is to combine two of these methods. As was mentioned above, the Weissenberg and de Jong-Bouman techniques need a crystal axis orientation normal to the X-ray beam direction. This automatically implies the orientation of both of the other axes in a plane containing the primary beam for all orthogonal crystal systems, and also for a monoclinic crystal if the monoclinic axis was chosen as the rotation axis. Therefore, only a further azimuthal setting is necessary to orient the crystal for a precession photograph. The principal experimental condition before adjusting the inclination angle is illustrated in Fig. 2.19a. Suppose a crystal with orthogonal axes has been set with the c-axis as the rotation axis. Then this axis is normal to the incident beam. The Weissenberg and de Jong-Bouman techniques then provide the experimentor with reciprocal lattice planes of type hk0, hk1, ... etc. The a- and b-axis and the primary beam are then in one plane and both axes can be brought into the X-ray beam direction by a simple rotation of the crystal around the c-axis. The precession photographs then record the 0kl, 1kl ... or the h0l, h1l, ... planes (Fig. 2.19b and c).

Hence only one crystal setting is necessary to obtain information about three mutual orthogonal reciprocal lattice planes, if a Weissenberg or de Jong-Bouman technique is combined with the precession technique. The resulting information is sufficient in most cases for a complete knowledge of all properties of the reciprocal lattice.

A camera designed to take both de Jong-Bouman and precession photographs has been developed recently by Wölfel [Wölfel, E.R., J. Appl. Cryst. 4, 297 (1971)]. This device, named the "Reciprocal Lattice Explorer" is a most convenient apparatus for combining different film techniques using only one crystal. We shall describe the principles of using this equipment when dealing with the application of film techniques in practice (2.3.3).

2.2 X-Rays

2.2.1 *Generation of X-Rays*

X-rays, first discovered by Wilhelm Conrad Roentgen on November 8, 1895, are electromagnetic radiation with wavelengths λ in a range $0.1 < \lambda < 100\,\text{Å}$ ($1\,\text{Å} = 10^{-10}\,\text{m}$). X-rays are produced when electrons of high speed hit a target material and are rapidly decelerated. Their kinetic energy

$$E = \frac{m}{2} v^2$$

is then transformed into X-radiation. Since the X-ray quanta absorb the incident energy not completely, their energies $E = (hc)/\lambda$ are spread over a large range and form a continuous spectrum of so-called "white radiation".

Another radiation type of high intensity is obtained if the incident electrons have sufficient energy to remove an electron from an inner orbital of a target atom. Since the excited atom is very unstable, it completes its inner orbital with an electron of an outer shell. The energy difference ΔE gained from the transition of an outer to an inner shell is transformed into X-radiation with a discrete wavelength λ corresponding

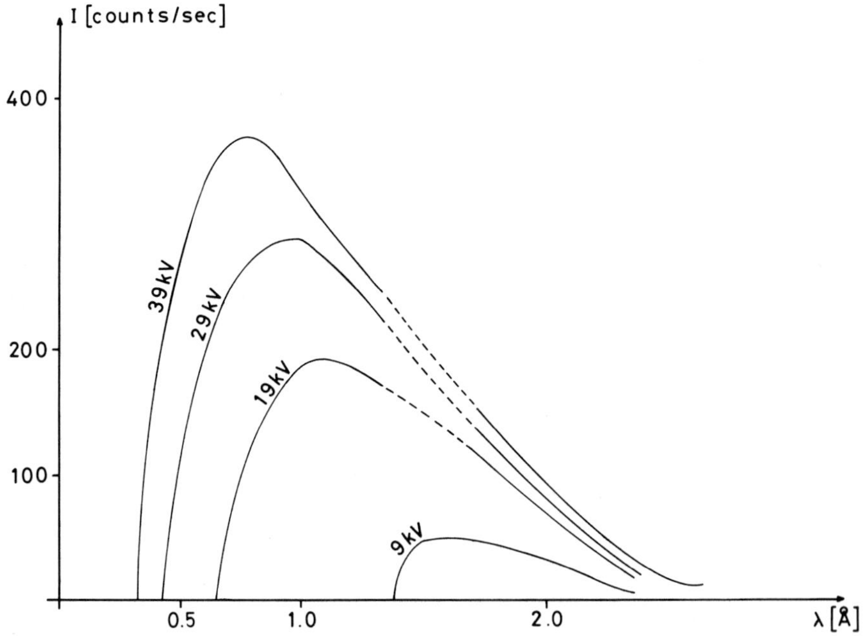

Fig. 2.20. Experimentally measured intensity distribution of white radiation plotted versus λ for various voltages V. Target material, Cu. The interval plotted by dotted lines is superimposed by the characteristic radiation.

to ΔE, which is named "characteristic" radiation:

$$\frac{hc}{\lambda} = \Delta E$$

Fig. 2.20 shows the intensity distribution I of white radiation plotted versus λ for various voltages V. There is a characteristic limit for a minimum wavelength λ_{min} for every voltage. This wavelength corresponds to the maximum energy gained if the complete kinetic energy

$$E = \frac{m}{2} v^2 = eV$$

of incident electrons is transformed into radiation energy (hc)/λ. Then we get

$$eV = h \frac{c}{\lambda_{min}}$$

or

$$\lambda_{min} = \frac{hc}{eV}$$

Taking the numerical values for the constants

$$h = 6.625 \times 10^{-34} \text{ J sec}$$
$$c = 2.998 \times 10^{8} \text{ m sec}^{-1}$$
$$c = 1.602 \times 10^{-19} \text{ C}$$

we obtain

$$\lambda_{min} = \frac{12.4}{V[kV]} [Å] \tag{2.29}$$

as the minimum wavelength of white radiation to be obtained with a given voltage V. Formula (2.29) is called the Duane-Hunt-law. Another feature, which can be derived from Fig. 2.20, is that the position of intensity maximum shifts toward smaller wavelengths with increasing voltage. For a fixed voltage the maximum intensity is at a λ-value of about 1.5 times λ_{min}.

A good approximation for the white radiation intensity I_W in terms of V, the electronic current i and the atomic number Z of target material is given by

$$I_W = cV^2 iZ \tag{2.30}$$

where c is a constant.

For single crystal diffraction experiments, we use a fixed value for λ, and are more interested in the monochromatic lines of characteristic radiation than in the continuous radiation. To understand why these lines appear and what types of lines can appear, we have to consider the quantum mechanical model of discrete atomic energy levels.

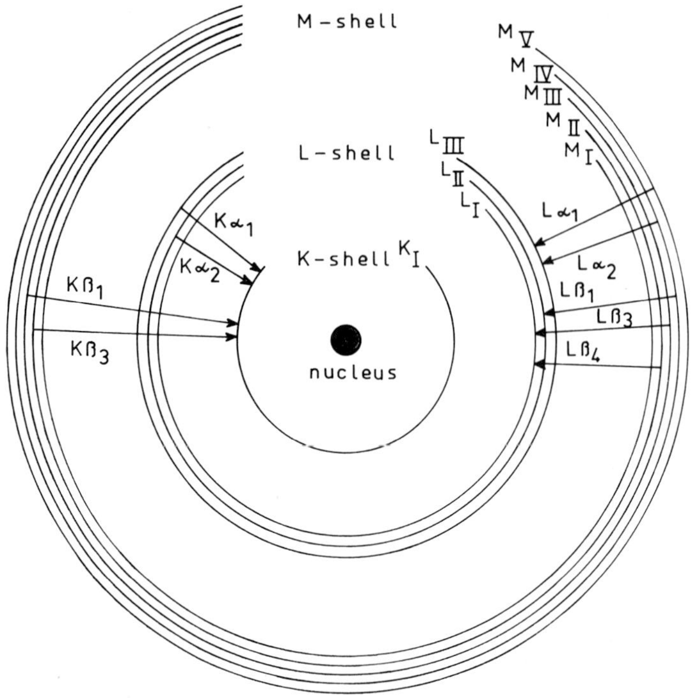

Fig. 2.21. Energy levels and transitions causing characteristic X-ray lines.

In Fig. 2.21, the three inner electronic shells are displayed, named K, L, M, corresponding to the principal quantum numbers n = 1, 2, 3. Every shell is separated into 2n-1 discrete energy levels with respect to the three further quantum numbers l, m and s. So we get one level for the K-shell, three for the L-shell, five for the M-shell, etc. Every transition of an outer to an inner level causes a characteristic line. However, there are "selection rules" which make some transitions impossible, for instance that from the inner L-shell to the K-shell.

The nomenclature of characteristic lines is defined in the following way: The radiation type is defined by the capital letter of the accepting shell. If the donor shell is the neighbouring shell, the radiation is said to be α-radiation; if it is the next shell, it is called β-radiation (although for historical reasons, there are some exceptions). The most important radiation types for the purpose of X-ray diffraction are the $K\alpha_1$ and $K\alpha_2$ and the $K\beta$-radiation. $K\alpha_1$ and $K\alpha_2$ wavelengths result from the two possible transitions of the L_{II} and L_{III}-level to the K-shell. Since the energy difference between the two L-levels is small, the wavelengths of $K\alpha_1$- and $K\alpha_2$-radiation are close together. Therefore, under some experimental conditions, the two lines are not separated and coalesce in one line, the $K\alpha$-line. The same holds for the $K\beta$-line. Actually, we have a doublet of the β_1 and β_3 line due to the $M_{III} \rightarrow K$ and $M_{II} \rightarrow K$ transition, but in practice they are not separated, and are therefore always used as

$K\beta$-radiation. Although a large number of further lines exists, they are of no practical importance in crystal structure analysis because of their weak intensities. Since an electron jump to the K-shell is more likely from the L-shell than from the M-shell, the $K\alpha$-line has the greatest intensity.

For the wavelength λ of a characteristic line, Balmer's formula holds

$$\frac{1}{\lambda} = R(Z-\sigma)^2 \left(\frac{1}{n^2} - \frac{1}{m^2}\right) \tag{2.31}$$

with $R = 1.097 \times 10^5\ \text{cm}^{-1}$ = Rydberg constant, n, m are the principal quantum numbers of participating shells, and σ is a screening constant taking repulsion of other electrons into account. For $n = 1$ and $m \to \infty$ we get a special wavelength λ_K with the corresponding energy

$$\Delta E_K = h \frac{c}{\lambda_K}$$

necessary to remove an electron from the K-shell to infinity. It follows from (2.29) that a minimum voltage V_K of

$$V_K = \frac{12.4}{\lambda_K} \tag{2.32}$$

is necessary for the excitation of λ_K. This voltage, V_K, is called the excitation potential of the K-series.

For fixed n and m, the factor

$$R\left(\frac{1}{n^2} - \frac{1}{m^2}\right)$$

is a constant p and we get

$$\frac{1}{\lambda} = p(Z-\sigma)^2 \tag{2.33}$$

This is Moseley's law. It expresses the property whereby the characteristic radiation wavelength decreases with the reciprocal second power of the atomic number Z. Since $1/\lambda$ is proportional to the radiation energy and since radiation is said to become harder with increasing energy, we can state that the degree of hardness of a radiation increases with the second power of Z. For the intensity I_K of K-radiation, a good approximation is given by

$$I_K = L(V - V_K)^{1.5}\, i \tag{2.34}$$

where L has a particular value for every type of line, depending on the transition probability of the corresponding pair of shells. For the $K\alpha_1 - K\alpha_2$ doublet, the intensity ratio is approximately

$$\frac{I(K\alpha_1)}{I(K\alpha_2)} = 2 \tag{2.35}$$

The ratio of integral Kα-radiation to integral Kβ-radiation is approximately

$$\frac{I(K\alpha)}{I(K\beta)} = 5 \tag{2.36}$$

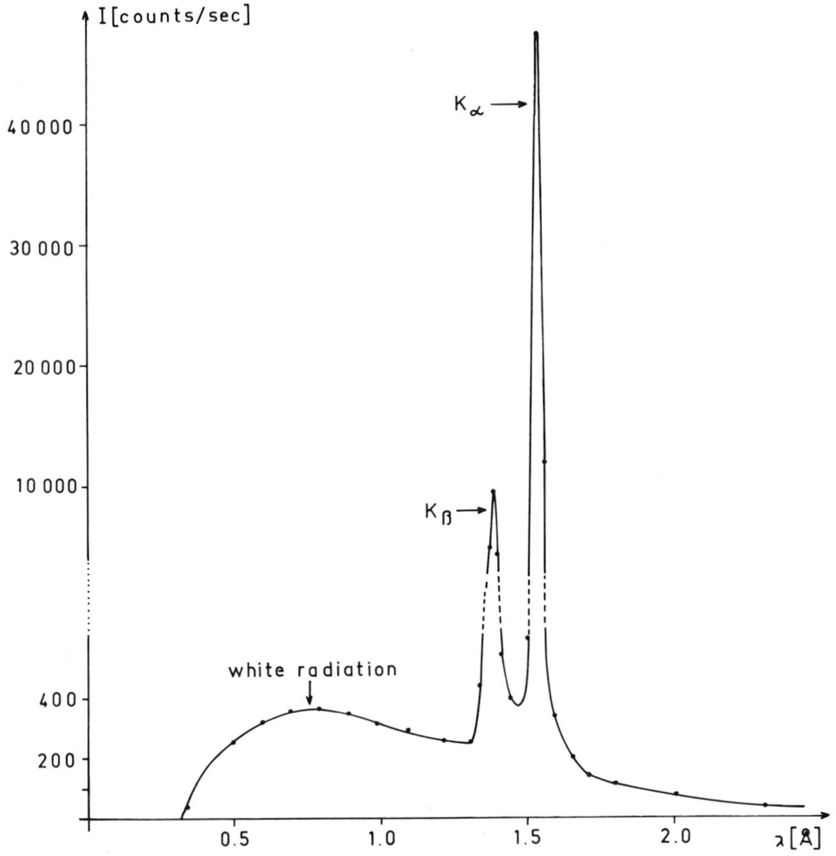

Fig. 2.22. Complete spectrum of white and characteristic radiation from a copper X-ray tube, taken experimentally at 39 kV.

Fig. 2.22 shows a complete intensity diagram of the continuous and characteristic Kα- and Kβ-radiation. Since we intend to use the monochromatic radiation, we try to reduce, or better, to remove, all but the Kα-radiation. The optimum ratio, Q, of characteristic to white radiation, in terms of V, is given by setting its first derivative equal to zero, where, from the ratio of (2.34) and (2.30),

$$Q = \frac{I_K}{I_W} = \frac{L}{c}\frac{(V - V_K)^{1.5} i}{V^2 i Z} = L' \frac{(V - V_K)^{1.5}}{V^2}$$

Then

$$0 = \frac{dQ}{dV} = L' \frac{1.5(V - V_K)^{0.5} V^2 - 2V(V - V_K)^{1.5}}{V^4}$$

It follows that

$$1.5\,V = 2(V - V_K)$$

or

$$V = 4V_K \tag{2.37}$$

From (2.37) we see that we have to apply a voltage four times V_K to get a best ratio of I_K and I_W. However, the white and β-radiation are not removed. This is done, at least in parts, by absorption, discussed in 2.2.2.

2.2.2 *Absorption*

X-rays are attenuated when passing through matter. This loss of radiation is due to two effects, photoelectrical absorption and scattering. If an X-ray beam of primary intensity I_0 passes a sample of homogenous material of thickness dx, the loss dI at the element dx is proportional to I_0 and dx.

$$dI = -\mu I_0\, dx$$

By integration, we get

$$I = I_0 e^{-\mu x} \tag{2.38}$$

an expression called Beer's law. The factor μ, having the dimension of a reciprocal length, is called the "linear absorption coefficient". Taking the two sources of absorption into account, μ is sometimes decomposed into a sum of $\mu = s + t$, with s as the scattering and t as the absorption coefficient. For our purposes, it is sufficient to operate with μ only.

Absorption is an additive atomic property of matter, which means that its magnitude does not depend on the physical or chemical state of the atoms. The absorption coefficient of any kind of matter can therefore be calculated by a simple addition, if the absorption coefficients of contributing elements are known.

In addition to the linear absorption coefficient, two further types of absorption coefficients are used. The first, denoted by μ_m, is the ratio of μ and the density ϱ, and is called the "mass absorption coefficient",

$$\mu_m = \frac{\mu}{\varrho} \tag{2.39}$$

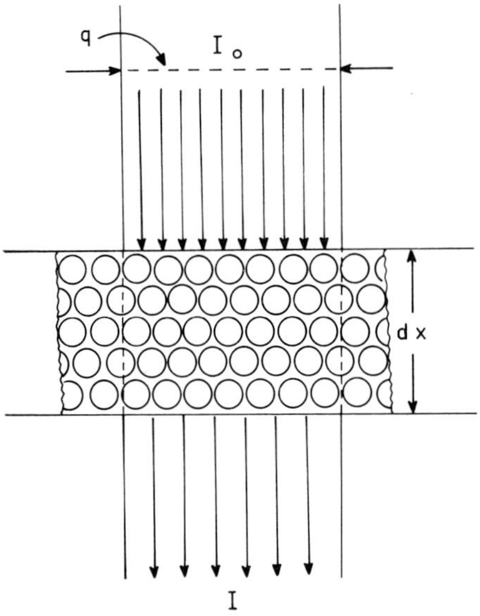

Fig. 2.23. Attenuation of an X-ray beam by a layer of atoms.

The second, called the atomic absorption coefficient, is useful in taking the atomic properties of absorption into account. It is denoted μ_a and defined by the following consideration (see Fig. 2.23). If an X-ray beam of cross-section q passes through a sample of matter with thickness dx, the attenuation dI is proportional to the number dN of atoms passed:

$$dI = -\mu_a I_0 dN$$

The volume element $dV = q dx$ may contain dM atoms, then $dN = dM/q$. With L being Avogadro's number, A the atomic weight, and ϱ the density, we get

$$dM = \frac{dm L}{A}$$

if $dm = \varrho dV$ is the mass of dV. Then

$$dN = \frac{dm L}{qA} = \frac{\varrho dV L}{qA} = \frac{\varrho q dx L}{qA} = \frac{\varrho L}{A} dx$$

hence $dI = -\mu_a \frac{\varrho L}{A} I_0 dx$, or by integration

$$I = I_0 \exp\left(-\mu_a \frac{\varrho L}{A} x\right)$$

Comparing this result with (2.38) and (2.39), we get

$$\mu = \mu_a \frac{\varrho L}{A}$$

$$\mu_m = \mu_a \frac{L}{A} \tag{2.40}$$

$$\mu_a = \frac{A}{L} \frac{\mu}{\varrho}$$

Since μ has the dimension of a reciprocal length, the dimensions of μ_m and μ_a are $cm^2\, g^{-1}$ and cm^2. The diension of μ_a is that of a "cross-section" and therefore μ_a is sometimes called the "atomic cross-section for absorption". Since absorption depends only on the atomic properties of matter, the molecular absorption coefficient μ_{mol} is

$$\mu_{mol} = \sum_i N_i \mu_{a_i} = \frac{1}{L} \sum_i N_i \left(\frac{\mu}{\varrho}\right)_i A_i$$

if the molecule is composed of N_i atoms of type i. An analogue to (2.40), $\mu_{m,mol}$ is defined by

$$\mu_{m,mol} = \mu_{mol}(L/W)$$

if W is the molecular weight. Then we get for an arbitrary compound

$$\frac{\mu}{\varrho} = \mu_{m,mol} = \sum_i N_i \frac{A_i}{W} \left(\frac{\mu}{\varrho}\right)_i$$

Setting $g_i = \frac{N_i A_i}{W}$, we get

$$\frac{\mu}{\varrho} = \sum_i g_i \left(\frac{\mu}{\varrho}\right)_i \tag{2.41}$$

with g_i being the mass fraction of element i contributing to the compound.

In terms of μ_a for an absorber of volume V, we get for μ, with (2.40):

$$\mu = \mu_{mol} \frac{\varrho L}{W} = \mu_{mol} \frac{mL}{VW}$$

With $M = \frac{mL}{W}$ being the number of molecules in the volume V, we have $n_i = M N_i$ as the number of atoms of type i in the volume V. Then

$$\mu = \frac{1}{V} \sum_i n_i \mu_{a_i} \tag{2.42}$$

If $V = V_c$ is the unit cell volume of a single crystal and n is the number of molecules in the unit cell, we get

$$\mu = \frac{n}{V_c} \sum_i n_i \mu_{a_i} \tag{2.43}$$

where the summation is taken over the atoms of one molecule. The quantities μ_a and $\left(\frac{\mu}{\varrho}\right)$ are dependent only on the absorbing material and are tabulated in the "International Tables for X-Ray Crystallography" (1968), Vol. 3, pp. 162, Birmingham: Kynoch Press, for the wavelengths most frequently used and for nearly all elements.

Let us, for example, calculate μ for the three structures we wish to determine. All quantities needed are given in Table 2.1. Since in crystal structure analysis only CuKα or MoKα radiation are of interest, we calculate μ for these radiation types only (cell volume V_c and n are taken from the result obtained in 3.3). The sum corresponding to equation (2.43) reads for KAMTRA.

$$\mu(\mathrm{CuK}\alpha) = (4/629.1)\,(4 \times 9.1 + 5 \times 0.07 + 6 \times 30.5 + 1 \times 925) \times 10^{24} \times 10^{-23}$$

Note the two different powers of 10. The factor 10^{24} derives from the cell volume, being usually given in Å^3, so that it must be transformed by 10^{-24} into cms. The factor 10^{-23} results from the μ_a's. We then get

$$\mu(\mathrm{CuK}\alpha) = (4/629.1)\,(36.4 + 0.35 + 183 + 925) \times 10 = 72.8\,[\mathrm{cm}^{-1}]$$

By similar calculations we get all further linear absorption coefficients listed in the last column of Table 2.1. As expected, μ is always significantly smaller for MoKα radiation. The magnitude of μ will be an important consideration when the choice of radiation for intensity measurement is made.

Table 2.1. Numerical Values of Atomic Absorption Coefficients Needed for the Calculation of μ for the Test Structures.

			$\mu_a \times 10^{23}$ [cm^2]		μ [cm^{-1}]	
Compound	V_c [Å]	n	CuKα	MoKα	CuKα	MoKα
KAMTRA ($C_4H_5O_6K$)	629.1	4	C: 9.17 H: 0.07 O: 30.5 K: 925	C: 1.25 H: 0.06 O: 3.49 K: 102	72.8	8.2
NITROS ($NH_4[S_4N_5O]$)	762.8	4	N: 17.5 H: 0.07 S: 474 O: 30.5	N: 2.13 H: 0.06 S: 50.8 O: 3.49	106.5	11.8
SUCROS ($C_{12}H_{22}O_{11}$)	715.9	2	C: 9.17 H: 0.07 O: 30.5	C: 1.25 H: 0.06 O: 3.49	12.5	1.5

For a given element, $\mu_a \left(\text{as well as } \frac{\mu}{\varrho}\right)$ depends on the wavelength λ. Fig. 2.24 shows the characteristic distribution of $\frac{\mu}{\varrho}$ plotted versus λ. Sharp discontinuities occur at particular values for λ. These discontinuities are called absorption edges, found once in the K-series, and three times in the L-series.

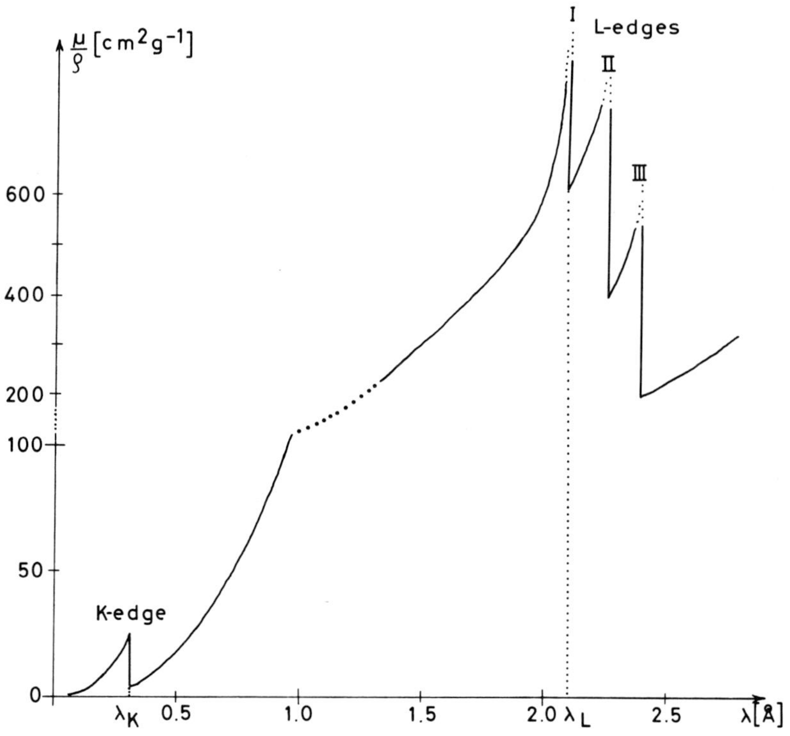

Fig. 2.24. Schematic representation of μ/ϱ plotted versus λ.

An explanation for the rapid increase of μ_a when approaching the absorption edge from the long-wave side can be given by considering the energy level model in Fig. 2.21. When the X-ray quanta have an energy less than the energy necessary to remove an electron, say from the K-shell, their photoelectrical absorption is possible – more or less probable – with respect to outer shells. If the critical energy is E_K and λ reaches the corresponding magnitude $\lambda_K = \frac{hc}{E_K}$, photoelectrical absorption with respect to a further shell, the K-shell, can take place. Therefore μ_a increases discontinuously at λ_K. It is clear that absorption edges are present for every level, hence we have one absorption edge for the K-shell, three for the L-shell, and five or more for the outer shells. From this explanation of absorption edges, it follows that the wavelength of the edge is equal to λ_K defined in (2.32) and the result of (2.37) can now be explained as follows. The wavelength λ_K corresponding to the excitation potential V_K is equal to the wavelength of K-absorption edge (of course, the same holds for all other energy levels) and can be determined experimentally from this property. If we calculate V_K from (2.32) we get an optimum ratio of characteristic to white radiation with a voltage $V = 4V_K$.

For the absorption discontinuities it is very difficult to give an equation relating μ_a to λ or Z. Between the absorption edges a good approximation is given by

$$\mu_a = c\lambda^3 Z^3 \qquad c = \text{const.} \tag{2.44}$$

(2.44) tells us that absorption can be reduced by using hard radition and that hard radiation is essentially necessary if the material to be examined is composed of elements with large atomic number.

The large variation of absorption near the K-absorption edge is a useful property for filtering out Kβ-radiation. Fig. 2.25 shows the λ-dependence of the intensity distribution of copper (Z = 29) radiation with the mass absorption coefficient of nickel (Z = 28). The absorption edge of nickel is situated between the CuKβ and the Kα lines, attenuating Kβ radiation much more than the Kα. Experimental investigations

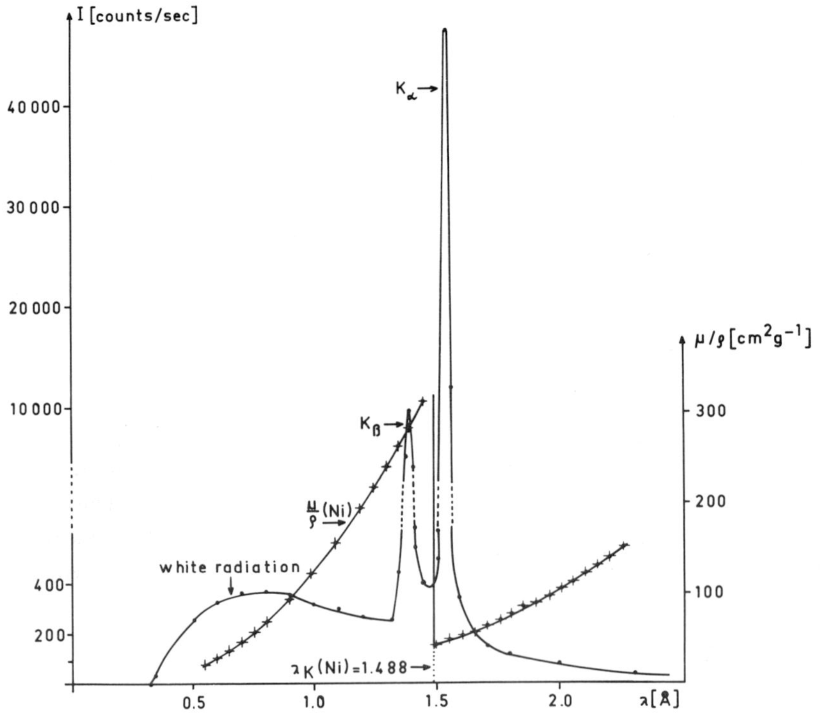

Fig. 2.25. Intensity of the Cu radiation and the mass absorption coefficient of nickel, both plotted versus λ.

show that the use of a nickel filter with thickness of 0.015 mm changes the ratio of $I(K\alpha):I(K\beta)$ to

$$\frac{I(K\alpha)}{I(K\beta)} \approx 100 \tag{2.45}$$

Comparing (2.45) to (2.36), we see that the ratio of $I(K\alpha)$ to $I(K\beta)$ is increased by a factor of 20. However, this improvement is accompanied by a loss of almost 50% of the $K\alpha$ radiation.

It is common practice to use as a filter material, an element whose atomic number is one or two less than that of the target element, since the absorption edge of the filter then lies between the $K\beta$ and $K\alpha$ line. A summary of radiation types most frequently used and their filter materials is given in Table 2.2 (from International Tables for X-Ray Crystallography (1968), Vol. 3, Birmingham: The Kynoch Press)

Table 2.2. K Wavelengths and β-Filters for some Commonly Used X-ray Tube Target Elements.

Target Element	β-filter	Wavelength [Å]			
		$K\alpha_1$	$K\alpha_2$	$K\alpha^*$	$K\beta_1$
Cr	V	2.28962	2.29351	2.29092	2.08480
Fe	Mn	1.93597	1.93991	1.93728	1.75653
Co	Fe	1.78892	1.79278	1.79021	1.62075
Cu	Ni	1.54051	1.54433	1.54178	1.39217
Mo	Zr	0.70926	0.71354	0.71069	0.63225
Ag	Pd	0.55936	0.56378	0.56083	0.49701

2.2.3 *X-Ray Tubes*

A number of X-ray tubes of different types are available commercially today. Although we shall not discuss construction details (if the reader is interested, he may read the instruction manual of the producer), some of their principal properties can be described. Fig. 2.26 shows a general representation of an X-ray tube. Electrons from the heated cathode are accelerated by high voltage towards the anode. The X-rays emitted from the anode target leave the tube by a window which is normally of beryllium. Since the intensity maximum of X-rays is found to be at a take-off angle of approximately 10°, windows are positioned at such a distance from the target to realize this favorable angle.

The conversion of the incident energy of the electron beam into X-radiation is a very inefficient process. Less than 1% is transformed into radiation, the remaining energy appears as heat which has to be dissipated by water-cooling the X-ray tube.

When deciding which type of X-ray tube to use, the practical crystallographer must be aware of three features. The first is the choice of target material and the radiation

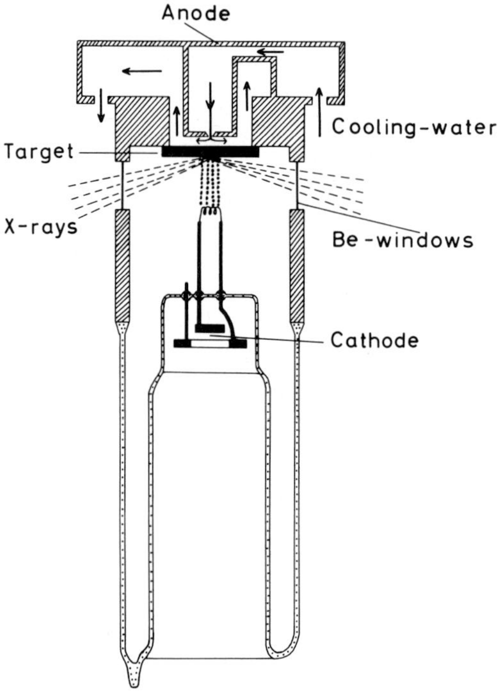

Fig. 2.26. Schematic representation of a commercial X-ray tube.

filter. Tubes with Cr, Fe, Cu, Mo, and Ag targets are commercially avaiable. For single crystal diffractometry, Cu and Mo tubes are adequate for 99% of all problems. For rotation and Weissenberg photographs, a Cu tube is most useful, while for precession and de Jong-Bouman photographs, a Mo tube is generally preferred, especially for crystals with small unit cell dimensions. The second and third important properties of an X-ray tube are the size of the focal point and the power, since the X-ray intensity depends on these two parameters.

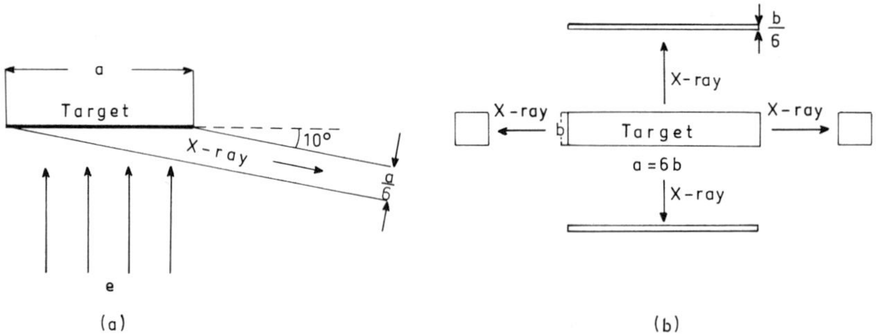

Fig. 2.27. (a) Reduction of X-ray beam diameter to one sixth of target length at a 10° take-off angle (since $\sin 10° \approx {}^1/_6$);
(b) target dimensions and focus size.

The shape of the target is generally rectangular with dimensions $a:b=6:1$. At a take-off angle of 10° we then get a square focus of size $b \times b$ (see Fig. 2.27) in the direction parallel to the long dimension of the target, and a line focus, size $6b \times b/6$ in the perpendicular direction. The four tube windows therefore provide two line and two square foci from each X-ray tube. For single crystal diffraction, only the square focus is used, which means that two single crystal diffraction instruments can be aligned with one X-ray tube. The two line foci can be used for powder diffraction cameras. Since circular collimators are used to limit the incident beam, we always have a circular X-ray beam with a diameter less than b at the crystal. Because the diffracted intensity depends on the crystal volume, great care has to be taken that the largest dimensions of the crystal are less than the diameter of the primary beam at the crystal. Since the edge of the primary beam is sometimes less homogeneous than the center, the crystal size should, in fact, be significantly smaller than the diameter of the X-ray beam.

Since X-ray collimators generally have diameters between 0.5 to 1.0 mm, the size of single crystals used in crystal structure analysis are always smaller than 1 mm. A useful estimate of an optimum crystal size is obtained by application of Beer's law, equation 2.38. Assuming to have a crystal of volume $\approx x^3$, we have $I_0 \sim V \approx x^3$ and

$$I \sim x^3 e^{-\mu x}$$

A maximum I is obtained in terms of x by setting its first derivative equal to zero

$$\frac{dI}{dx} \sim -\mu x^3 e^{-\mu x} + 3x^2 e^{-\mu x} = 0$$

Hence

$$x = 3/\mu \tag{2.46}$$

However, equation 2.46 is applicable only if $\mu \gg 30\ \text{cm}^{-1}$. Otherwise the collimator size is the more restrictive limit.

The power dissipation of modern X-ray tubes varies from 1 KW to 2.5 or 3 KW with a stationary anode. Since the optimum voltage for a given target material does not depend on the tube's power, but on equation (2.37), the full capacity can be utilized by choosing the current as high as possible. If we have, as an example, a copper tube with a power of 1.75 KW, voltage V and current i have to be chosen as follows. V_K is about 9 KV; from (2.37) we obtain a best voltage $V = 4 \times V_K = 36$ KV. Then i must be less than 48.6 mA to avoid an overcharge of the tube. Setting $i = 48$ mA we have the best condition for the optimum intensity of X-radiation.

Notice that the intensity obtained from a tube depends on the power as well as on the focus size. It is evident that for a given crystal size, you do not gain any intensity in the diffraction experiment if you change the tube to another with double power, if the focus size is simultaneously doubled. Only by improving the ratio of power to focus size can a higher intensity be obtained.

2.3 Practicing Film Techniques

2.3.1 *Choice of Experimental Conditions*

The aim of these preliminary investigations is to get information about the unit-cell dimensions, about the crystal symmetry, and about the crystal quality and its stability to X-rays. At this stage, no intensity measurements are made, since it is assumed that an automatic four-circle counter diffractometer is available for this purpose. Up to the middle sixties, diffraction intensities were measured by film techniques, but since then nearly all single crystal investigations are carried out with diffractometers. These instruments run automatically and measure intensities more precisely than film methods (with the exception of proteins with very large unit cell dimensions).

Some investigators maintain that taking film exposures is superfluous and that all experiments can be done by diffractometer only. Although this is true in principle, we believe that it is frequently easier to use both film and diffractometer methods. We consider that the geometrical properties of the reciprocal lattice are better determined from photographs. Abnormal crystal properties, such as disorder, twinning, or crystal splitting, which influence the diffraction intensities, are recognized far more easily by film methods than when using only a diffractometer. If film cameras are available, we always recommend that these are used in the preliminary study of the crystals.

When choosing a single crystal for film techniques we must consider three properties.

Fig. 2.28. Choice of some crystals. The specimen suitable for single crystal measurements are marked by arrows.

(1) The crystal must be a real single crystal. No more than one individual should be selected. A polarization microscope should be used to observe whether the crystal is twinned (see Fig. 2.28).

(2) The crystal should be as large as possible within the limitations given by the primary beam diameter and the $\sim 1/\mu$ limitation discussed above. A large crystal will usually shorten the time of exposure.

(3) Since film techniques are generally used to determine the crystal setting, it is useful to select a crystal with well-formed edges and faces if possible. Needle-shaped crystals are the most favorable for a fast setting, but not suitable for diffractometry because of the difficulty of enclosing *all* the crystal within the X-ray beam.

The crystal shape depends on the compound to be investigated and not all crystals are needle-shaped. But if you have one, use this directions as the rotation axis for film exposures. Frequently, the longest elongation of the crystal will coincide with the shortest lattice constant. This is understandable if we consider a simple model as shown below.

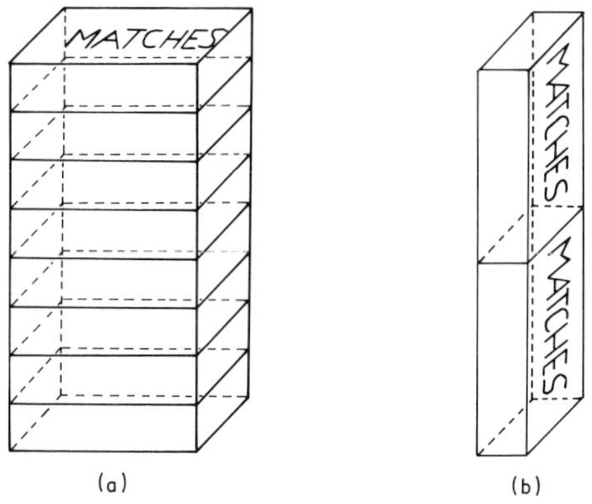

Fig. 2.29. Several ways of stacking match boxes. Method (a) is the most stable.

Suppose we want to stack match boxes (see Fig. 2.29). The most stable way is that shown in Fig. 2.29a, in which there is the maximum area between two match boxes and therefore the best stability. This direction of stacking is the shortest dimension of the box.

Having chosen a good crystal it has to be mounted on a "goniometer head", which is shown in Fig. 2.30. Two arcs A and A′ allow the crystal axis to be turned about $\pm 20°$ in two perpendicular planes. Two sledges B and B′ provide further alignment by allowing the crystal axis to be centered. The crystal is mounted by an adhesive on a glass fiber of length approximately 10 mm. A preliminary crystal setting on the camera should be done optically, followed by a precise alignment using X-rays, by means of a "setting photograph". Several techniques for setting photographs are

Fig. 2.30. A commercial goniometer head (Fa. Huber, Germany).

known. Most of them are based on the method of Bunn [Bunn, C.W., "Chemical Crystallography", (1945), pp. 173–175, Oxford: Clarendon Press]. Here we shall deal with a modification of Bunn's procedure by Dragsdorf [Dragsdorf, R.D. Acta Cryst. 6, 220 (1953)]. With misaligned crystals, the layers of the rotation photograph are no longer projected on the film as lines, but as curves. The aim of the alignment procedure is to obtain values for the corrections of goniometer head arcs from the displacements from linearity of the layer lines. The method of Dragsdorf requires two "oscillation photographs", taken in the same way as rotation photographs, but with a restriction of crystal rotation to a small oscillation range, for instance $\Delta = \pm 10$ or 15°. The detailed experimental conditions are shown in Fig. 2.31a and b.

When starting the exposure, the goniometer head is positioned with one of its arcs, A', perpendicular to the primary beam direction. The orientation of the film is indicated by a mark M. The cylindrical film holder is shifted slightly in the direction of the goniometer head so that the reflections of the zero-layer will not hit the film at its center, but are recorded unsymmetrically. With this orientation, a first oscillation photograph is taken. If the investigator looks from the crystal into the direction of the primary beam, he will see the *left* half of the film exposed (Fig. 2.31a). The time of exposure depends on the crystal size, but it is usually 15–20 min.

The second oscillation photograph is taken on the *same* film, after the crystal was turned by 180° and after the film holder was shifted to the right by a distance C (away from the goniometer head) so that now the right half of the film is exposed (Fig. 2.31b). If the crystal is misaligned, the traces of zero-layer lines will be curvillinear on the unrolled film. At an angle θ, the differences A and B between the zero layers of the

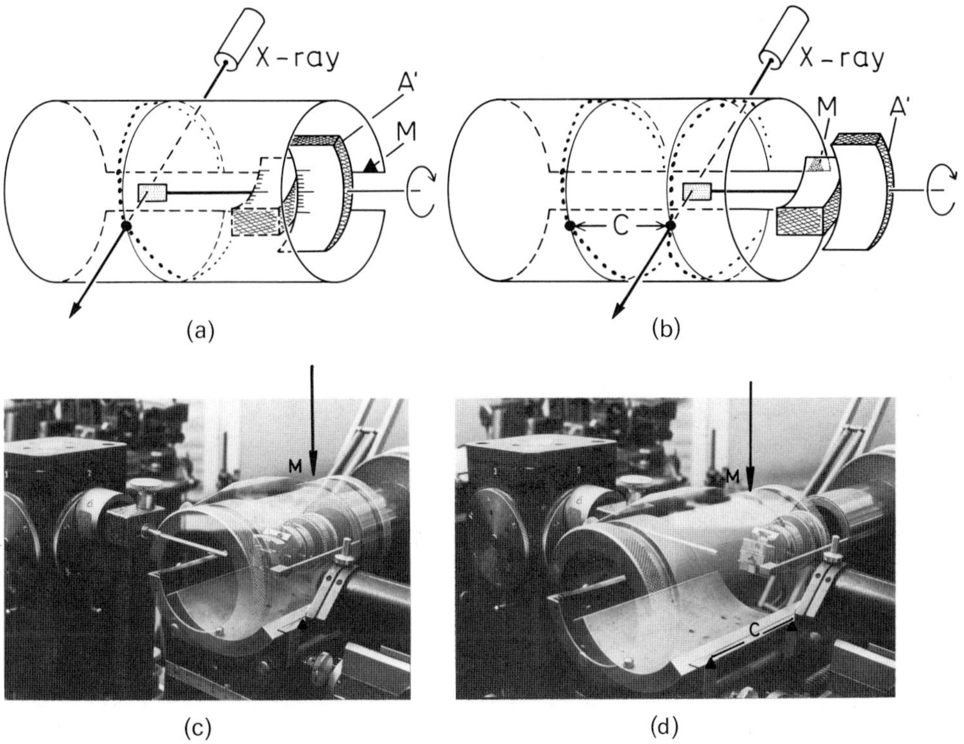

Fig. 2.31. Principle of taking a pair of setting photographs after Dragsdorf, (a) and (b); and its experimental realization, (c) and (d).

two exposures are measured on the upper (A) and on the lower (B) part of the film (see Fig. 2.32).

Assuming that the crystal is accurately centered, the misalignment is due to an inclination of the crystal axis against the rotation axis with the angular components Δi_1 and Δi_2 (in degrees) in the planes of the large and the small goniometer head arc, respectively. The displacement of the diffraction spots from the equator, in terms of Δi_1 and Δi_2, for small inclinations, is

$$\begin{aligned} -\Delta i_1 \sin 2\theta &= \tfrac{1}{2}(A+B) - C \qquad \text{(large arc)} \\ -\Delta i_2 (1-\cos 2\theta) &= \tfrac{1}{2}(A-B) \qquad \text{(small arc)} \end{aligned} \tag{2.47a}$$

where the camera radius R is assumed to be $R = 180/(2\pi) = 28.65$ mm. If A, B, C are measured in mm, we get Δi_1 and Δi_2 in degrees. The trigonometric factors in (2.47a) vanish if A and B are taken for $\theta = 45°$. Then we obtain

$$\begin{aligned} -\Delta i_1 &= \tfrac{1}{2}(A+B) - C \\ -\Delta i_2 &= \tfrac{1}{2}(A-B) \end{aligned} \tag{2.47b}$$

[Hendershot, O. P., Rev. Sci. Instrum. 8, 436 (1937)].

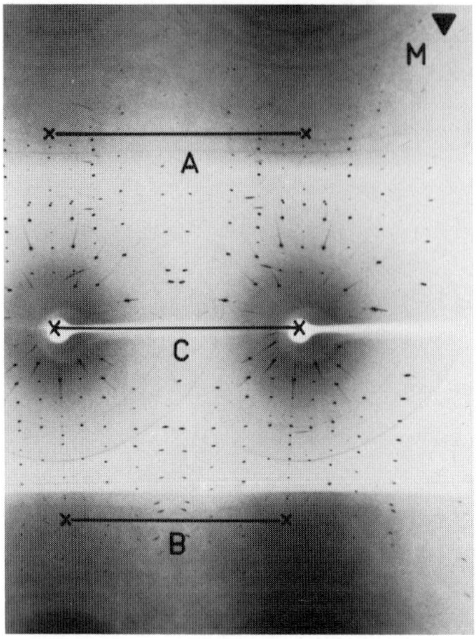

Fig. 2.32. Interpretation of a pair of setting photographs.

The sign of the corrections is obtained as follows. If Δi_k (k = 1, 2) is positive, the correction has to be applied in an anticlockwise direction. The signs of formulae (2.47a and b) are defined for a goniometer head arc scale orientation as indicated, for instance, in Fig. 2.31a. If the scale of the larger arc faces the experimentor, the scale of the small arc is assumed to point upwards. For goniometer heads having an opposite orientation of scales, the sign of Δi_2 becomes "+". Fig. 2.32 shows the first setting photograph of a SUCROS crystal (oscillation angle $\pm 8°$, time of exposure 20 min each). We find A = 62.0, B = 53.8, C = 58.6 (all taken at $\theta = 45°$), and from (2.47b) we get

$$\Delta i_1 = +0.7° \text{ and } \Delta i_2 = -4.1°$$

The DRAGSDORF method can be applied to one oscillation photograph. In this case A and B are measured as deviations from the theoretical equator direction. For a precise determination of that direction, the primary beam stop is removed for a short time while the film holder is translated perpendicular to the primary beam. In this way a black line is produced on the film representing the normal to the equator direction (see Fig. 2.33). The corrections are now given by

$$\begin{aligned} \Delta i_1 &= A + B \\ \Delta i_2 &= A - B \end{aligned} \qquad (2.48)$$

Both these methods provide only approximate corrections, so that one or more additional setting photographs may be necessary. The last method requires one

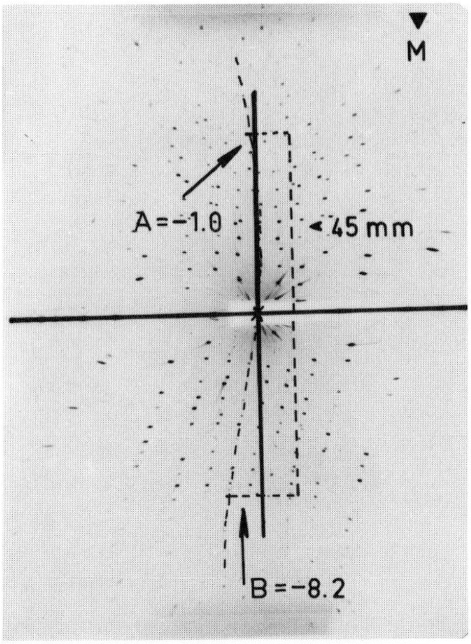

Fig. 2.33. One exposure modification of the Dragsdorf method. With the actual values A = −1.0, B = −8.2, the corrections obtained from eq. (2.48) are $\Delta i_1 = -9.2°$ and $\Delta i_2 = 7.2°$.

exposure but seems to be less precise and converges less rapidly. When the crystal is set, the photographs described in 2.1 can be made.

2.3.2 *Rotation and Weissenberg Photographs of KAMTRA and SUCROS*

Single crystals of KAMTRA are shown in Fig. 2.34. We have chosen a colorless needle-shaped crystal having dimensions 0.2 × 0.3 × 0.8 mm. All film exposures will be made with a 1 KW copper X-ray tube and nickel filter. The voltage is 35 KV, with a current of 28 mA.

With a needle axis the preliminary optical setting is nearly perfect. The first oscillation photograph (oscillation angle ±7°, time of exposure, 30 min) indicates that only small corrections are necessary. With this relatively large crystal, an exposure time of one hour is sufficient for the rotation photograph. If smaller crystals are used, several hours of exposure may be necessary.

The zero-level Weissenberg photograph can be started immediately after the rotation photograph. Several hours of exposure time is needed for the Weissenberg technique, depending on the crystal size and its scattering power. The following timetable can be kept. Crystal selection, setting and rotation photographs can be done

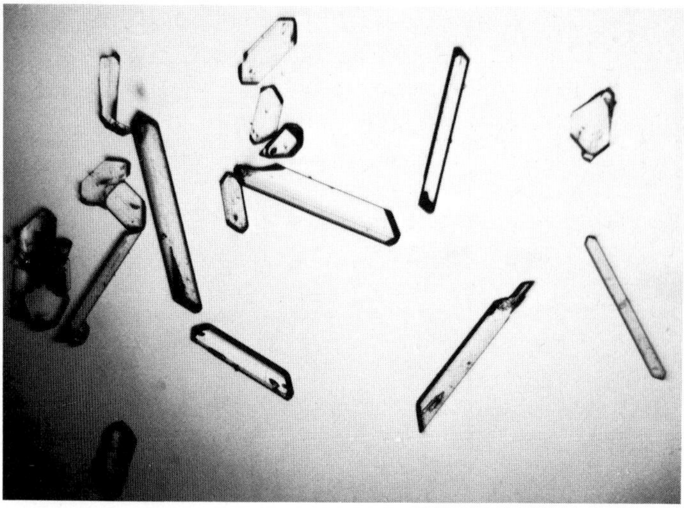

Fig. 2.34. Single crystals of KAMTRA.

within one day by starting the work in the morning. The first Weissenberg photograph can then be started in the late afternoon, and finished and processed the next morning.

To prepare upper level Weissenberg photographs by the equi-inclination technique, the inclination angle φ_n and the displacement constant t_n have to be calculated. From

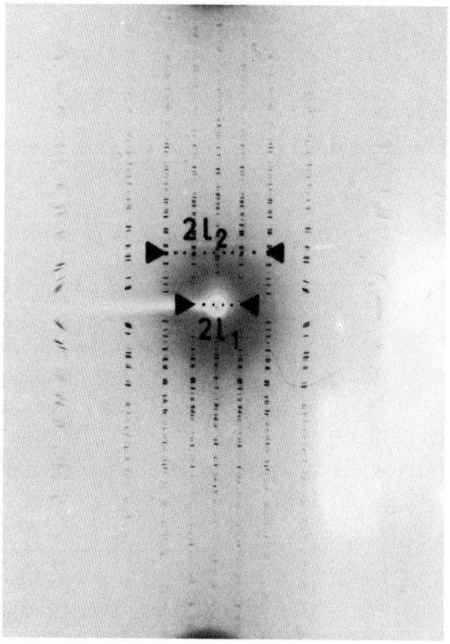

Fig. 2.35. Rotation photograph of KAMTRA.

the rotation photograph (Fig. 2.35) we can measure l_n for formula (2.1). To improve the accuracy, we measure $2l_n$ and divide by 2R, being 57.3 mm for our camera type. Application of (2.11) and (2.12) leads to the desired quantities given in Table 2.3. If the rotation axis is long, i.e. > 10 Å, higher layer-lines than two should be used to reduce the error caused by the relatively bad precision of the $2l_n$ values.

Table 2.3. Device Setting Constants for KAMTRA (Weissenberg Technique).

n	$2l_n$ [mm]	D [Å]	φ_n [°]	t_n [mm]
1	11.7	7.71	5.7	2.5
2	25.1	7.69	11.6	5.0

The zero first, and second level Weissenberg photographs taken for KAMTRA are shown in Fig. 2.36. These three levels give us almost all the information about the dimensions and symmetry of the crystal lattice. The type of layer-line levels that are recorded with one crystal setting is given by (2.3).

If the needle axis of the crystal is the direction of the shortest lattice constant, say c, then the quantity D from the rotation photograph is equal to c and the layer-line levels contain the hk0, hk1, hk2, ... reflections. Special features of Weissenberg photographs should be noted. The first is that the upper and lower half of each photograph look very similar. As we shall see in 3.3.2, the information obtained from both halves is generally identical, and we need only discuss the upper half.

Consider the two lines of reflections A and B on the zero-layer photograph shown in Fig. 2.36a. On either side of the line A we can always observe pairs of reflections having equal intensity. We have marked five reflections (No. 1 to 5) on line A and 10 reflections (No. 6 to 15) next to it (see Fig. 2.36b). Characterizing the intensities by very strong (vs), strong (s), medium (m), weak (w), and very weak (vw), we get an estimated intensity distribution given in Table 2.4.

It is apparent that the primed reflections have the same intensity as the unprimed ones with the equal number. The same is true for line B. Reflection intensities for No. 16 to 30 are equal to those of No. 16′ to 30′ and this is true for the other reflections which are not marked. Looking at the lines A′ and B′ on the first layer or at A″ and B″ on the second layer, we find the same symmetric behavior of intensities.

Table 2.4. Rough Intensities of Reflections No. 1 to 30 of KAMTRA, Zero Layer-line.

No.	1	2	3	4	5	6	7	8	9	10	11	12	13	14	15
Int.	s	s	s	s	vs	vs	vw	s	s	vw	s	w	m	w	s

No.	16	17	18	19	20	21	22	23	24	25	26	27	28	29	30
Int.	s	vs	m	w	vw	vs	s	s	w	s	vw	vw	s	vw	s

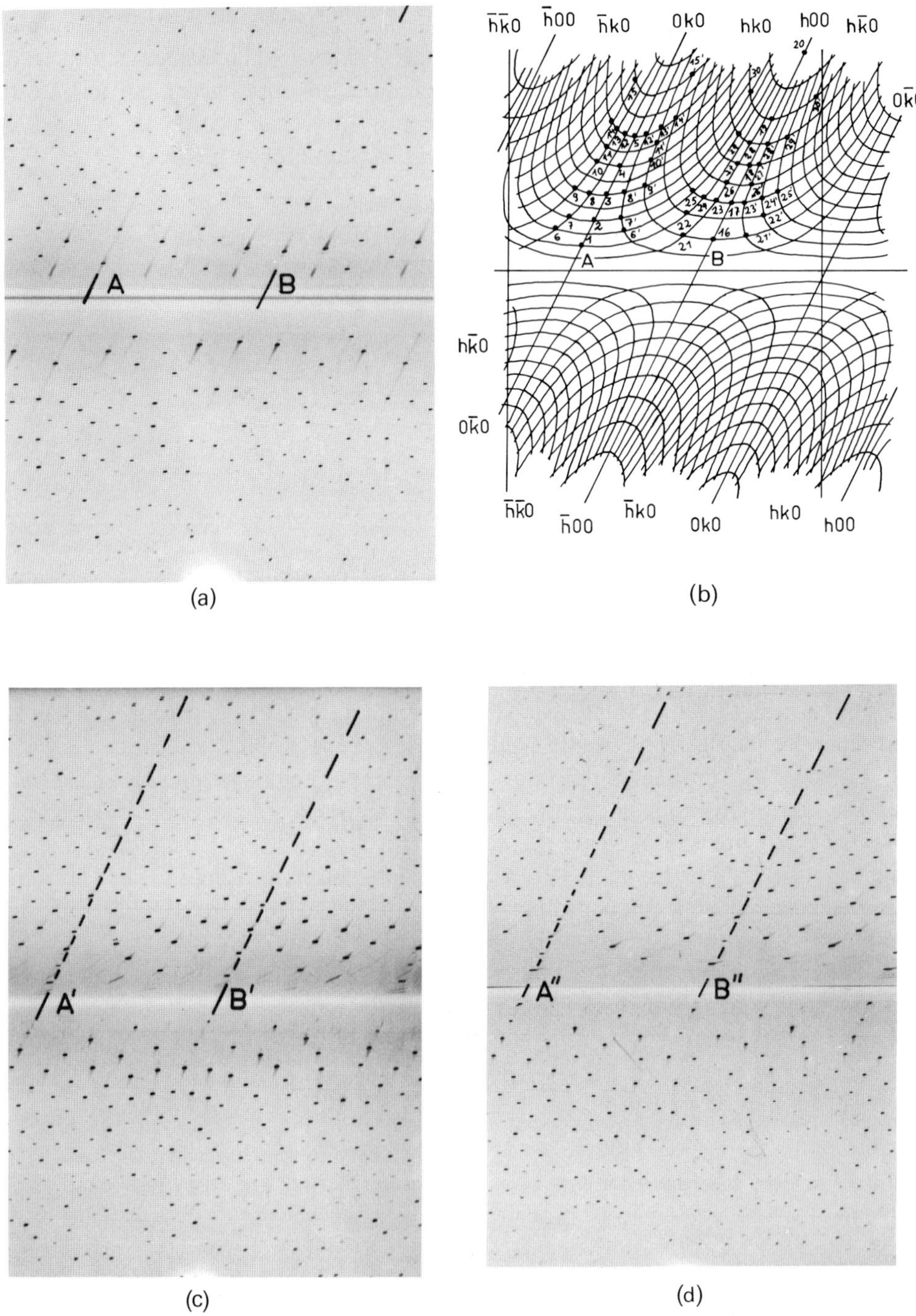

Fig. 2.36. Weissenberg exposures of KAMTRA. (a) zero-layer; (b) schematic representation with various reflections marked; (c) first layer Weissenberg exposure; (d) second layer Weissenberg exposure.

The lines of reflections A and B in all layers are therefore symmetry lines. This is the first experimental result concerning the symmetric properties of the crystal.

A further feature should be noted. The reflections on the zero and first layer appear at nearly the same positions. This is best seen by placing the two photographs together. There are, however, two exceptions. The reflection lines A′ and B′ seem to contain twice as many reflections, or in other words, on lines A and B every second reflection is systematically absent. On the second layer, however, the number of reflections on lines A″ and B″ is the same as on A′ and B′.

Now we have two important results. The intensity distribution shows a symmetry property and there are special groups of reflections which are systematically absent. We can now examine the corresponding photographs of the two other compounds to see whether similar results are observed.

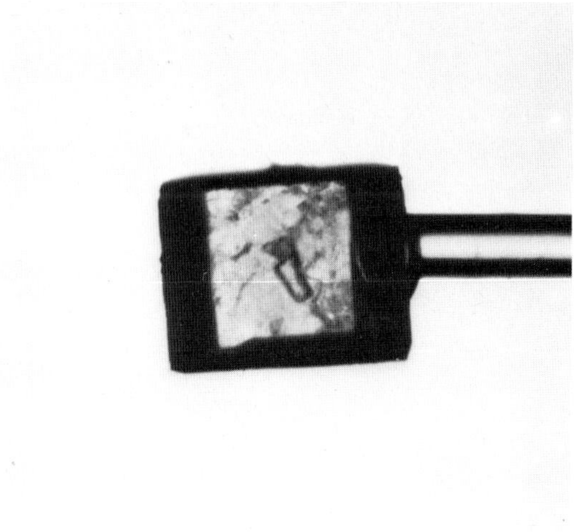

Fig. 2.37. A SUCROS single crystal mounted on a goniometer head.

The next example for the application of film techniques concerns sucrose. Since crystals of this well-known compound are more plate-shaped, we now have two, rather than one, prominent axis. We mount a crystal with dimensions 0.4 × 0.4 × 0.2 mm on a goniometer head, as shown in Fig. 2.37, and select the same conditions for X-ray diffraction as for KAMTRA. From the setting photograph shown in Fig. 2.32, the

Table 2.5. Instrument Setting Constants for SUCROS (Weissenberg Technique).

n	$2l_n$ [mm]	D [Å]	φ_n [°]	t_n [mm]
1	10.2	8.80	5.0	2.2
2	21.5	8.78	10.1	4.4

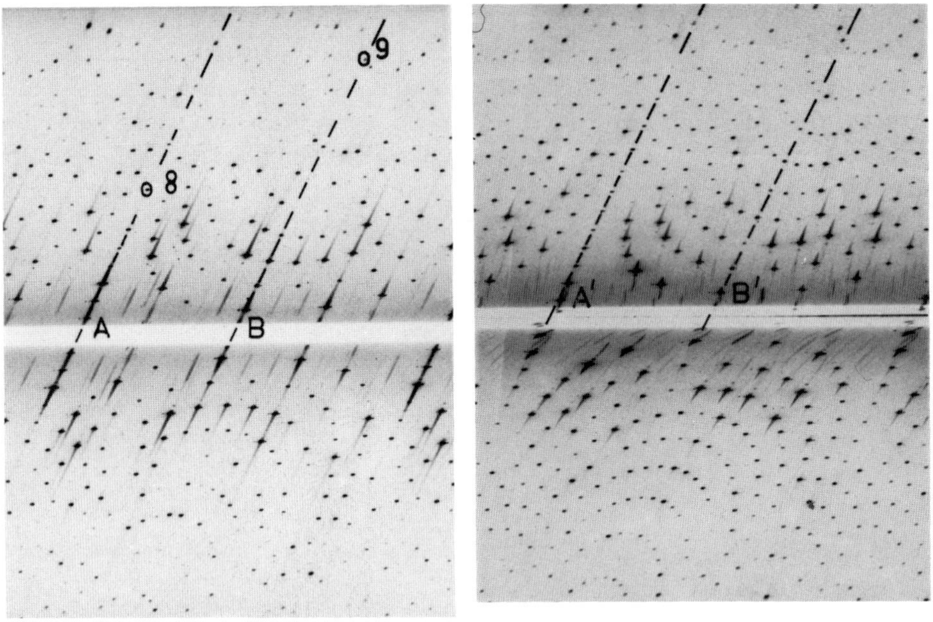

(a) (b)

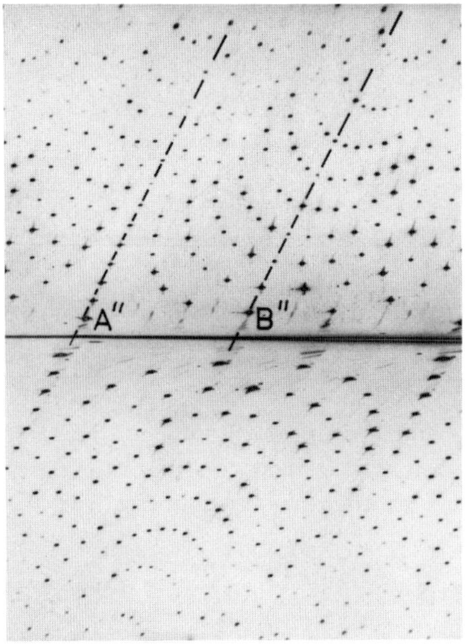

(c)

Fig. 2.38. Weissenberg exposures of SUCROS: (a) zero layer; (b) first layer; (c) second layer.

corrections derived are precise enough to obtain a good rotation photograph. The instrument setting constants needed for the Weissenberg technique are given in Table 2.5, the resulting photographs are shown in Fig. 2.38.

On the zero and upper layers we see again the lines of reflections marked A and B, A′ and B′, A″ and B″. In contrast to the KAMTRA layers, however, they are not lines of symmetry, neither on the zero nor in the upper layers. Furthermore, none of these lines or any other reflection groups show systematically absent reflections. An explanation of these observations will be given in Section 3.3.4.

2.3.3 *De Jong-Bouman and Precession Photographs of NITROS and SUCROS*

To obtain undistorted representations of the reciprocal lattice layers we make use either of the de Jong-Bouman or the precession method. However, as already mentioned in 2.1.4, it is a good practice to combine these two methods, which is possible with the "Lattice Explorer" instrument developed by Wölfel. Only minor changes in the instrument are necessary to pass from one technique to the other. Therefore this apparatus is very convenient for the examination of a crystal lattice.

We can operate the "Explorer" on the same CuX-ray tube as the Weissenberg camera using both quadratic foci (see 2.2.3), although MoKα radition would be more appropriate for these methods. Since we use these photographs only for space group determination, however, the information obtained with CuKα radiation is sufficient.

One of the yellow prismatic crystals of NITROS, which is our example of an inorganic structure, is set on the goniometer head, so that the normals to the larger crystal faces are oriented perpendicular to the axis of rotation. For a precise setting of the crystal on the "Explorer" there is an optical system similar to that of an optical goniometer. This setting by optical methods can be very precise if the crystal has well-shaped faces, as for NITROS and SUCROS. (For details of this optical setting, see the instruction manual of the "Explorer", Fa. Stoe & Cie, Darmstadt, Germany.) To take photographs of de Jong-Bouman type on the "Explorer", some fixed camera adjustments are used. The crystal to film axis distance d_F (taken along the primary beam) is 75 mm (note that this quantity is generally fixed at 60 mm for precession instruments). The annular screen has a fixed radius $r_s = 30$ mm. It is recommended that a constant $\varphi_n = 45°$ is used for de Jong-Bouman exposures (to get Laue cones with constant opening angles of $\varphi_n = 90°$); the screen to crystal distance can be fixed also to $d_s = 30$ mm (see 2.1.4, formula 2.15). With these camera settings, we take a zero-layer exposure of NITROS, together with a cone-axis photograph which is obtained by placing a second film on the front side of the screen. The exposure time is 90 min for the cone-axis photograph and approximately 5 hours for the zero-layer photograph. The results are shown in Fig. 2.39a and b.

An undistorted lattice representation is recorded on the de Jong-Bouman photograph (Fig. 2.39b). The "layer-line circles" are recorded on the cone-axis exposure (Fig. 2.39a), each circle representing the Laue cone of one layer. The zero-layer is

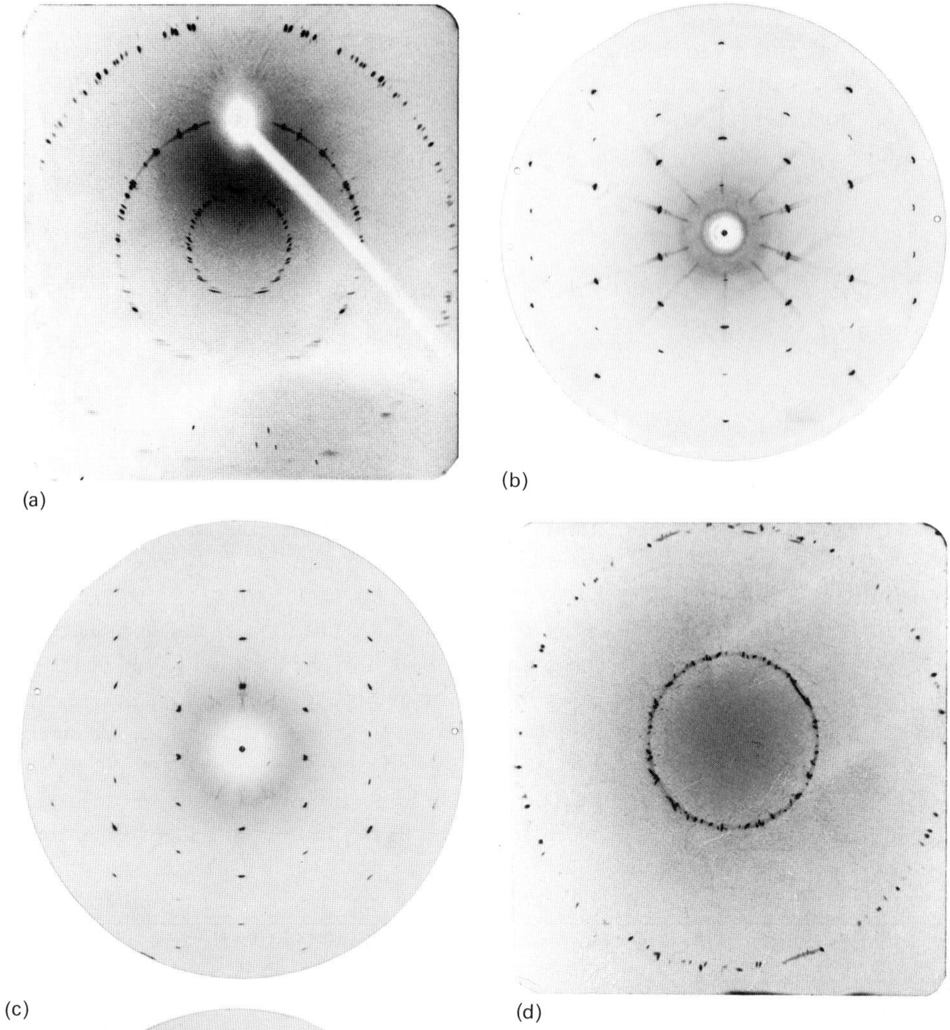

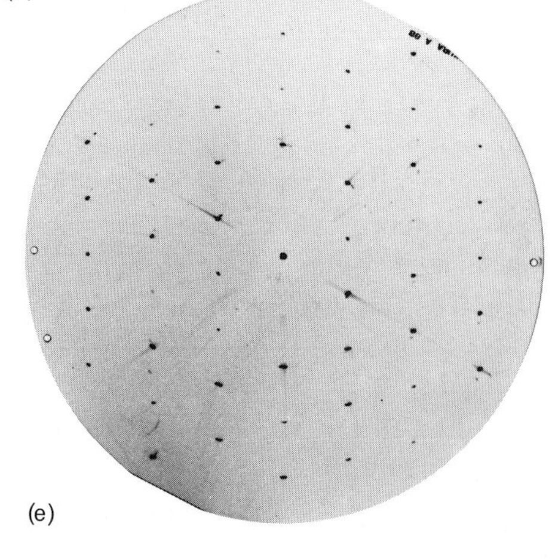

Fig. 2.39. Film exposures of NITROS:
(a) cone axis, de Jong-Bouman setting; (b) zero layer, de Jong-Bouman setting;
(c) first layer, de Jong-Bouman setting;
(d) cone axis, precession setting;
(e) zero layer, precession setting.

that circle which is intersected by the shadow of the primary beam stop. Because of the relatively small lattice constant and the use of CuKα radiation, only the zero and the plus and minus first layer cone are observed on the film.

From the radii of the zero and first layer circle, we obtain the lattice layer distance, by application of (2.19) and (2.18). We find $2\ r_0 = 56.5$ mm and $2\ r_1 = 23.3$ mm. With $\varphi_0 = \mu_0 = 45°$ we get $\varphi_1 = 22.41°$ (from 2.19) and then ($\lambda = 1.5418$ Å) $1/D = 0.141$ Å^{-1} or $D = 7.09$ Å (from 2.18). We will use D for the reciprocal spacing of the lattice planes, since at this time we do not know whether the crystal rotation axis coincides with one basis vector. From D, we can calculate the camera settings for the first layer. To obtain $\varphi_1 = 45°$ we enter $\varphi_0 = 45°$ in (2.20) and get $\mu_1 = 60.7°$. Note that for the "Explorer", the inclination angle is defined relative to the normal direction of the incident beam, so that the apparatus has to be adjusted to an angle $90° - 60.7° = 29.3°$. The displacement t_1 was calculated from (2.21) to be 16.3 mm. Note that the film holder shift has to be directed away from the crystal. With the setting described above, we get the first layer of NITROS (with exposure time 5 hours) shown in Fig. 2.39c.

We can also obtain a precession photograph by making an azimuthal crystal-setting so that the largest crystal faces are in the plane of the primary beam. A zero-layer precession photograph together with the corresponding cone-axis exposure is obtained using an inclination angle $\mu_0 = \varphi_0 = 23.5°$, which is that recommended by the manufacturer for use with a 75 mm camera. With an annular screen having a radius $r_s = 20$ mm we get a crystal to screen distance $d_s = 46.0$ mm. The results are shown in Fig. 2.39d and e.

The lattice layer spacing can be obtained from the radii of the zero and first layer circle (note that in this case, $r_0 < r_1$) and we could therefore calculate the settings for the first layer ($2\ r_0 = 34.7$ mm, $2\ r_1 = 89.2$ mm). However, for the space group determination of NITROS, this first layer is unnecessary (as we shall see in 3.3.3). From (2.19) and (2.23) we get $1/D = 0.162$ Å^{-1} ($D = 6.16$ Å).

Although we have already taken Weissenberg exposures of SUCROS, which provide us with much of the information required on the crystal lattice, we shall also describe the application of de Jong-Bouman and precession techniques to this compound. We can then compare the results, mainly with respect to the amount of information obtained.

We start again with a simultaneous recording of a zero-layer de Jong-Bouman exposure and a corresponding cone-axis photograph (Fig. 2.40a and b). From the layer-line circles $2\ r_0 = 50.9$ mm and $2\ r_1 = 27.1$ mm ($\varphi_0 = \mu_0 = 45°$) we get $\varphi_1 = 28.03°$ and $D = 8.78$ Å (formulae 2.19 and 2.18). It follows that $\mu_1 = 57.9°$ and $t_1 = 13.2$ mm ($d_F = 75$ mm) as instrumental constants for the first layer, to give the photograph shown in Fig. 2.40c.

For the precession photograph, we use the same instrumental constants as for NITROS, i.e. $\mu_0 = 23.5°$, a screen with diameter 40 mm, and a screen distance $d_s = 46.0$ mm. The zero-layer and the cone-axis photographs are shown in Fig. 2.40d and e.

To demonstrate the upper level precession technique, lat us record the first layer of SUCROS. The radii of zero and first layer circles are found to be $2\ r_0 = 36.8$ mm,

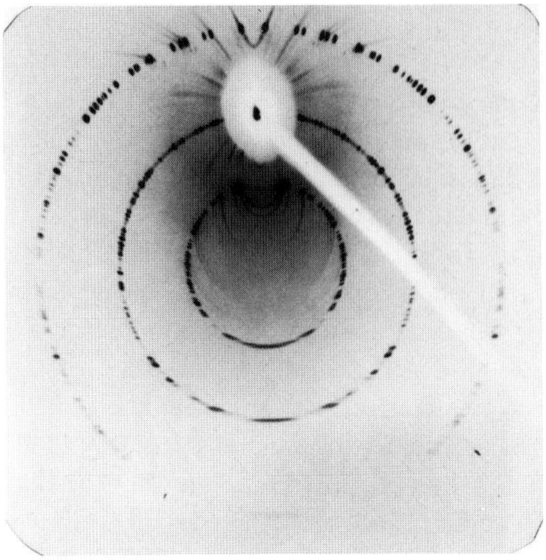

(a)

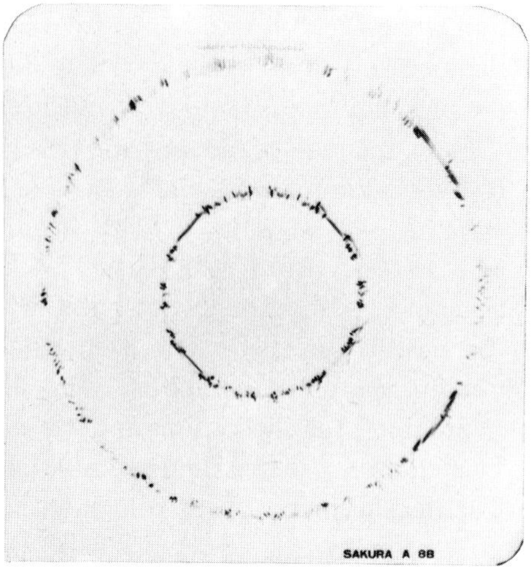

(d)

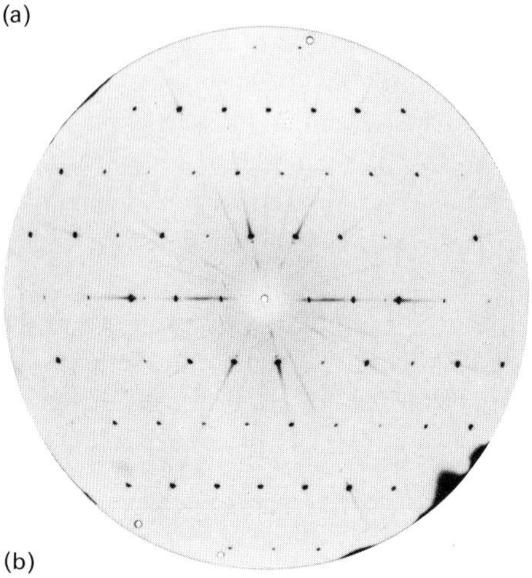

(b)

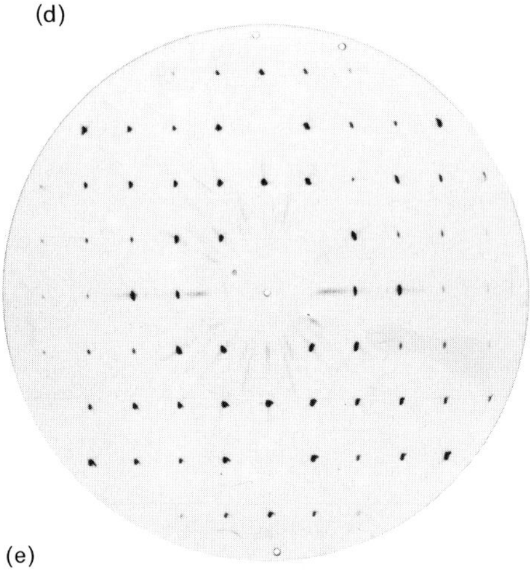

(e)

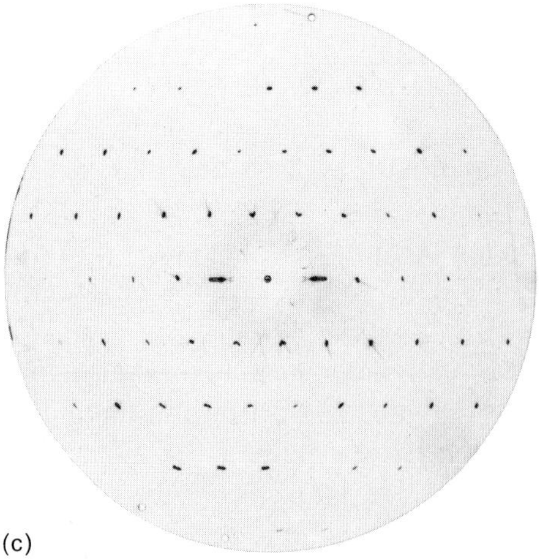

(c)

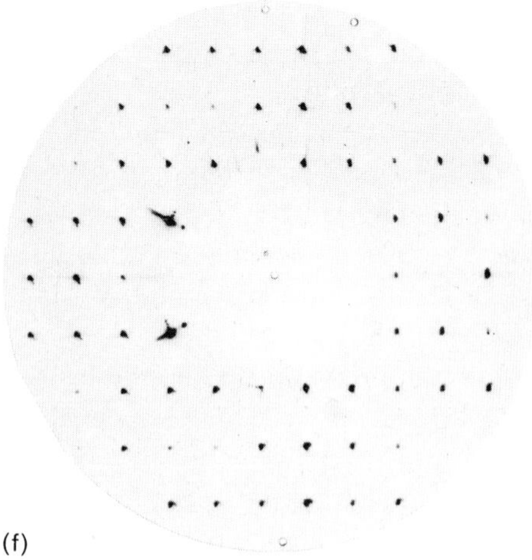

(f)

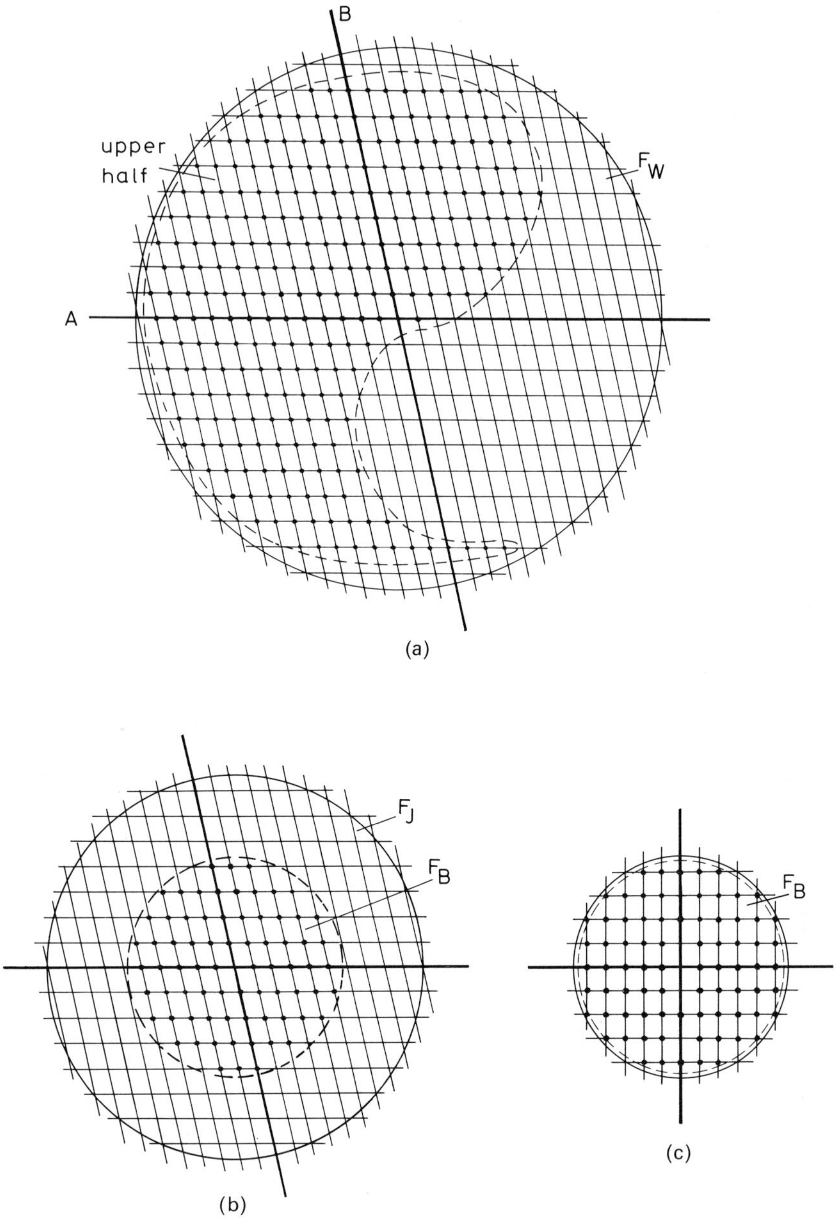

Fig. 2.41. Areas of reciprocal layers covered by the different film techniques (SUCROS, h 0 l plane (a) and (b); h k 0 plane (c).

◀ *Fig. 2.40.* Undistorted exposures of SUCROS:
(a) cone axis, de Jong-Bouman setting; (b) zero layer, de Jong-Bouman setting;
(c) first layer, de Jong-Bouman setting; (d) cone axis, precession setting;
(e) zero layer, precession setting; (f) first layer, precession setting.

$2\, r_1 = 81.5$ mm. With $\varphi_0 = \mu_0 = 23.5°$, we get (from 2.19) $\varphi_1 = 43.9°$ and then (from 2.23) $1/D = 0.128$ Å^{-1}, $D = 7.84$ Å.

The precession angle for the first layer could, in principle, be chosen arbitrarily. However, to avoid collision between the screenholder and the X-ray tube, the screen to crystal distance must be at least 35 mm. Since d_s increases with increasing screen radius and decreasing φ_1, we have to take a larger screen radius and a smaller precession angle for the first layer. If we decide to choose a 60 mm screen ($r_s = 30$ mm) and $\mu_1 = 16.5°$, we get $d_s = 35.3$ mm, which is large enough to avoid a collision. The film displacement is only a function of D, and from (2.25a), $u_1 = 14.8$ mm (with $d_F = 75$ mm).

The first-layer precession photograph is shown in Fig. 2.40f. This exposure with the remarkably big shadowed area in its center was already shown in Fig. 2.18 as an example of this property. From the experimental conditions we can calculate $r_{shad,B}$ and $r'_{shad,B}$ from (2.28). We get $r_{shad,B} = 0.236$ Å^{-1} and the result $r'_{shad,B} = 27.2$ mm agrees with the radius found on Fig. 2.18.

We have now exposures of all three types present for SUCROS and the investigator my compare the results. In Fig. 2.41 the zero layer areas which can be recorded by the three methods are represented schematically. As was shown already by equation 2.26, the area sizes differ remarkably. If the areas which are actually on the films are compared (see Fig. 2.38a, 2.40b, 2.40e and the ranges marked by dashed lines in Fig. 2.41) it comes out that for the Weissenberg and the precession method almost all possible reflections were recorded. Since there is only one film cassette supplied with the "Explorer" and its radius was chosen in accordance with the precession technique, de Jong-Bouman exposures can only be taken with a film of that size so that the areas covered when using the "Explorer" are in fact the same for both techniques. So we can summarize. The Weissenberg exposure (Fig. 2.38a) shows almost all reflections within the r_W-limit, the de Jong-Bouman (Fig. 2.40b) and the precession exposure (Fig. 2.40e) have recorded only the reflection below the r_B-limit.

3 Crystal Symmetry

3.1 Symmetry Operations in a Crystal Lattice

3.1.1 *Introduction*

The main result of Section 1.2.3 was the fact, known as the Laue condition, that diffraction intensities from single crystals are spread discretely in space. The intensities are collected over the finite range of the limiting sphere with radius $2/\lambda$. For a given crystal with unit cell of volume V, the number of reflections which can be recorded is a fixed number given by equation (1.46).

Since investigators are interested in the minimum number of necessary measurements, we must consider the way that the symmetry properties of the crystal can reduce the amount of experimental data required and the number of atomic parameters to be determined.

Fig. 3.1 illustrates various two-dimensional patterns around a "baby-head" motif. In Fig. 3.1a we have a pattern where the only symmetry is periodicity. The unit cell is the smallest non-periodic volume element consisting of a one-face motif. In Fig. 3.1b, however, the situation is quite different. Here the motif is present in four different orientations, two showing their front side and two their back side. Here the identical motif reappears first after having passed another one in both directions. The lattice constants have twice the magnitude of those of Fig. 3.1a, hence the unit cell has the four-fold volume.

Nevertheless, both patterns consist of the same motif. Obviously the structure of crystal (a) is completely known if the structure of this motif and the lattice constants are known. For crystal (b) we have, in principle, two possibilities. We could regard all four orientations as four different motifs, in which case the crystal structure would be known from the determination of all four motifs and the lattice constants. However, it is simpler to determine the one motif, which together with the knowledge of the symmetry and the lattice constants, completely describes the crystal structure. In example (b) at least two symmetry operations are present. Once you get the back of the head by rotation of the baby's head about the axis marked "2" by 180^0. The lower part of the unit cell is obtained by reflection on a plane marked "m", perpendicular to the plane of the drawing. If the symmetry of (b) occurs in a crystal, as shown in Fig. 3.1c, the advantage of using this symmetry can be expressed numerically. We have four molecules in the unit cell, of wich only one is symmetry-independent. The crystal structure is described by the positions of the five atoms of this one molecule, the two symmetry

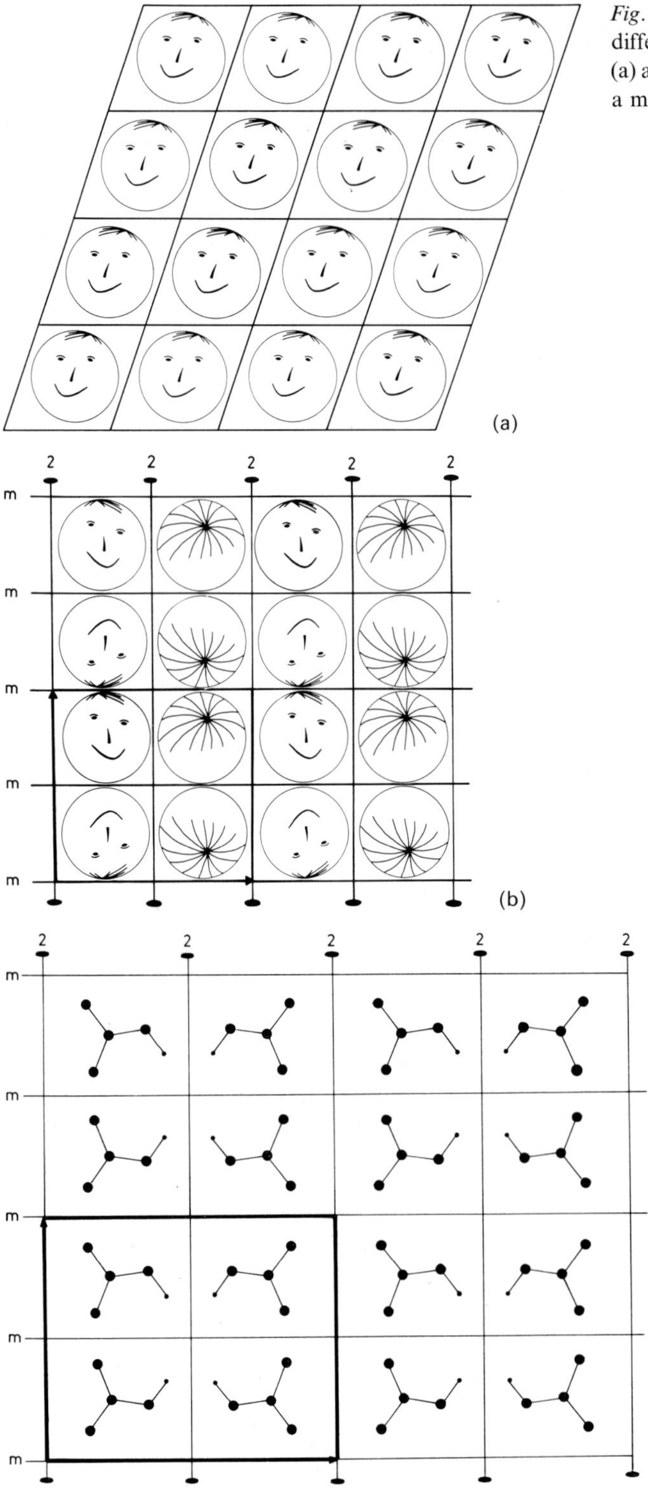

Fig. 3.1. Baby head motif in two different periodic arrangements, (a) and (b). (c) symmetry of (b) in a molecular crystal.

operations, and the cell constants. If the symmetry were neglected, the positions of all 20 atoms in the unit cell would have to be determined.

The asymmetric part of the unit cell, which is the whole cell in crystal 3.1(a), a quarter cell or one motif in (b), and one molecule in (c), is said to be the *asymmetric unit*. By taking crystal symmetry into account, we can restrict ourselves to the determination of the asymmetric unit.

Since the diffraction experiment can provide some information about the crystal symmetry, an important task will be to find out the relationship between the diffraction intensities and the crystal symmetry. We shall do this by considering all possible crystal symmetry operations and their effect on the intensity distribution.

3.1.2 *Basic Symmetry Operations*

Before proceeding further, we have to define the symmetry notation. We shall denote a *crystal symmetry operation* as one which leaves the crystal lattice indistinguishable from that initially present. The geometrical objects related by a symmetry operation are said to be the corresponding *symmetry elements*. From the example in Fig. 3.1, we illustrate two important crystal symmetry operations; a rotation about an axis, and a reflection on a plane. The symmetry elements belonging to those operations are a rotation axis and a mirror plane.

Further symmetry operations are all translations by one or more periodicity vectors. Their main difference from the rotation and reflection symmetries is the absence of a "fixed point". A point in space is called a "fixed point" if it is left invariant with the execution of a symmetry operation.

We shall first deal with those operations having a fixed point, referred to as *point group symmetry operations*. The only crystal symmetry operations with fixed points are

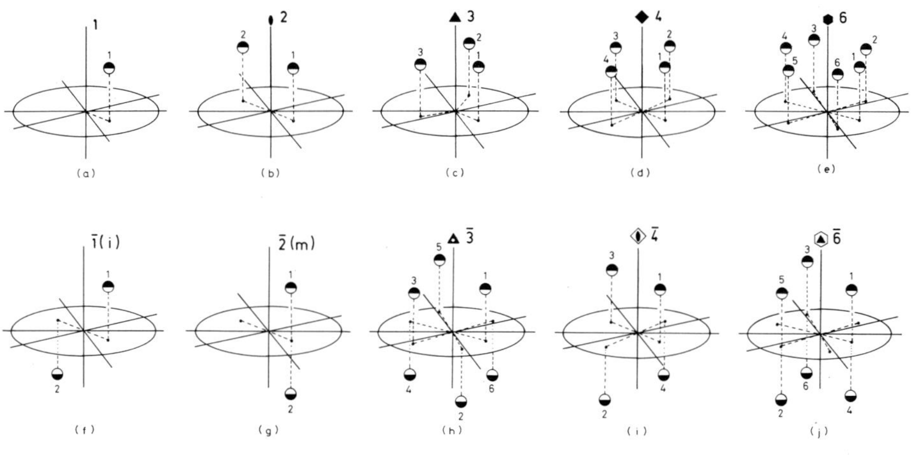

Fig. 3.2. The ten basic symmetry operations.

reflections, rotations and their combinations. The fixed points are all points on the corresponding symmetry elements.

A rotation axis is said to be an n-fold axis if the rotation angle $\varphi = 360^0/n$. In the examples in Fig. 3.1b and 3.1c, we had rotations about $\varphi = 180^0 = 360^0/2$, hence we had a two-fold axis. Crystal symmetry operations cannot be chosen arbitrarily. They must be consistent with three-dimensional periodicity, and therefore only 1-, 2-, 3-, 4-, and 6-fold axes are possible (Fig. 3.2). Proof of this fundamental proposition can be given geometrically, and is illustrated in Fig. 3.3. Suppose that N_1 is the origin of an n-fold axis perpendicular to the plane of the paper. From the periodicity of the crystal, we must have the same n-fold axis at N_2 if **a** is one lattice vector. If φ is the rotation angle, we get, by application of the rotation axis at N_2, a further lattice point at N_3, which is also the origin of an n-fold axis. This axis then produces the next lattice point at N_4 which, in accordance with the translational concept of the crystal lattice, must be at a distance d from N_1 wich is an integer multiple of $a = |\mathbf{a}|$. From trigonometry, we get

$$a = d + 2a\cos\varphi$$

If $d = ma$, it follows that

$$a = ma + 2a\cos\varphi$$

or

$$1 - m = 2\cos\varphi \tag{3.1}$$

Since $|\cos\varphi| \leqq 1$, the integer $M = 1 - m$ must satisfy the inequality

$$|M| \leqq 2$$

Then M can only have the five values ± 2, ± 1, 0. With these values we get from (3.1) the five possible rotation angles which are specific for 1-, 2-, 3-, 4-, and 6-fold axes (see Table 3.1). [Analytical proof of this proposition can be found in the book by Burzlaff and Zimmerman, "Symmetrielehre", (1977), pp. 97ff, Stuttgart: Georg Thieme.]

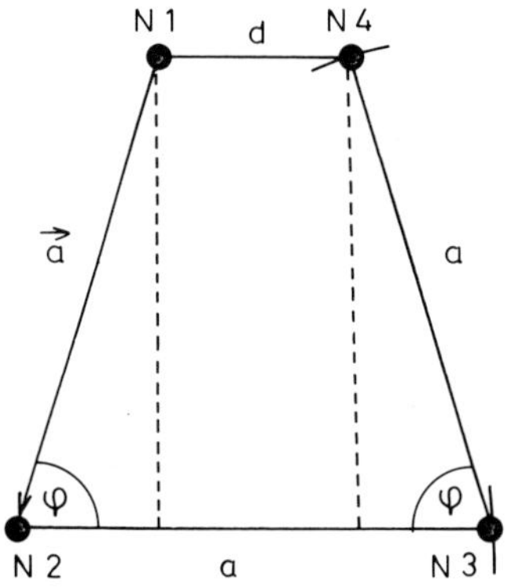

Fig. 3.3. Restriction on 1-, 2-, 3-, 4-, and 6-fold axes.

Table 3.1. Possible Rotation Axes.

M	m	φ [°]	$n = 360/\varphi$
−2	3	180	2
−1	2	120	3
0	1	90	4
1	0	60	6
2	−1	0	1

To get all basic symmetry operations, we could combine n-fold axes and mirror planes. This is the basis of the older Schoenflies notation. In crystallography, we use a combination of the n-fold axis and the "center of symmetry" or "inversion center". This operation, sometimes called "reflection on a point", (see Fig. 3.4) is equivalent to a rotation about 180^0 followed by reflection on a plane. The intersection of a two-fold axis with the mirror plane is the inversion center and the only fixed point of the operation. Motifs related by a center of symmetry are said to be centrosymmetric. Whether a crystal structure is centrosymmetric or not plays an important role in phase determination, as we shall see in Section 5.

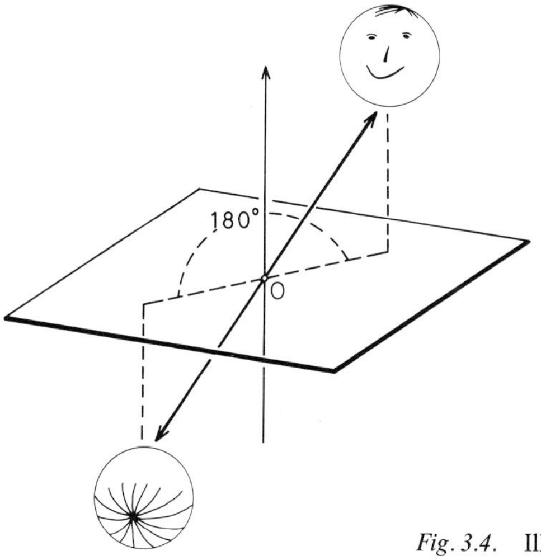

Fig. 3.4. Illustration of an inversion center.

The basic symmetry operations are summarized as follows. Crystal lattices can have 10 basic point group symmetry operations wich are the 1-, 2-, 3-, 4-, or 6-fold rotation axes, or rotation-inversion axes wich are combinations of these axes with a center of symmetry. The five possible n-fold axes were already shown in Fig. 3.2a-e. Now let us discuss the rotation-inversion axes. A one-fold axis followed by an inversion center is just the inversion center itself (Fig. 3.2f). A 2-fold inversion axis results in a well-known symmetry element, the mirror plane. Its operation is illustrated in Fig. 3.2g. A rotation about 180^0 of the original point 1, followed by an inversion, leads to the final position 2, a reflection of position 1.

Note that while a 2-fold inversion axis is identical to a mirror plane, the 3-, 4-, and 6-fold inversion axes illustrated in Figs. 3.2h, i and j, are new operations and not equivalent to any introduced previously. (However, a 6-fold inversion axis can also be described by a 3-fold axis combined with a mirror plane, $\bar{6} = 3/m$.)

A descriptor is assigned to every symmetry operation. N-fold axes are denoted by 1, 2, 3, 4, 6; the three and more fold inversion axes by $\bar{3}$, $\bar{4}$, $\bar{6}$. The 1- and 2-fold inversion axes, wich are identical to a center of symmetry and to a mirror plane, are generally denoted $\bar{1}$ and m, although sometimes the symbols i for the inversion center and $\bar{2}$ for the mirror plane are used.

3.1.3 *Crystal Classes and Related Coordinate Systems*

The examples in Fig.3.1 show that a motif does not necessarily contain only *one* symmetry element, but in general more will be present. The rule that every symmetry operation transforms a crystal to a position which is indistinguishable from the initial state is also true for a combination of two symmetry operations. It can be shown that the set of all possible symmetry operations of a motif has the properties of a mathematical group. For this reason, a systematic discussion of crystallographic symmetry theory is provided by group theoretical methods.

With these methods, it can be shown that due to the translational restrictions in a crystal, the combination of basic symmetry operations is restricted to 32 possibilities. These 32 symmetry groups, each containing a finite number of basic symmetry operations, are called the *crystal classes*.

A crystal class which is commonly found is that named 2/m (say two over m), consisting of a 2-fold axis and a mirror plane oriented perpendicular to the axis. For illustration of this symmetry, let us again use the "baby head" model. Before starting any symmetry operation it may have position 1 (see Fig. 3.5). From group theoretical aspects it is necessary to include the identity as one symmetry operation in every crystal class (we had already included the identity as the 1-fold axis in the basic elements). Now let identity be symmetry element I, the 2-fold axes, II, and the mirror plane, III. Then we get position 2 by application of II and position 3 and 4 by application of the

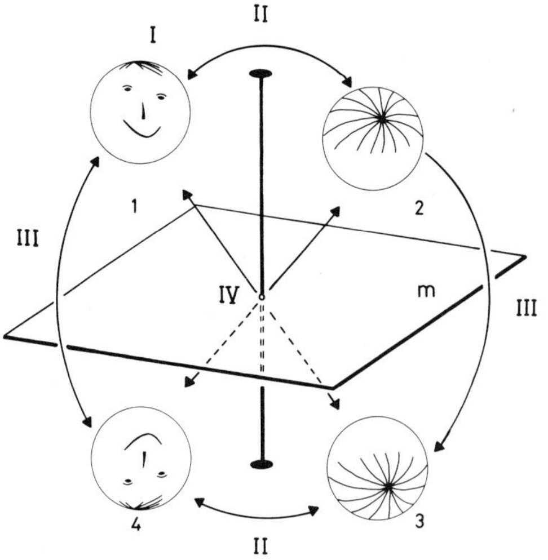

Fig. 3.5. Symmetry elements in the crystal class 2/m.

reflection. These two symmetry operations automatically produce a further one, an inversion. We have an additional symmetry element, IV, a center of symmetry at the intersection of the 2-fold axis and the mirror plane. Thus the crystal class 2/m contains four symmetry elements:

(1) the identity,

(2) the 2-fold axis,

(3) the mirror plane,

(4) the center of symmetry.

The symmetry in this crystal class is said to be 4-fold, since every point is present in four equivalent positions. Looking again at our starting example in Fig. 3.1b, we see that this was a crystal belonging to crystal class 2/m.

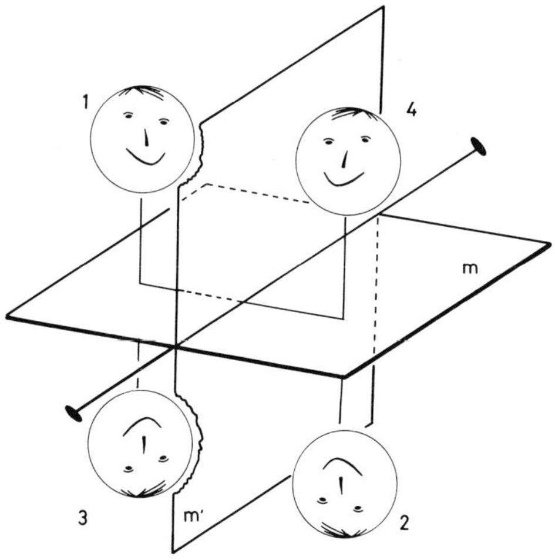

Fig. 3.6. A 2-fold axis in a mirror plane (crystal class mm2).

Another possibility of combining a 2-fold axis and a mirror plane is obtained by having the axis in the mirror plane. The complete set of symmetric positions for this combination is shown in Fig. 3.6. Let 1 be the indentical position. By rotation about the 2-fold axis, we get position 2. Reflection of 1 and 2 on m leads to positions 3 and 4. A second mirror plane m′ is produced, perpendicular to the first and having the 2-fold axis as the intersection of the two mirror planes. It follows that a crystal class 2m defined by a 2-fold axis and a parallel mirror plane is identical to a crystal class mm2, defined by two perpendicular mirror planes and a 2-fold axis.

What happens if the 2-axis is replaced by a $\bar{2}$-axis in the examples discussed above? A crystal class $\bar{2}$/m is nonsense, of course, since $\bar{2} = \mathrm{m}$. In principle, $\bar{2}$m is possible, but it is the same crystal class as 2m. Examining Fig. 3.6, application of $\bar{2} = \mathrm{m}'$ on 1 leads to 4. Then the reflection of 1 and 4 on m (having $\bar{2}$ in its plane) gives 2 and 3, and we get

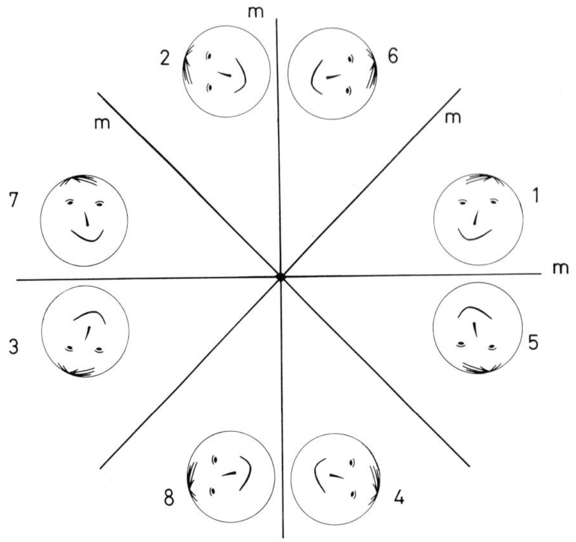

Fig. 3.7. A 4-fold axis in a mirror plane (projection down the 4-fold axis).

the same result as for 2m. Now we have the equality of three combinations: 2m = mm2 = $\bar{2}$m.

A further example of possible combinations of basic symmetry operations is the crystal class 4mm, resulting from a 4-fold axis and a mirror plane containing the axis (see Fig. 3.7, which shows the projection perpendicular to the axis and the plane). If the baby's head is initially present at position 1, the 4-fold axes will produce positions 2, 3, and 4. Reflection of these four positions on the mirror plane leads to the positions 5 to 8. Again the initial symmetry elements, the 4-fold axis and the mirror plane, produce further mirror planes, one perpendicular to the first and two at angles of 45^0. The symmetry in this crystal class is 8-fold, since every point occurs in 8 equivalent positions.

Table 3.2 gives the 32 crystal classes in the notation of Hermann and Mauguin, wich is recommended by the International Union of Crystallography.

Table 3.2. The 32 Crystal Classes.

Crystal system	No.	Crystal classes
Triclinic	2	1; $\bar{1}$
Monoclinic	3	2; m; 2/m
Orthorhombic	3	222; mm2; mmm
Tetragonal	7	4; $\bar{4}$; 4/m; 422; 4mm; $\bar{4}$2m; 4/mmm
Hexagonal	7	6; $\bar{6}$; 6/m; 622, 6mm; $\bar{6}$m2, 6/mmm
Trigonal	5	3; $\bar{3}$; 32; 3m; $\bar{3}$m
Cubic	5	23; m3; 432; $\bar{4}$3m; m3m

As previously noted, the symbols 1, 2, 3, 4, 6 are used for the corresponding n-fold axes; the symbols $\bar{1}, \bar{2}, \bar{3}, \bar{4}, \bar{6}$ for the inversion axes, and m for mirror planes. A slash (/) between an axis and a mirror symbol indicates that the two symmetry elements are perpendicular to each other. If an axis is situated parallel to a mirror plane, a separation character between their symbols is omitted. So, as already shown in the examples, 2m indicates that the 2-fold axis is parallel to m, while 2/m is the symbol for a 2-fold axis perpendicular to m.

The crystal classes are related to the basic symmetry operations which were characterized by the property of having a fixed point in space. Before expanding the symmetry concept by the addition of translational elements, we have to introduce the mathematical representation of symmetry operations and the implications of crystal symmetry to the selection of unit cell vectors. In terms of unit cell vectors **a**, **b**, **c**, every point P in space represented by the vector **r** can be uniquely written as

$$\mathbf{r} = x\mathbf{a} + y\mathbf{b} + z\mathbf{c}$$

The quantitities x,y, z, having no physical dimensions, are said to be the *fractional coordinates* of the point P. If P is situated inside the chosen unit cell, its fractional coordinates have numerical values between 0 and 1.

If P is transformed by any of the basic symmetry operations to an equivalent point P′, the question arises, what mathematical operation produces its vector **r**′? The solution is very simple. Every basic symmetry operation can be represented by 3x3 matrix, the matrix elements depending on the choice of the unit cell vectors. This can be illustrated with an example. In Fig. 3.8a we have a molecular crystal with no symmetry but the identity, 1, and in Fig. 3.8b, we have the molecular crystal of Fig. 3.1c in the crystal class 2/m. In Section 1.2.3, we had defined the unit cell as the smallest nonperiodic unit. Now, by examination of the example in Fig. 3.8a, we find that more than one possible unit cell can be chosen, all having the same volume.

Obviously, none of the possible choices has any advantage over the others. It follows that the lattice constants a, b, c have no restriction, and the angles α, β, γ between the unit cell vectors have no specific values. A crystal having a unit cell of that type is said to belong to the triclinic crystal system. Since it can be shown that the same arbitrary choice of unit cell can be made if no symmetry except an inversion center is present, we can say that a triclinic crystal system is characterized by the absence of any symmetry except the identity and a center of symmetry.

For a crystal with symmetry 2/m (see Fig. 3.8b) we could also in principle choose several unit cells unless we had the problem of representing the symmetry operations by a simple arithmetic expression. Suppose the 2-fold axis is oriented as shown in Fig. 3.8b, and the mirror plane is perpendicular to the plane of the paper. Let us choose 0 as the origin of the unit cell and let **r** be the vector of one atom of the molecule. From the special symmetry in this crystal, one choice of unit cell vectors is necessarily the most favorable, that is, one vector in the direction of the 2-fold axis, the two others in the mirror plane. As shown in Fig. 3.8b, the representation of the symmetry transformed vectors is very simple. For **r**′, transformed via the 2-fold axis,

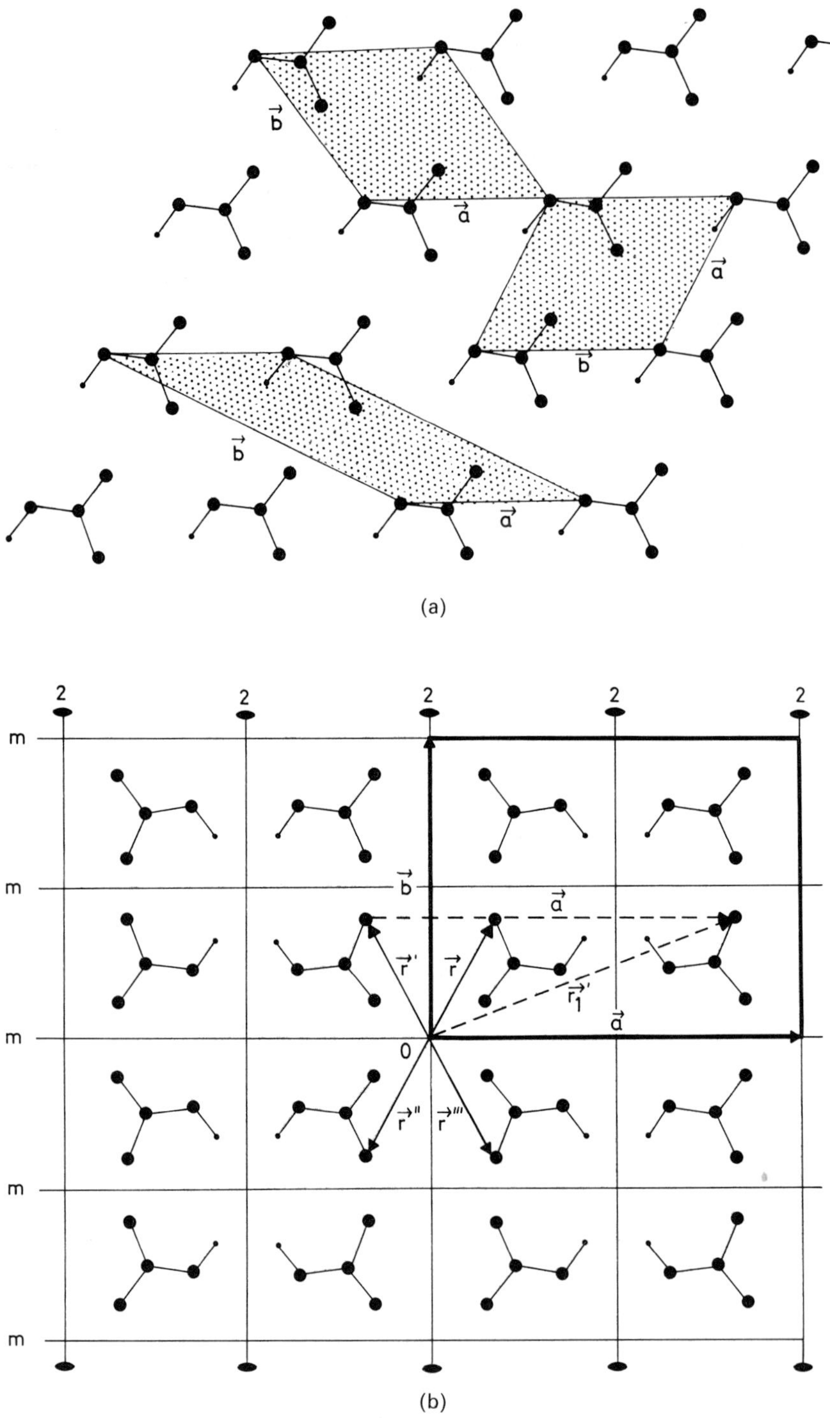

Fig. 3.8. Unit cell choices in a triclinic crystal (a) and in a crystal having the symmetry 2/m (b).

$$\mathbf{r}' = -x\mathbf{a} + y\mathbf{b} - z\mathbf{c}$$

for $\mathbf{r}''$, transformed via the center of symmetry,

$$\mathbf{r}'' = -x\mathbf{a} - y\mathbf{b} - z\mathbf{c}$$

and for $\mathbf{r}'''$, transformed via the mirror plane,

$$\mathbf{r}''' = x\mathbf{a} - y\mathbf{b} + z\mathbf{c}$$

(To obtain all symmetry related vectors in the same unit cell as $\mathbf{r}$, only a pure translation by one of the unit cell vectors is necessary, as illustrated in Fig. 3.8b for $\mathbf{r}'$ and $\mathbf{r}'_1$.)

Using the representation of a vector by a column (see 1.1.4), with

$$\mathbf{r} = \begin{pmatrix} x \\ y \\ z \end{pmatrix},$$

we get,

$$\mathbf{r}' = \begin{pmatrix} -x \\ y \\ -z \end{pmatrix}, \quad \mathbf{r}'' = \begin{pmatrix} -x \\ -y \\ -z \end{pmatrix}, \quad \mathbf{r}''' = \begin{pmatrix} x \\ -y \\ z \end{pmatrix}$$

or (from the definition of a matrix product),

$$\mathbf{r}' = \begin{pmatrix} -1 & 0 & 0 \\ 0 & 1 & 0 \\ 0 & 0 & -1 \end{pmatrix} \mathbf{r}, \quad \mathbf{r}'' = \begin{pmatrix} -1 & 0 & 0 \\ 0 & -1 & 0 \\ 0 & 0 & -1 \end{pmatrix} \mathbf{r},$$

$$\mathbf{r}''' = \begin{pmatrix} 1 & 0 & 0 \\ 0 & -1 & 0 \\ 0 & 0 & 1 \end{pmatrix} \mathbf{r}$$

With these three matrices we have a mathematical representation of all symmetry operations in this crystal class, if we add the unit matrix for the identical operation. We can identify the symmetry operations by the 3×3 matrices

$$S(1) = \begin{pmatrix} 1 & 0 & 0 \\ 0 & 1 & 0 \\ 0 & 0 & 1 \end{pmatrix} \qquad S(2) = \begin{pmatrix} -1 & 0 & 0 \\ 0 & 1 & 0 \\ 0 & 0 & -1 \end{pmatrix}$$

$$S(i) = \begin{pmatrix} -1 & 0 & 0 \\ 0 & -1 & 0 \\ 0 & 0 & -1 \end{pmatrix} \qquad S(m) = \begin{pmatrix} 1 & 0 & 0 \\ 0 & -1 & 0 \\ 0 & 0 & 1 \end{pmatrix}$$

and this matrix representation permits the analytical calculation of symmetry related positions. While another choice of unit cell vectors would also have resulted in a matrix representation, this is the most simple and the most suitable for this kind of symmetry.

It is therefore customary for all crystallographers to choose the unit cell vectors as described above.

From this special choice it follows that the lattice constants have some restrictions. Since two vectors have to be in a plane perpendicular to the third vector, two of the angles are restricted to 90°. Usually the axis in the direction of the 2-fold axis, called the "unique axis" is designated b. Then only β differs from 90° and $\alpha = \gamma = 90°$. A crystal with lattice constants of that kind is said to belong to the *monoclinic* system. The system is monoclinic when only one 2-fold axis or m-symmetry plane is present.

With the other crystal classes, other restrictions due to symmetry are observed. We shall not explain this in detail, but only present the results.

Table 3.3. Crystal Systems.

Name	Conditions for Lattice Constants
Triclinic	no restrictions for a, b, c, α, β, γ
Monoclinic	no restrictions for a, b, c, β; $\alpha = \gamma = 90°$ (2nd setting) (no restrictions for a, b, c, γ; $\alpha = \beta = 90°$ (1st setting)
Orthorhombic	no restrictions for a, b, c; $\alpha = \beta = \gamma = 90°$
Tetragonal	no restrictions for a and c, but a = b and $\alpha = \beta = \gamma = 90°$
Hexagonal	no restrictions for a and c, but a = b, $\alpha = \beta = 90°$, $\gamma = 120°$
Trigonal (rhombohedral axes)	no restriction for a, but a = b = c, no restriction for α, but $\alpha = \beta = \gamma$
Cubic	no restriction for a, but a = b = c, $\alpha = \beta = \gamma = 90°$

It was shown that seven base systems are sufficient do describe all possible crystal symmetries. These seven systems, which are nothing more than special coordinate systems, are shown in Table 3.3, together with the conditions for the lattice constants. In Table 3.2, which shows the crystal classes, an additional column indicates the crystal system for every crystal class.

In all but the trigonal system, the base system (i.e. the coordinate system) and the crystal system have the same notation. In the trigonal system the axes having the properties a = b = c and $\alpha = \beta = \gamma$ (not necessarily equal to 90°) are named "rhombohedral". It can be shown that the symmetry of a crystal belonging to the trigonal system can also be described in the hexagonal system. It would seem to be superfluous to have this system, but for special purposes (see 3.1.4) it is desirable to use either a trigonal or a hexagonal system.

There are several conventions which are used with these crystal systems. In the monoclinic system, the unique axis is generally denoted "b". Then the non-right angle must be β. This choice is called the "second setting", in contrast to the choice of c as the unique axis, which is denoted the "first setting", but is used less frequently.

In all other systems the axis of highest symmetry is named c, as in the tetragonal and hexagonal system. The reciprocal lattice constants as introduced in 1.2.3 are restricted in the same way as for the direct space lattice. If we calculate, for instance, the reciprocals in a monoclinic system, we get, by the definition in 1.1.4,

$$V = \mathbf{abc} = \mathbf{b}(\mathbf{c} \times \mathbf{a}) = abc \sin\beta$$

$$a^* = \frac{bc \sin\alpha}{V} = \frac{1}{a \sin\beta}$$

$$b^* = \frac{ca \sin\beta}{V} = \frac{1}{b}$$

$$c^* = \frac{ab \sin\gamma}{V} = \frac{1}{c \sin\beta}$$

$$\cos\alpha^* = \frac{0-0}{\sin\beta} = 0, \text{ hence } \alpha^* = 90^\circ$$

$$\cos\beta^* = -\cos\beta, \text{ hence } \beta = 180^\circ - \beta$$

$$\cos\gamma^* = \frac{0-0}{\sin\beta} = 0, \text{ hence } \gamma^* = 90^\circ$$

The result of these calculations for all seven crystal systems is given in Table 3.4.

With these different crystal systems we can complete the representation of the basic symmetry operations by 3×3 matrices, as we have done with the identity, inversion center, 2-fold axis and mirror plane.

For a 3-fold axis, the trigonal system is usually preferred. The base vectors, which are equal in length and all include the same angle, form a unit cell as shown in Fig. 3.9a, the 3-fold axis is in the body diagonal of the unit cell.

In principle, another system would be possible, and for some symmetry combinations with 3- and 6-fold axes, a trigonal or a hexagonal system can be chosen. The orientation of the two types of unit cells is shown in Fig. 3.9b. The hexagonal c-axis coincides with the body diagonal of the rhombohedral cell. The transformation between the base system **a**, **b**, **c** (rhombohedral) and **A**, **B**, **C** (hexagonal), is given by

$$\mathbf{A} = \mathbf{a} - \mathbf{b}$$
$$\mathbf{B} = \mathbf{b} - \mathbf{c}$$
$$\mathbf{C} = \mathbf{a} + \mathbf{b} + \mathbf{c}$$

Table 3.4. Cell Volume and Reciprocal Lattice Constants in Terms of Direct Cell Constants for Various Crystal Systems.

Crystal System	Non-restricted-Cell Constants	Volume	Reciprocal Cell Constants
Triclinic	a, b, c α, β, γ	V = (see below)	$a^* = \frac{bc \sin\alpha}{V}$; $b^* = \frac{ca \sin\beta}{V}$ $c^* = \frac{ab \sin\gamma}{V}$; $\alpha^*, \beta^*, \gamma^*$ as given by equation (1.13) in 1.1.4
Monoclinic	a, b, c, β	$V = abc \sin\beta$	$a^* = \frac{1}{a \sin\beta}$; $b^* = \frac{1}{b}$; $c^* = \frac{1}{c \sin\beta}$; $\beta^* = 180° - \beta$
Orthorhombic	a, b, c	$V = abc$	$a^* = \frac{1}{a}$; $b^* = \frac{1}{b}$; $c^* = \frac{1}{c}$
Tetragonal	a, c	$V = a^2 c$	$a^* = \frac{1}{a}$; $c^* = \frac{1}{c}$
Hexagonal	a, c	$V = \frac{a^2 c \sqrt{3}}{2}$	$a^* = \frac{2}{a\sqrt{3}}$; $c^* = \frac{1}{c}$; $(\gamma^* = 60°)$
Trigonal (rhombohedral axes)	a, α	V = (see below)	$\cos\frac{\alpha^*}{2} = \frac{1}{2\cos\frac{\alpha}{2}}$; $a^* = \frac{1}{a \sin\alpha \sin\alpha^*}$
Cubic	a	$V = a^3$	$a^* = \frac{1}{a}$

In the triclinic system, $V = abc\sqrt{1 - \cos^2\alpha - \cos^2\beta - \cos^2\gamma + 2\cos\alpha\cos\beta\cos\gamma}$

In the trigonal system, $V = a^3\sqrt{1 - 3\cos^2\alpha + 2\cos^3\alpha}$

or in matrix notation

$$\begin{pmatrix} \mathbf{A} \\ \mathbf{B} \\ \mathbf{C} \end{pmatrix} = \begin{pmatrix} 1 & -1 & 0 \\ 0 & 1 & -1 \\ 1 & 1 & 1 \end{pmatrix} \begin{pmatrix} \mathbf{a} \\ \mathbf{b} \\ \mathbf{c} \end{pmatrix}$$

(The transformation given above is said to belong to an "obverse" orientation of the rhombohedral axes relative to the hexagonal axes. Another orientation may be chosen, denoted as "reverse orientation" by $\mathbf{A} = \mathbf{a} - \mathbf{c}$, $\mathbf{B} = \mathbf{b} - \mathbf{a}$, $\mathbf{C} = \mathbf{a} + \mathbf{b} + \mathbf{c}$. By convention, the obverse orientation is standard.)

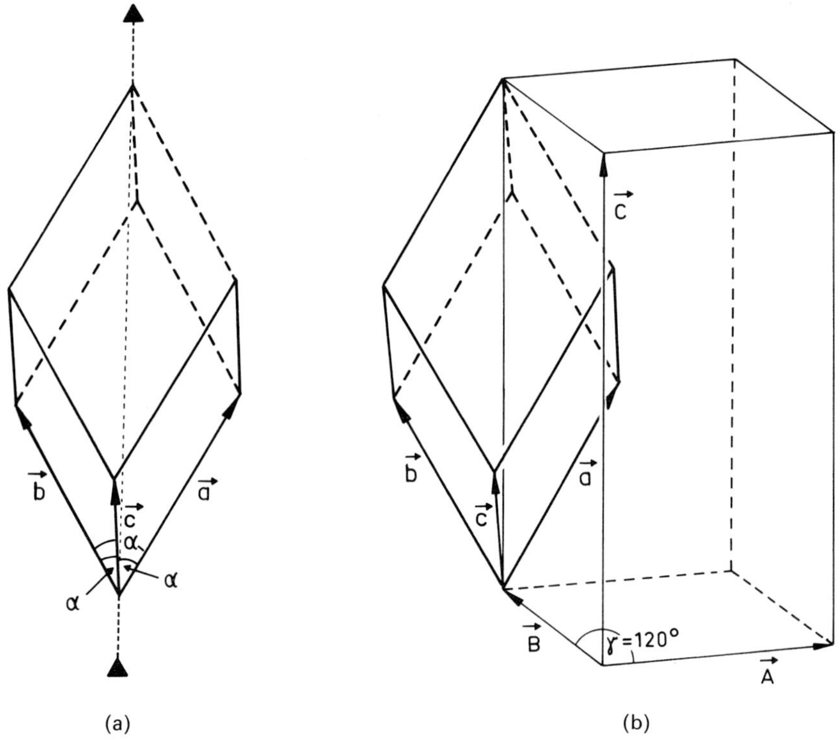

Fig. 3.9. (a) Unit cell defined by rhombohedral base vectors, (b) Relation between rhombohedral and hexagonal cell.

The transformation of all other quantities such as reciprocal base vectors, indices, etc., can be done as in 1.1.5.

Fig. 3.10 illustrates the transformation by a 3-fold axis. If our motif has initially a position vector **r** with components (x, y, z), we get, after rotation about 120°, a position vector **r**′ with components (x′, y′, z′) in the rhombohedral cell. From Fig. 3.10 we find

$$x' = z$$
$$y' = x$$
$$z' = y$$

The matrix representation is then

$$S(3) = \begin{pmatrix} 0 & 0 & 1 \\ 1 & 0 & 0 \\ 0 & 1 & 0 \end{pmatrix}$$

Since a 3-fold axis is a symmetry element which is responsible for more than one equivalent position, we have to take care of a further equivalent position **r**″. Its com-

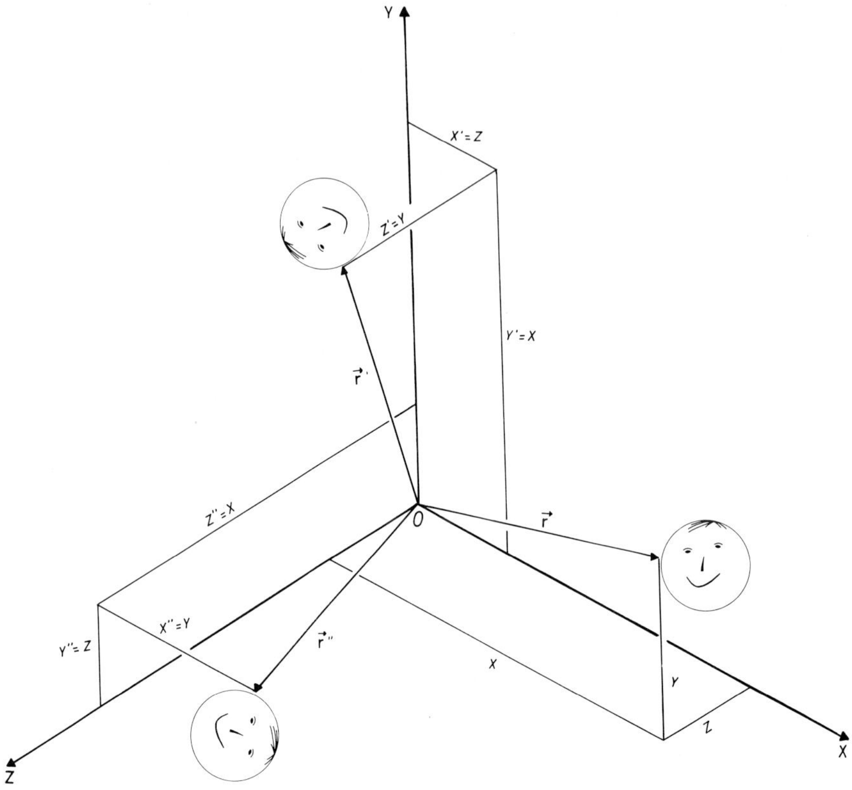

Fig. 3.10. Equivalent positions with respect to a 3-fold axis.

ponents are

$$x'' = y$$
$$y'' = z$$
$$z'' = x$$

obtained by application of S(3) to $\mathbf{r}'$ or by application of the second power of S(3) to $\mathbf{r}$,

$$\mathbf{r}'' = \mathrm{S}(3)\,\mathbf{r}' = \mathrm{S}(3)\,\mathrm{S}(3)\,\mathbf{r}$$

If we repeat this process once more, we get,

$$\mathbf{r}''' = \mathbf{r} = \mathrm{S}(3)^3\,\mathbf{r}$$

but that means that $\mathrm{S}(3)^3 = \mathrm{S}(1) =$ identity. Since all three equivalent positions can be calculated by the matrix product of a suitable power of S(3) and $\mathbf{r}$, we can say that S(3) is a complete representation of a 3-fold axis in the rhombohedral system.

It follows immediately from what has been deduced above that a 3-fold inversion axis (Fig. 3.2h) has the following six equivalent positions in the rhombohedral system:

(1) x, y, z; (2) −z, −x, −y; (3) y, z, x;
(4) −x, −y, −z; (5) z, x, y; (6) −y, −z, −x

(When dealin with equivalent positions, crystallographers prefer to write the column representation of a position vector as a row, with the fractional coordinates separated by a comma, as above.)

The matrix representation is

$$S(\bar{3}) = \begin{pmatrix} 0 & 0 & -1 \\ -1 & 0 & 0 \\ 0 & -1 & 0 \end{pmatrix}$$

With the first six powers of $S(\bar{3})$ all positions are obtained. Since there are close relations between 3- and 6-fold symmetry, let us proceed with 6-fold axes in a hexagonal system. The situation is illustrated in Fig. 3.11 (the c-axis is normal to the plane of the paper).

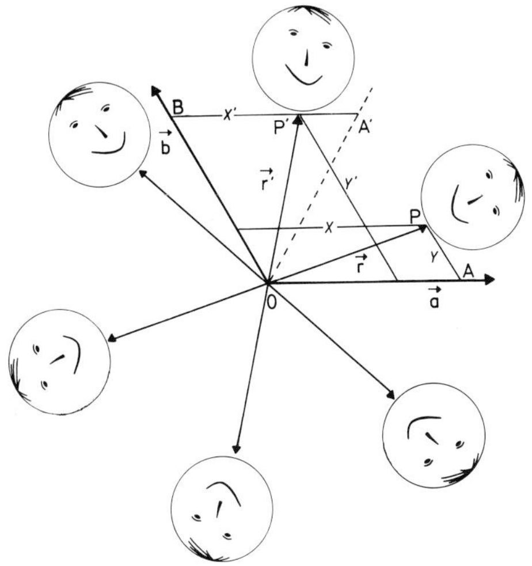

Fig. 3.11. Equivalent positions with respect to a 6-fold axis.

The dotted line OA′ is the bisector of the hexagonal a- and b-axis. By rotation of the complete triangle OAP it is transformed to OA′P′. Since the angles A′OB and OBA′ are 60°, the triangle A′OB is equilateral, hence we get $y' = x$ and $x' + y = x$. From a similar consideration we get all six equivalent positions:

(1) x, y, z; (2) x − y, x, z; (3) −y, x − y, z;
(4) −x, −y, z; (5) y − x, −x, z; (6) y, y − x, z

The corresponding matrix representation is then

$$S(6) = \begin{pmatrix} 1 & -1 & 0 \\ 1 & 0 & 0 \\ 0 & 0 & 1 \end{pmatrix}$$

The next basic operation for which the matrix representation has to be derived is the $\bar{6}$-axis. It has already been shown in Fig. 3.2j that a $\bar{6}$-axis is equivalent to a 3-fold axis with a perpendicular mirror plane, $\bar{3}/m$. Its representation in a hexagonal system follows immediately from that of a 6-fold axis:

(1) x, y, z;	(2) $y-x, -x, -z$;	(3) $-y, x-y, z$;
(4) $x, y, -z$;	(5) $y-x, -x, z$;	(6) $-y, x-y, -z$

The matrix representation is

$$S(\bar{6}) = \begin{pmatrix} -1 & 1 & 0 \\ -1 & 0 & 0 \\ 0 & 0 & -1 \end{pmatrix}$$

With the positions derived for a 6-fold axis, we are able to give the representation of a 3- and a $\bar{3}$-axis in a hexagonal system. The equivalent positions are (see Fig. 3.11 and 3.2):

(a) 3-fold axis

(1) x, y, z;	(2) $-y, x-y, z$;	(3) $y-x, -x, z$

Matrix representation:

$$S(3) = \begin{pmatrix} 0 & -1 & 0 \\ 1 & -1 & 0 \\ 0 & 0 & 1 \end{pmatrix}$$

(b) $\bar{3}$-axis

(1) x, y, z;	(2) $y, y-x, -z$;	(3) $y-x, -x, z$;
(4) $-x, -y, -z$;	(5) $-y, x-y, z$;	(6) $x-y, x, -z$

Matrix representation:

$$S(\bar{3}) = \begin{pmatrix} 0 & 1 & 0 \\ -1 & 1 & 0 \\ 0 & 0 & -1 \end{pmatrix}$$

Compare the results with the matrices obtained for a rhombohedral system!

For a 4-fold axis a tetragonal system is the most suitable, with the rotation axis having the direction of the tetragonal c-axis.

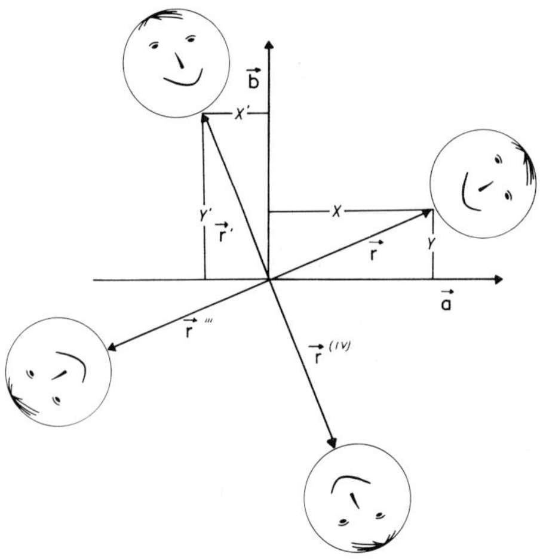

Fig. 3.12. Equivalent positions with respect to a 4-fold axis.

From Fig. 3.12 (4-fold axis normal to the plane of the paper), it follows immediately for the four equivalent positions:

(1) x, y, z; (2) −y, x, z; (3) −x, −y, z; (4) y, −x, z

The matrix representation is

$$S(4) = \begin{pmatrix} 0 & -1 & 0 \\ 1 & 0 & 0 \\ 0 & 0 & 1 \end{pmatrix}$$

Similarly we get, for the $\bar{4}$-axis, the four positions,

(1) x, y, z; (2) y, −x, −z; (3) −x, −y, z; (4) −y, x, −z

and the matrix representation

$$S(\bar{4}) = \begin{pmatrix} 0 & 1 & 0 \\ -1 & 0 & 0 \\ 0 & 0 & -1 \end{pmatrix}$$

3.1.4 *Translational Symmetry, Lattice Types and Space Groups*

The basic point group symmetry operations have to satisfy two conditions; first, to comply with the three-dimensional periodicity of crystals, and second, to have at least one "fixed point" in space. To these conditions; we can add a number of translational elements.

As shown in the last section, every basic symmetry operation has a matrix representation. The addition of a translational element can be expressed by a vector

$$\mathbf{t} = \begin{pmatrix} t_1 \\ t_2 \\ t_3 \end{pmatrix}$$

creating a new situation where we now have to operate with symmetry operations of the form

$$B = S + \mathbf{t} \tag{3.2}$$

where S is one of the 3×3 matrices derived in the last section and $\mathbf{t}$ is a translational vector which complies with the crystal periodicity. (Note that (3.2) has to be regarded as a formal operator to act on a position vector $\mathbf{r}$ so that the execution of B should lead to a vector $\mathbf{r}'$, given by $\mathbf{r}' = B\mathbf{r} = S\mathbf{r} + \mathbf{t}$. Only in this sense is (3.2) a reasonable expression. Regarded as a matrix equation, B would be undefinied since S and $\mathbf{t}$ are of different types.)

It can be shown that the set of all B operations forms a mathematical group, named the "group of motions", M. The complete symmetry of any motif can be described by a

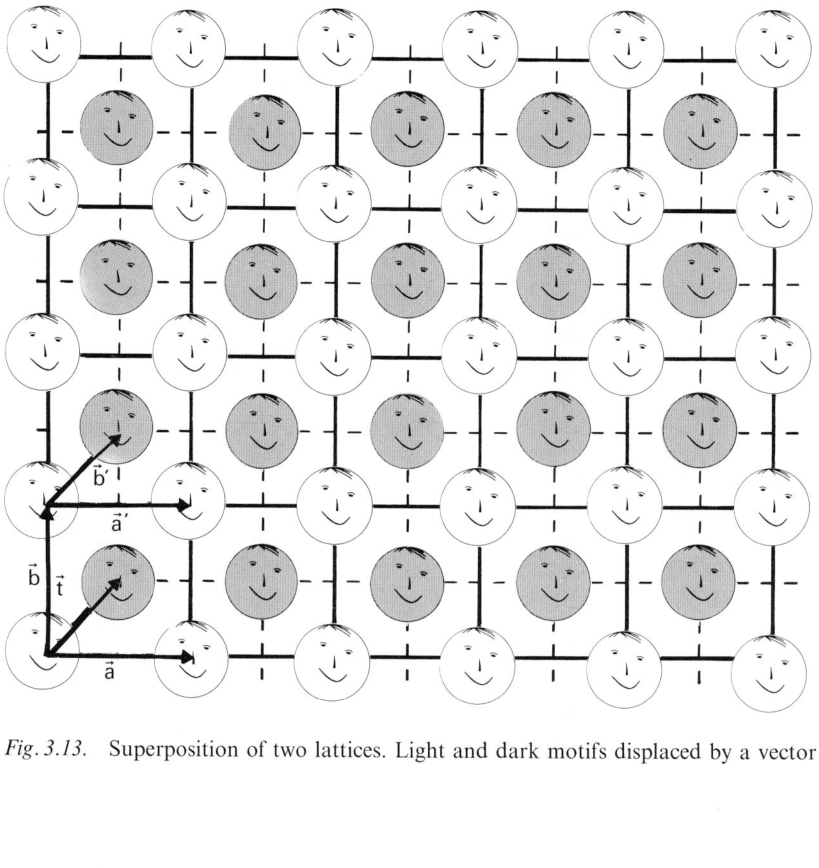

Fig. 3.13. Superposition of two lattices. Light and dark motifs displaced by a vector **t**.

special subgroup of M. No limitation can be given for the number of subgroups of M unless we restrict ourselves to the consideration of crystal symmetry. For this case, it was deduced by Schönflies and Fedorov, that the number of subgroups of M necessary to describe the symmetry of a crystal lattice is limited to 230. These 230 subgroups are called "space groups", and every crystal lattice has a symmetry which is described by one of the 230 space groups. Every space group is related to a crystal class, since it can be shown that the set of matrices in (3.2) needed for the description of its symmetry is equal to the set of matrices in one of the 32 crystal classes. In this sense, every space group belongs to a crystal class, or, we can say that every crystal class can be expanded to a number of space groups by addition of some translation vectors to its symmetry matrices.

The first and most simple possibility for expanding a crystal class is the addition of one overall translation vector to all symmetry matrices of a crystal class. The geometrical result of such an operation is illustrated in Fig. 3.13. Supposing this (two-dimensional) lattice is spanned by unit cell vectors **a** and **b**. We find, for every light motif, its equivalent shadowed motif, translated by a vector $\mathbf{t} = 1/2(\mathbf{a} + \mathbf{b})$. Thus every unit cell contains one additional shadowed motif which can be obtained from a light one by a pure translation operation. A lattice having this property is said to be centered; non-centered lattices are called primitive. As we see from Fig. 3.13, this centered lattice can be regarded as the sum of two identical primitive lattices, displaced against each other by the translation vector **t**; the two primitive lattices are given by the light and the shadowed motifs.

Note that in the example in Fig. 3.13, it is not absolutely necessary to regard the lattice as centered. If the unit cell vectors $\mathbf{a}'$ and $\mathbf{b}'$ are chosen, a primitive lattice is obtained. The reason for this is the low symmetry of this lattice, which has the identity as its only basic symmetry element. So it can be regarded as a (two-dimensional) triclinic lattice. Triclinic lattices permit an arbitrary choice of unit cells, so the construction of a centered lattice can be avoided.

The situation changes fundamentally for a lattice with higher symmetry. Consider, for the present, only the light motifs in Fig. 3.14a; we have again a lattice with symmetry 2/m. The $\mathbf{a} - \mathbf{c}$ plane is the mirror plane, and the 2-fold axis has the direction of **b**. The conventions of a monoclinic system require the choice of unit cell vectors as illustrated in Fig. 3.14a, with **b** as the monoclinic "unique axis" (second setting), and therefore the non-right angle is β (the direction of b has been chosen downward to get **a**, **b** and **c** in a right-handed system). As we have already seen, the matrices necessary to represent the lattice symmetry are the unit matrix E, S(2), S(m), and S(i). The addition of a vector $\mathbf{t} = 1/2\ (\mathbf{a} + \mathbf{b})$ to all these symmetry matrices leads to a centered lattice, resulting in the addition of the shadowed motifs to the light ones. Again we can regard this three-dimensional lattice as the sum of two primitive lattices displaced by the vector $\mathbf{t} = 1/2\ (\mathbf{a} + \mathbf{b})$, and again we have the possibility of avoiding the description of this lattice as centered by choosing another unit cell, for example with the vectors $\mathbf{a}'$, $\mathbf{b}'$, $\mathbf{c}'$. But this unit cell does not comply with the rules of a monoclinic system, since the two angles no longer have values of 90°. Now we have two possibilities. Either we

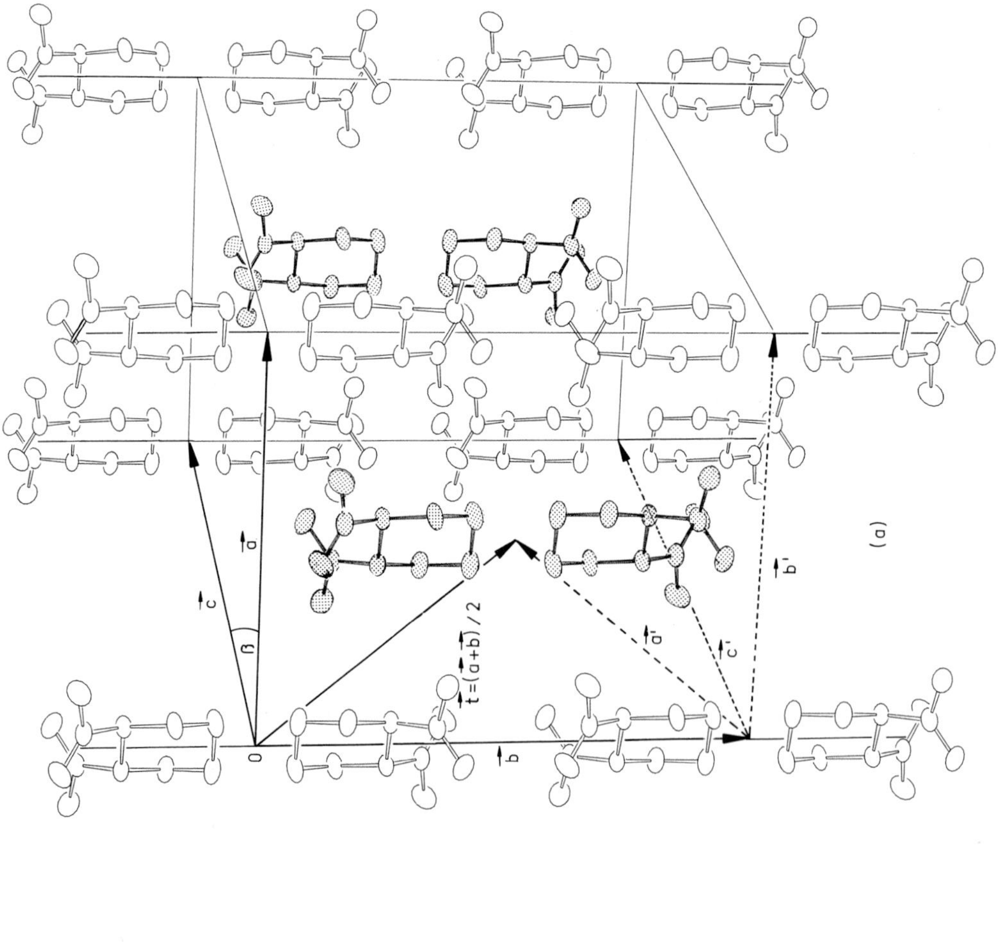

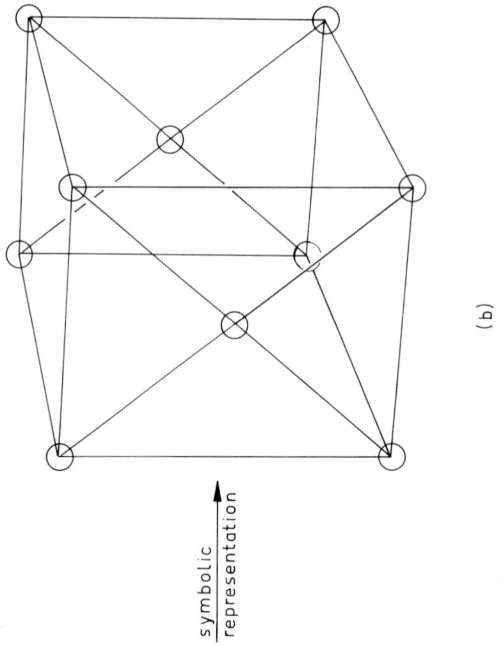

Fig. 3.14. (a) Centered lattice with symmetry 2/m, and (b) its schematic representation.

refrain from choosing the crystal system appropriate to the symmetry, or we include a centered lattice. Crystallographers use the latter, which means that in the example in Fig. 3.14a, the correct unit cell vectors are **a**, **b**, and **c**, resulting in a centered lattice.

The translation vector $\mathbf{t} = 1/2\,(\mathbf{a} + \mathbf{b})$, which is responsible for this centered lattice, points from the origin 0 of the unit cell to the center of the $\mathbf{a} - \mathbf{b}$ plane. Crystallographers call this plane the C-plane; the $\mathbf{a} - \mathbf{c}$ and $\mathbf{b} - \mathbf{c}$ planes are called the B- and A-planes, respectively. Lattices with centering of one plane are called face-centered, with a designation of the special face which is centered. That is, they are said to be A-, B-, or C-centered. The lattice in Fig. 3.14a is then C-centered. In a symbolic representation (see Fig. 3.14b) all motifs related by non-translational operations are collected in one single point in the eight equivalent corners of the unit cell, while the motifs obtained by pure translation are represented as a single point at the top of the translational vector.

Two more types of centering can be present in crystal lattices. A face-centered lattice like the monoclinic C-lattice is called doubly primitive, since it can be derived by two primitive lattices and an overall displacement vector transforming every point in one lattice to its equivalent in the second. Generalizing this concept, we can define centered lattices of higher multiplicity by n primitive lattices together with n − 1 displacement vectors. Fortunately, this general aspect need not be considered, since there are only two more centered lattice types present in crystals. These are those with all faces centered, designated F, and the body-centered lattices, desginated I. Their symbolic representations are given in Fig. 3.15. An F-centered lattice is quadruply primitive. In addition to every lattice point, we get three further points by translation vectors pointing to the centers of all faces. An I-lattice is doubly primitive, with one translation vector pointing to the body center of the unit cell.

The analytical representations of the displacement vectors are, for the F-lattice,

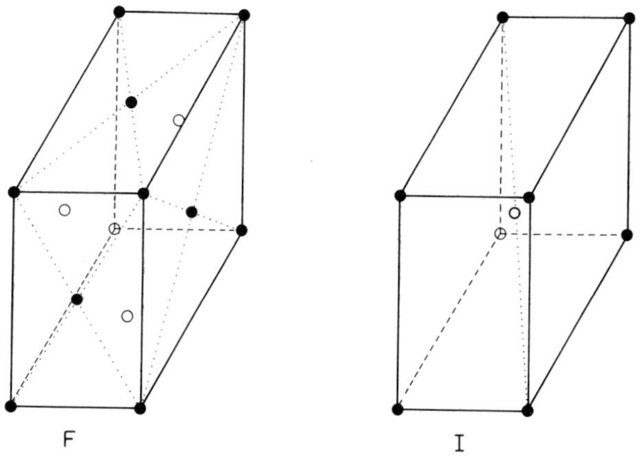

Fig. 3.15. Schematic representation of unit cells for F- and I-centered lattices.

$$\mathbf{t}_1 = \begin{pmatrix} 1/2 \\ 1/2 \\ 0 \end{pmatrix} \quad \mathbf{t}_2 = \begin{pmatrix} 0 \\ 1/2 \\ 1/2 \end{pmatrix} \quad \mathbf{t}_3 = \begin{pmatrix} 1/2 \\ 0 \\ 1/2 \end{pmatrix}$$

and for the I-lattice,

$$\mathbf{t} = \begin{pmatrix} 1/2 \\ 1/2 \\ 1/2 \end{pmatrix}$$

It should be noted here that a triply primitive lattice can be observed in a hexagonal lattice. But, it can be replaced by a primitive one, if the lattice is described in the rhombohedral system. With this reservation, we can summarize. The addition of an overall translation vector to the symmetry operations known so far leads to three possible types of centered lattices. Two of them, the face-centered and the body-centered lattices, are doubly primitive, and the last, called all face-centered is quadruply primitive.

A detailed study of the possibilities for centering in all crystal systems shows that crystals can have only 14 different types of lattices. Since Bravais (1850) was the first to discover this, these 14 lattice types are called the Bravais lattices. We show the symbolic representations of the 14 lattice types in Fig. 3.16, and give a brief discussion here. In the triclinic crystal system, only primitive lattices, denoted P, are necessary, as has already been demonstrated in the example in Fig. 3.13. It is conventional to avoid centered lattices whenever possible, unless the choice of a primitive cell is inconsistent with the orientation of the symmetry axes in the unit cell. Following this convention, triclinic cells will only be primitive.

In the monoclinic system, it is sufficient to describe all centering by two types, a primitive or a C-centered lattice. An A-centered lattice which is equivalent to C-centering is avoided by choosing the non-unique axis vectors **a** and **c** in a way that the centered face will be **a** — **b** instead of **b** — **c**. A B-centered monoclinic cell can be transformed into a primitive monoclinic cell of half the volume. This transformation is illustrated in Fig. 3.17.

If **a**, **b**, **c** are the unit cell vectors of the B-centered cell (unique axis b), we obtain three unit cell vectors **A**, **B**, **C**

$$\mathbf{A} = 1/2\,(\mathbf{a} - \mathbf{c})$$
$$\mathbf{B} = \mathbf{b}$$
$$\mathbf{C} = 1/2\,(\mathbf{a} + \mathbf{c})$$

by the transformation representing a primitive monoclinic lattice, since the angles between **A** and **B** and **C** and **B** are still 90°, and the angle between **A** and **C** is generally not a right angle. The problem of showing the redundancy of I- and F-lattices in the monoclinic system is left as an exercise to the reader.

The orthorhombic crystal system allows all four types of centered lattices. For the

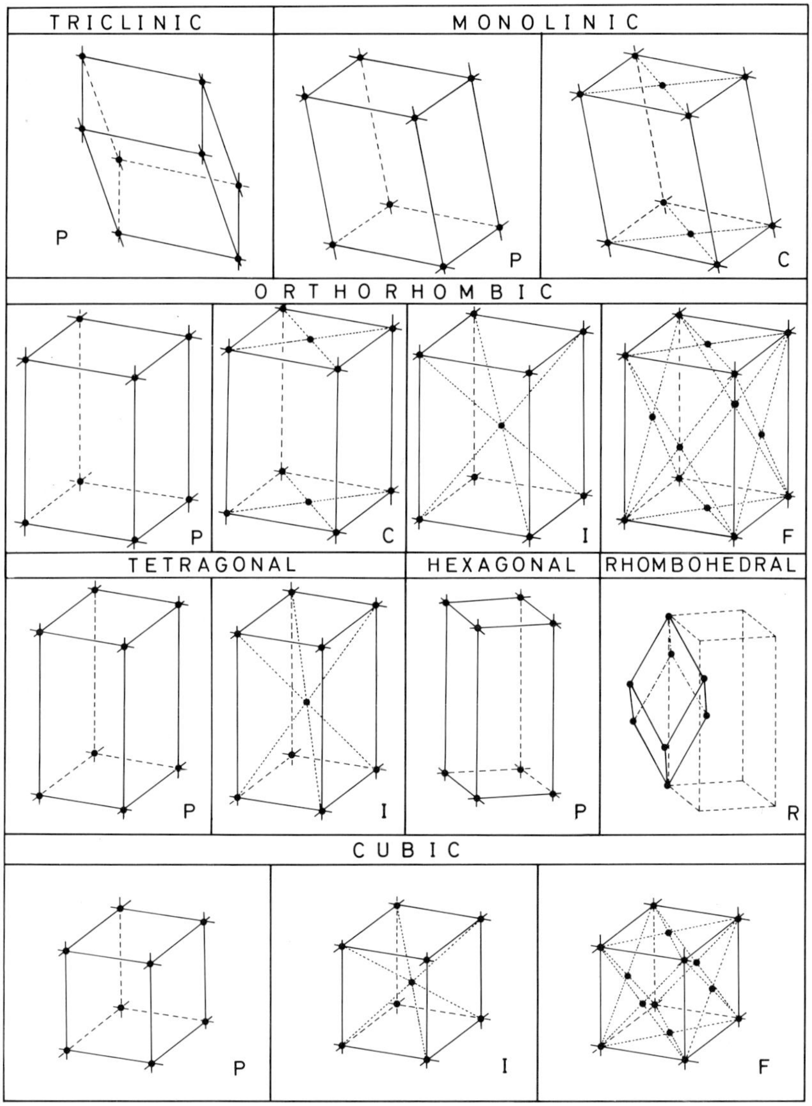

Fig. 3.16. The 14 Bravais lattice types.

face-centered lattice, the convention is the same as in the monoclinic case. The choice of unit cell vectors is such that the centered face is the $\mathbf{a}-\mathbf{b}$ plane.

In thc tetragonal system, there are only two lattice types, the primitive and the body-centered. One example of reduction of other types to one of these two representatives is demonstrated in Fig. 3.18. which shows the transformation between tetragonal F- and I-lattices.

In the hexagonal and trigonal system, only primitive lattices are present. However, it

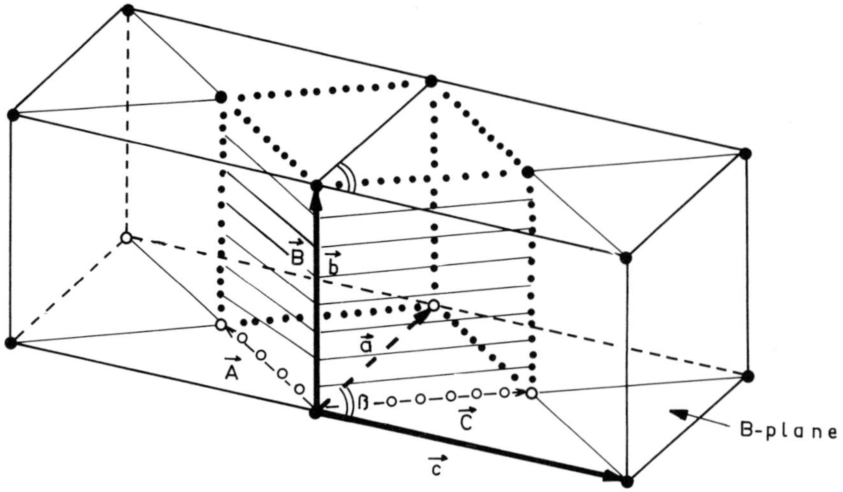

Fig. 3.17. Transformation of a monoclinic B-centered cell to a primitive cell.

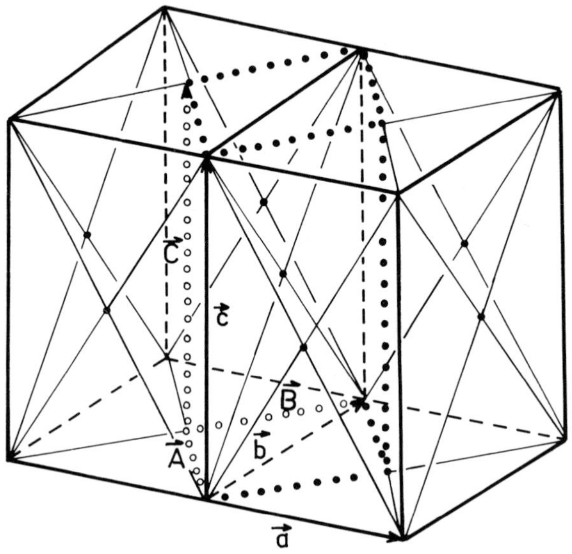

Fig. 3.18. Transformation of a tetragonal F-centered cell to an I-centered cell.

must be noted that a centering in the hexagonal system can only be avoided by transforming a triply primitive hexagonal lattice into a trigonal system. Fig. 3.19 shows a lattice of that type. We have three types of lattice points; those drawn by full circles having components in the c-direction of 0 or 1, those marked horizontally having z-components 1/3, and those marked vertically with z-components 2/3. The translation vectors in the sense of (3.2) are

$$\mathbf{t}_1 = (1/3)\,\mathbf{a} + (2/3)\,\mathbf{b} + (2/3)\,\mathbf{c}$$
$$\mathbf{t}_2 = (2/3)\,\mathbf{a} + (1/3)\,\mathbf{b} + (1/3)\,\mathbf{c}$$

or, written as columns

$$\mathbf{t}_1 = \begin{pmatrix} 1/3 \\ 2/3 \\ 2/3 \end{pmatrix}; \qquad \mathbf{t}_2 = \begin{pmatrix} 2/3 \\ 1/3 \\ 1/3 \end{pmatrix}$$

The transformation into the rhombohedral cell allows the description of this lattice as primitive. However, note that here the centering does not disappear by choosing another cell in the same crystal system, but only by changing to another system.

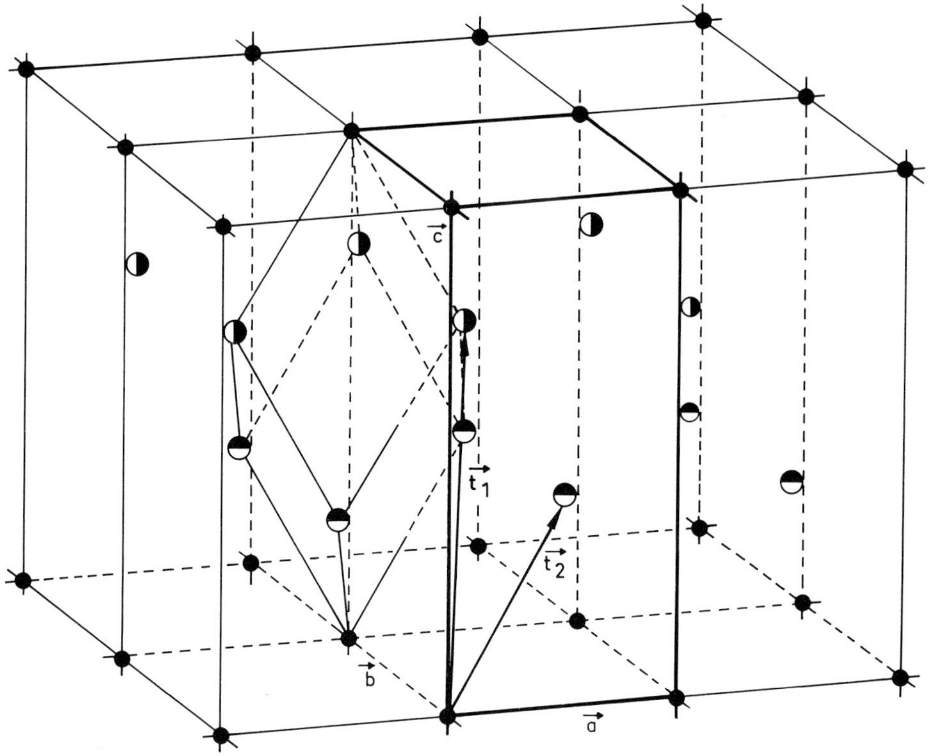

Fig. 3.19. Unit cell transformation of a triply primitive hexagonal lattice to a primitive rhombohedral lattice.

The notation of lattice types in the hexagonal and trigonal system is as follows. Primitive lattices in both systems are indicated by a P. A lattice which is triply primitive in the hexagonal system, but primitive in the rhombohedral system, is denoted R.

Finally, the cubic system allows P-, I-, and F-lattices. A face-centered lattice is impossible, since all directions are equivalent because of the symmetry in this system.

The last subject to be discussed is the consequence of adding an individual translation vector to a basic symmetry operation. This can be done briefly, since the results are easy to formulate. In a crystal lattice there are two possibilities of combining a basic symmetry operation with an individual translation vector. The first is the addition of a displacement to a reflection, the displacement vector necessarily having a direction parallel to the mirror plane. A symmetry element of that kind is designated a "glide plane". The second is the addition of a displacement to a rotation, the displacement vector necessarily having a direction parallel to the rotation axis. A symmetry element of that kind is called a "screw axis".

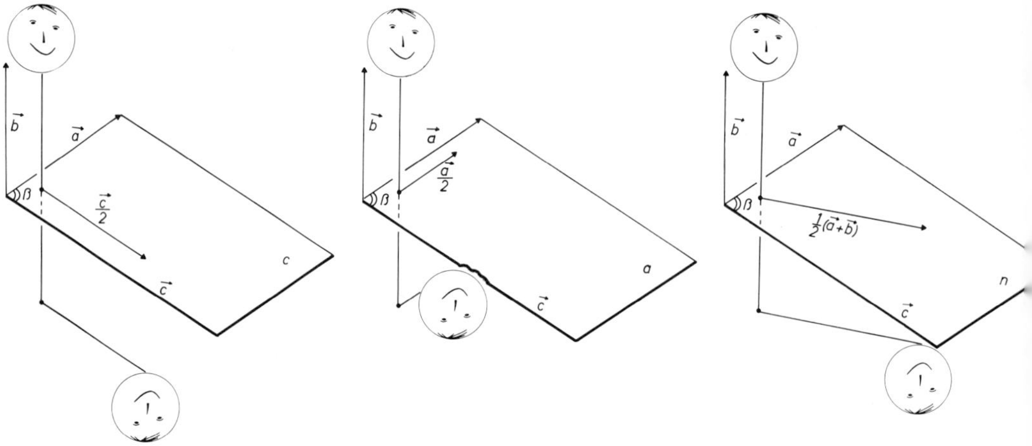

Fig. 3.20. Glide planes in a monoclinic cell.

We have seen that the components of the overall displacement vectors could not have arbitrary values, but were restricted to integer fractions of the unit cell vectors. The same holds for glide planes and screw axes. Usually glide planes are parallel to one of the unit cell planes, either $\mathbf{a}-\mathbf{b}$, $\mathbf{a}-\mathbf{c}$ or $\mathbf{b}-\mathbf{c}$ (there are only a few exceptions in the high symmetric crystal systems). In Fig. 3.20, we have drawn a glide plane parallel to the $\mathbf{a}-\mathbf{c}$ plane of a monoclinic cell. Since we have already pointed out that the glide vector $\mathbf{t}$ must be parallel to the mirror plane, it follows that

$$\mathbf{t}=\alpha\mathbf{a}+\beta\mathbf{c}$$

α and β can have only the values 0 and 1/2, and we have three possibilities

$$\mathbf{t}(\mathrm{a}) \quad =(1/2)\,\mathbf{a} \tag{3.3a}$$

$$\mathbf{t}(\mathrm{c}) \quad =(1/2)\,\mathbf{c} \tag{3.3b}$$

$$\mathbf{t}(\mathrm{a,c})=1/2\,(\mathbf{a}+\mathbf{c}) \tag{3.3c}$$

Glide planes are called axial if the glide component is parallel to an axial direction; they are called diagonal if the glide component is diagonal. In the axial case, they are designated by the character indicating the glide direction. In the diagonal case, they are designated by n. The three possible glide planes are shown in Fig. 3.20; two axial of types a (with $\mathbf{t} = \mathbf{t}(a)$) and c (with $\mathbf{t} = \mathbf{t}(c)$), and one diagonal of type n (with $\mathbf{t} = \mathbf{t}(a, c)$).

There is a further type of glide plane known as the diamond glide d, since its name derives from its occurrence in the diamond structure. The translation vector is given by a quarter of a diagonal, i.e., it is of the form $1/4(\mathbf{a}+\mathbf{b})$, $1/4(\mathbf{a}+\mathbf{c})$, $1/4(\mathbf{b}+\mathbf{c})$, or $1/4(\mathbf{a}+\mathbf{b}+\mathbf{c})$.

Screw axes exist for every allowed rotation axis. Their classification is very simple. If the repetition vector in the axis direction is $\mathbf{R}$, we can derive from every n-fold axis $n-1$ screw axes, having the translational components $(1/n)\mathbf{R}$, $(2/n)\mathbf{R}$, ..., $((n-1)/n)\ \mathbf{R}$ ($n = 2, 3, 4, 6$. This proposition is also valid for $n = 1$, but it has no practical sense.). The symbolic notation is n_m if the axis is n-fold and the translation vector is $(m/n)\mathbf{R}$. Fig. 3.21 shows all possible screw axes. Two properties should be pointed out:

(1) Usually the rotation axis has the direction of one unit cell vector, say $\mathbf{c}$. Then for a n_m-screw axis, the translation vector is $\mathbf{t} = (m/n)\mathbf{c}$ or, written as a column,

$$\mathbf{t} = \begin{pmatrix} 0 \\ 0 \\ m/n \end{pmatrix} \qquad (3.4)$$

It is possible for 3-, 4-, and 6-fold screw axes that with multiple applications of the srew axis operation, the translation component exceeds the repetition period. In a 3_2-axis, for example, the first rotation about 120° is accompanied by a translation of $(2/3)\mathbf{c}$, producing the point 2. The second rotation about 120° needs a further translation of $(2/3)\mathbf{c}$; that means 3′ is already translated by $(4/3)\mathbf{c}$, which is $\mathbf{c} + (1/3)\mathbf{c}$. Since $\mathbf{c}$ is the repetition period, 3′ is the equivalent of point 3 (having $\mathbf{t} = (1/3)\mathbf{c}$) in the initial cell. The same situation is observed for all other screw axes for which this problem arises.

(2) Let us compare all equivalent points produced by a 3_1- and a 3_2-axis. It is clear that the points of the 3_1-axis are related by an *anticlockwise* rotation of 120° each, plus translation by 1/3 in the $\mathbf{c}$-direction. The points produced by the 3_2-axis operation can be regarded as rotated by 120° *clockwise*, and than translated by $(1/3)\mathbf{c}$. Thus the two axes differ only in their direction of rotation. In other words, the motifs produced by these two screw axes are mirror-image related; they behave like left- and right-hand screws.

Motifs having these properties are said to be enantiomorphous. Enantiomorphy plays an important role when dealing with the structures of optically active compounds. An enantiomorphic relationship exists also for the 4_1- and 4_3-axis and for two pairs of 6-fold screw axes, the 6_1–6_5 pair and the 6_2–6_4 pair.

Let us summarize the crystal symmetry concepts (Fig. 3.22).

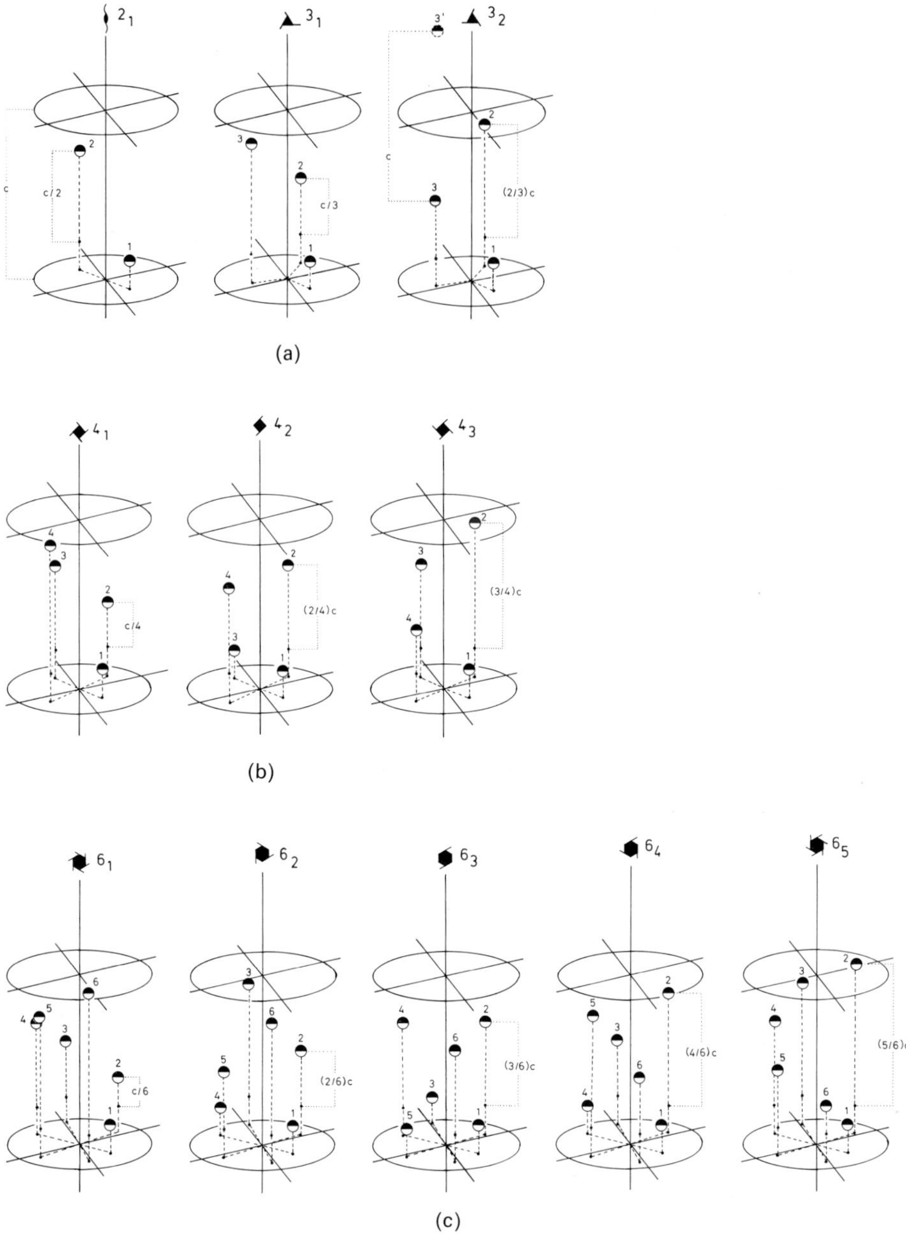

Fig. 3.21. 2-, 3-, 4- and 6-fold screw axes.

(1) From the five rotation axes possible in a crystal lattice, we get, by addition of their corresponding inversion axes, the 10 basic symmetry operations.

(2) All 10 basic operations can be expressed analytically by a 3×3 matrix representation.

(3) With the restrictions resulting from the three-dimensional periodicity of crystals, 32 combinations of the basic operations are possible. These 32 sets of basis symmetry operations, each forming a mathematical group, are the so-called 32 "crystal classes". Each crystal class can be represented by a finite set of 3×3 matrices corresponding to the basic operations actually present in this class.

(4) The introduction of translational elements results in the fourteen possible Bravais lattices on one hand and on the other hand in two new types of symmetry elements, the glide planes and the screw axes. Every symmetry element of this expanded set can be represented by a 3×3 matrix and a three-dimensional column vector. 230 subsets, each forming a mathematical group, can be deduced to describe the complete symmetry of a crystal lattice. These groups are called the 230 "space groups". Since the set of matrices needed in the matrix-vector description in one space group is identical to the set of matrices of one crystal class, it can be said that every space group belongs to a crystal class, or that every crystal class has an expansion to a certain number of space groups.

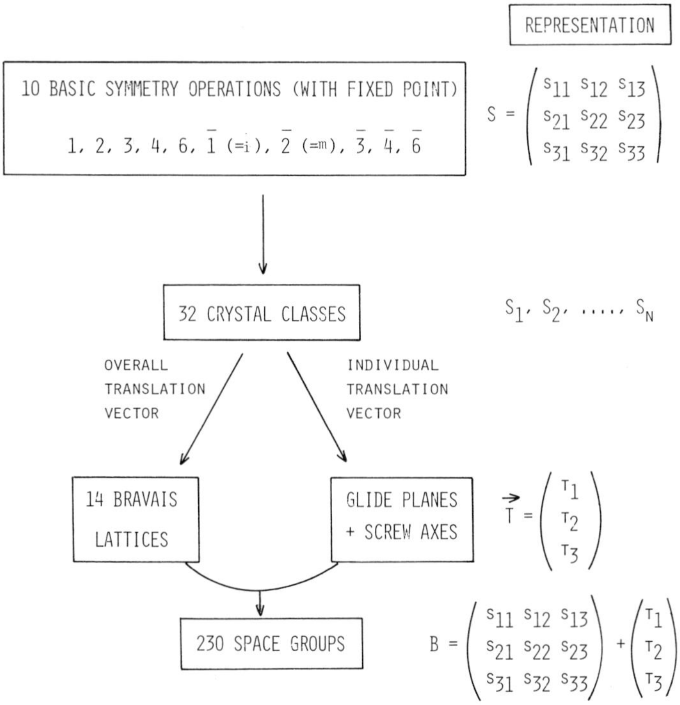

Fig. 3.22. Crystal symmetry derivation scheme.

For a better understanding of the derivation of space groups, let us discuss in detail the expansion of the monoclinic crystal class 2/m to its six possible space groups. Let us consider again the symmetry elements of this crystal class:

(1) the identity, 1,

$$S(1) = \begin{pmatrix} 1 & 0 & 0 \\ 0 & 1 & 0 \\ 0 & 0 & 1 \end{pmatrix}$$

(2) the 2-fold axis, 2,

$$S(2) = \begin{pmatrix} -1 & 0 & 0 \\ 0 & 1 & 0 \\ 0 & 0 & -1 \end{pmatrix}$$

(3) the mirror plane, m, perpendicular to the 2-fold axis,

$$S(m) = \begin{pmatrix} 1 & 0 & 0 \\ 0 & -1 & 0 \\ 0 & 0 & 1 \end{pmatrix}$$

(4) the inversion center, i,

$$S(i) = \begin{pmatrix} -1 & 0 & 0 \\ 0 & -1 & 0 \\ 0 & 0 & -1 \end{pmatrix}$$

All matrix representations are valid provided that a monoclinic system was chosen with the monoclinic b-axis having the direction of the 2-fold axis.

The first space group derived is a very simple one, obtained by addition of no translational elements, i.e., the first space group contains the same symmetry elements as the crystal class itself. It is called P2/m. The first capital character always denotes the Bravais lattice type, which is primitive in this case. Its matrix vector representation is that of the crystal class with all translational vectors equal to the null vector. Hence it follows for the four equivalent positions of an arbitrary vector $\mathbf{r} = (x, y, z)$,

(1) x, y, z; (2) $-x, y, -z$; (3) $x, -y, z$; (4) $-x, -y, -z$

We get new space groups by replacing the 2-fold axis with a screw axis, or the mirror plane with a glide plane (or both), and we have the opportunity of introducing a C-centering. No other centering is allowed in the monoclinic system.

Let us do these expansions step by step. The 2-fold axis can be replaced only by one screw axis, a 2_1-axis. This leads to the space group $P2_1/m$. The matrix-vector representation for the screw axis is now

$$B(2_1) = \begin{pmatrix} -1 & 0 & 0 \\ 0 & 1 & 0 \\ 0 & 0 & -1 \end{pmatrix} + \begin{pmatrix} 0 \\ 1/2 \\ 0 \end{pmatrix}$$

Its application to an arbitrary position vector $\mathbf{r} = (x, y, z)$ leads to

$$\begin{pmatrix} -1 & 0 & 0 \\ 0 & 1 & 0 \\ 0 & 0 & -1 \end{pmatrix} \begin{pmatrix} x \\ y \\ z \end{pmatrix} + \begin{pmatrix} 0 \\ 1/2 \\ 0 \end{pmatrix} = \begin{pmatrix} -x \\ 1/2 + y \\ -z \end{pmatrix}$$

The simple replacement of the 2-fold axis by the 2_1-screw axis has one important consequence in the choice of the unit cell origin. In the space group P2/m it can be seen from the four equivalent positions that the origin is located at the intersection of the 2-fold axis and the mirror plane. This point is also identical to a center of symmetry. The distribution of all symmetry elements in this space group is shown in Fig. 3.23a. In addition to the 2-fold axis at $x = 0, z = 0$, further axes are produced at $x = 1/2, z = 0$, $x = 0, z = 1/2$, and $x = 1/2, z = 1/2$. Similarly, a mirror plane at $y = 1/2$ results from the existence of mirror planes at $y = 0$ and $y = 1$.

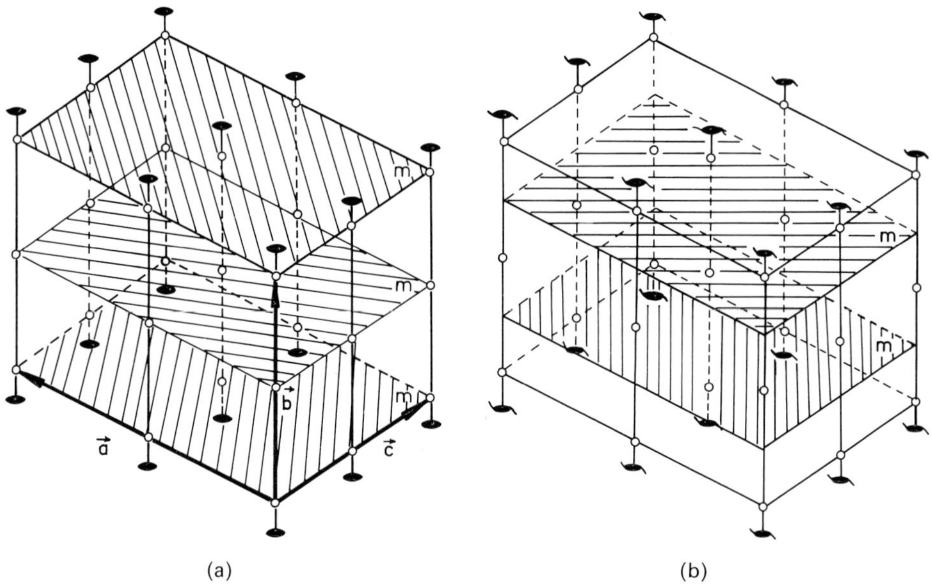

Fig. 3.23. Distribution of all symmetry elements in a unit cell: (a) space group P2/m; (b) space group $P2_1/m$.

We have already pointed out that the subsequent execution of two symmetry operations results in one symmetry operation of that space group. For example, execution of a reflection on m, followed by an inversion, is equal to a rotation about the 2-fold axis. This general property of all symmetry operations in every space group can also be examined by the matrix-vector representation. Every combination of two representation leads to one representation already present in the space group. This rule would not be true, if, in the space group $P2_1/m$, the origin were at the intersection of the screw axis and the mirror plane. This can easily be seen by subsequent application of an inversion after a 2_1-screw operation. We get

$$\begin{pmatrix} -1 & 0 & 0 \\ 0 & -1 & 0 \\ 0 & 0 & -1 \end{pmatrix} \left[\begin{pmatrix} -1 & 0 & 0 \\ 0 & 1 & 0 \\ 0 & 0 & -1 \end{pmatrix} + \begin{pmatrix} 0 \\ 1/2 \\ 0 \end{pmatrix} \right] = \begin{pmatrix} 1 & 0 & 0 \\ 0 & -1 & 0 \\ 0 & 0 & 1 \end{pmatrix} + \begin{pmatrix} 0 \\ -1/2 \\ 0 \end{pmatrix}$$

The translation component $-1/2$ is equal to $+1/2$, since $+1/2 = -1/2 + 1$ and the addition of a complete translation period is always allowed for crystals. So finally we get the result

$$\begin{pmatrix} 1 & 0 & 0 \\ 0 & -1 & 0 \\ 0 & 0 & 1 \end{pmatrix} + \begin{pmatrix} 0 \\ 1/2 \\ 0 \end{pmatrix}$$

However, this representation is that of a reflection at $y = 1/4$, rather than at $y = 0$. To avoid this problem, we shift the origin by 1/4 in the y-direction. This has the further advantage that the origin and the inversion center are still identical. With this choice of origin, the mirror plane lies at $y = 1/4$ (see Fig. 3.23b) and we have the complete representation of all symmetry elements

$$B(1) = \begin{pmatrix} 1 & 0 & 0 \\ 0 & 1 & 0 \\ 0 & 0 & 1 \end{pmatrix} + \begin{pmatrix} 0 \\ 0 \\ 0 \end{pmatrix}$$

$$B(2_1) = \begin{pmatrix} -1 & 0 & 0 \\ 0 & 1 & 0 \\ 0 & 0 & -1 \end{pmatrix} + \begin{pmatrix} 0 \\ 1/2 \\ 0 \end{pmatrix}$$

$$B(m) = \begin{pmatrix} 1 & 0 & 0 \\ 0 & -1 & 0 \\ 0 & 0 & 1 \end{pmatrix} + \begin{pmatrix} 0 \\ 1/2 \\ 0 \end{pmatrix}$$

$$B(i) = \begin{pmatrix} -1 & 0 & 0 \\ 0 & -1 & 0 \\ 0 & 0 & -1 \end{pmatrix} + \begin{pmatrix} 0 \\ 0 \\ 0 \end{pmatrix}$$

The application of these four representations to a position vector leads to the four equivalent positions

(1) x, y, z; (2) $-x, 1/2 + y, -z$; (3) $x, 1/2 - y, z$; (4) $-x, -y, -z$

The next space group, C2/m, is obtained by replacing the primitive lattice by a face-centered one, or in other words, by addition of the overall translation vector

$$\mathbf{t} = \begin{pmatrix} 1/2 \\ 1/2 \\ 0 \end{pmatrix}$$

to the symmetry elements of 2/m. We get, ultimately, the following eight equivalent positions

(1) x, y, z; (2) −x, y, −z; (3) x, −y, z; (4) −x, −y, −z;
(5) x + 1/2, y + 1/2, z; (6) −x + 1/2, y + 1/2, −z; (7) x + 1/2, −y + 1/2, z;
(8) −x + 1/2, −y + 1/2, −z

We would get eight matrix-vector representations, the first four identical to those of P2/m and the second four produced from the first four by addition of the translation vector. This representation can be specified by the primitive operations together with the translation vector:

(0, 0, 0; 1/2, 1/2, 0) +

(1) x, y, z; (2) −x, y, −z; (3) x, −y, z; (4) −x, −y, −z.

Now we can proceed with the replacement of the mirror plane by a glide plane. Since the standard choice of glide direction is c, we get the space group P2/c. To keep the origin identical with the inversion center, the 2-fold axis is shifted to x = 0, z = 1/4, and we get the four equivalent positions

(1) x, y, z; (2) −x, y, 1/2 − z; (3) x, −y, 1/2 + z; (4) −x, −y, −z.

If one of the two other glide directions is chosen (see Fig. 3.20), symmetry operation (3) must be changed. In the case of an a-glide plane, we get

(3)′ 1/2 + x, −y, z

This change causes the 2-fold axis to be situated now at x = 1/4 and z = 0, and symmetry operation (2) must also be changed,

(2)′ 1/2 − x, y, −z

For an n-glide plane, we get

(3)″ 1/2 + x, −y, 1/2 + z

and the 2-fold axis must be situated at x = 1/4, z = 1/4. Then (2) reads

(2)″ 1/2 − x, y, 1/2 − z.

In the case of an a- or an n-glide plane, the space group would be called P2/a or P2/n, but together with P2/c all three are regarded as one space group, since they all contain the same symmetry operations.

Replacing both the mirror plane and the 2-fold axis by a glide plane and a screw axis, we get the well-known space group $P2_1/c$. This is by far the space group most frequently present, especially for organic structures. Like the mirror plane in $P2_1/m$, the glide plane is positioned at y = 1/4. For the usual glide direction c, the 2-fold axis ist at x = 0, z = 1/4 and we have as matrix-vector representations of the four symmetry elements,

$$(1)\quad \begin{pmatrix} 1 & 0 & 0 \\ 0 & 1 & 0 \\ 0 & 0 & 1 \end{pmatrix} + \begin{pmatrix} 0 \\ 0 \\ 0 \end{pmatrix};$$

$$(2)\quad \begin{pmatrix} -1 & 0 & 0 \\ 0 & 1 & 0 \\ 0 & 0 & -1 \end{pmatrix} + \begin{pmatrix} 0 \\ 1/2 \\ 1/2 \end{pmatrix};$$

$$(3)\quad \begin{pmatrix} 1 & 0 & 0 \\ 0 & -1 & 0 \\ 0 & 0 & 1 \end{pmatrix} + \begin{pmatrix} 0 \\ 1/2 \\ 1/2 \end{pmatrix};$$

$$(4)\quad \begin{pmatrix} -1 & 0 & 0 \\ 0 & -1 & 0 \\ 0 & 0 & -1 \end{pmatrix} + \begin{pmatrix} 0 \\ 0 \\ 0 \end{pmatrix}$$

Application of these four representations to an arbitrary position vector $\mathbf{r} = (x, y, z)$ gives the four equivalent vectors

$$(1)\quad \begin{pmatrix} 1 & 0 & 0 \\ 0 & 1 & 0 \\ 0 & 0 & 1 \end{pmatrix} \begin{pmatrix} x \\ y \\ z \end{pmatrix} + \begin{pmatrix} 0 \\ 0 \\ 0 \end{pmatrix} = \begin{pmatrix} x \\ y \\ z \end{pmatrix}$$

$$(2)\quad \begin{pmatrix} -1 & 0 & 0 \\ 0 & 1 & 0 \\ 0 & 0 & -1 \end{pmatrix} \begin{pmatrix} x \\ y \\ z \end{pmatrix} + \begin{pmatrix} 0 \\ 1/2 \\ 1/2 \end{pmatrix} = \begin{pmatrix} -x \\ 1/2 + y \\ 1/2 - z \end{pmatrix}$$

$$(3)\quad \begin{pmatrix} 1 & 0 & 0 \\ 0 & -1 & 0 \\ 0 & 0 & 1 \end{pmatrix} \begin{pmatrix} x \\ y \\ z \end{pmatrix} + \begin{pmatrix} 0 \\ 1/2 \\ 1/2 \end{pmatrix} = \begin{pmatrix} x \\ 1/2 - y \\ 1/2 + z \end{pmatrix}$$

$$(4)\quad \begin{pmatrix} -1 & 0 & 0 \\ 0 & -1 & 0 \\ 0 & 0 & -1 \end{pmatrix} \begin{pmatrix} x \\ y \\ z \end{pmatrix} + \begin{pmatrix} 0 \\ 0 \\ 0 \end{pmatrix} = \begin{pmatrix} -x \\ -y \\ -z \end{pmatrix}$$

In crystallography, these vectors are written as the equivalent positions

(1) x, y, z; (2) $-x, 1/2 + y, 1/2 - z$; (3) $x, 1/2 - y, 1/2 + z$; (4) $-x, -y, -z$.

Because of the great frequency of space group $P2_1/c$, everyone working on crystal structures should be well experienced with its symmetry. It is evident that instead of a c-glide plane, an a- or n-glide plane can be chosen in this space group. Then we get as symbols,

$P2_1/a$ and (2)′ $1/2 - x, 1/2 + y, -z$
(3)′ $1/2 + x, 1/2 - y, z$

or $P2_1/n$ and (2)″ $1/2-x, 1/2+y, 1/2-z$
(3)″ $1/2+x, 1/2-y, 1/2+z$

The last space group, derived form the crystal class 2/m and designated C2/c, consists of a C-centered lattice, a 2-fold axis and a c-glide plane. As in C2/m, we write the eight equivalent positions in the form

$(0, 0, 0;\ 1/2, 1/2, 0)+$

(1) x, y, z; (2) $-x, y, 1/2-z$; (3) $x, -y, 1/2+z$; (4) $-x, -y, -z$.

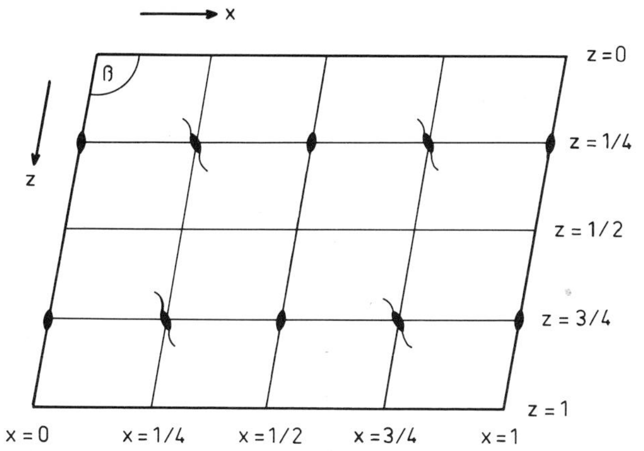

Fig. 3.24. Distribution of 2- and 2_1-axes in a unit cell, space group C2/c.

Note that the presence of a 2-fold axis, together with a C-centering, automatically produces 2_1-screw axes (see Fig. 3.24). As indicated by symmetry operation (2), the 2-fold axes are at $x=0$, $z=1/4$, and the screw axes are then at $x=1/4$ and $z=1/4$. Therefore this space group could also be written as $C2_1/c$. The same is true for the space group C2/m, which is equivalent to $C2_1/m$. All the possibilities of replacing one or more basic symmetry elements in 2/m by those with translational elements have now been utilized. We have derived the six space groups belonging to the crystal class 2/m, and no more can be derived from this class. Before finishing the description of crystal symmetry, it should be noted that questions concerned with crystal symmetry and space group representation are discussed in detail in the "International Tables for X-Ray Crystallography," Vol. I, of which a new issue will be distributed in 1980. Illustrated descriptions of each of the 230 space groups are given, with all the symbols and official abbreviations used for every symmetry operation. Everyone working on crystal structure analysis makes use of the "Tables". They are a tool no crystallographer should dispense with. All questions concerning symmetry which are not discussed here are surely described in the Tables.

3.2 Crystal Symmetry and Related Intensity Symmetry

As shown in Section 1.2.3, the advantage of single crystal diffraction is the production of discrete intensities which can be uniquely related to the integer lattice planes. Using this provision the basic formulae of diffraction theory can be transformed into expressions which are more suitable for numerical calculations.

3.2.1 *Representation of ϱ and F as Fourier Series*

First we shall transform the formulae for the electron density function $\varrho(\mathbf{x})$ and the structure factor F ($\mathbf{b}$) as derived in 1.2.2 for arbitrary materials into expressions which take into account the special properties of single crystals. ϱ and F are related by Fourier transforms, i.e.,

$$F(\mathbf{h}) = \int_V \varrho(\mathbf{r})\, e^{2\pi i \mathbf{h}\mathbf{r}}\, dV \tag{3.5}$$

$$\varrho(\mathbf{r}) = \int_{V^*} F(\mathbf{h})\, e^{-2\pi i \mathbf{h}\mathbf{r}}\, dV^* \tag{3.6}$$

with

$$\mathbf{h} = h\mathbf{a}^* + k\mathbf{b}^* + l\mathbf{c}^* \qquad h, k, l \text{ are integers} \tag{3.7}$$

and

$$\mathbf{r} = x\mathbf{a} + y\mathbf{b} + z\mathbf{c} \tag{3.8}$$

A few remarks must be added to these formulae. The argument of F is designated $\mathbf{h}$ instead of $\mathbf{b}$ as in 1.2.2, which indicates that F now has to be considered only for lattice vectors satisfying condition (3.7). Base vectors $\mathbf{a}$, $\mathbf{b}$, $\mathbf{c}$ and $\mathbf{a}^*$, $\mathbf{b}^*$, $\mathbf{c}^*$ as used in (3.7) and (3.8) are no longer chosen arbitrarily, but will have to be in agreement with the conditions of one of the seven crystal systems given in Table 3.3.

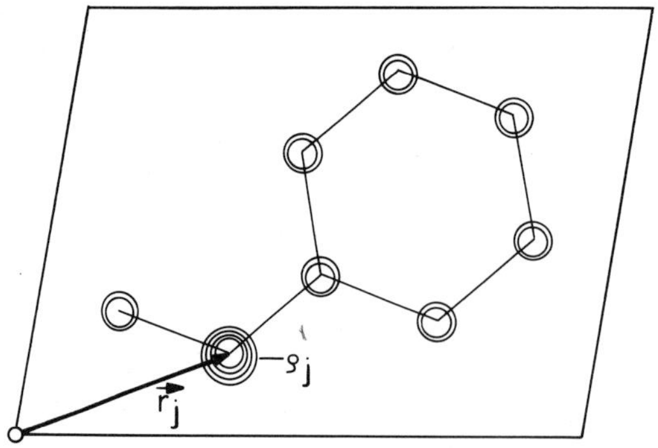

Fig. 3.25. Representation of $\varrho(\mathbf{r})$ in terms of the ϱ_j's.

$\varrho(\mathbf{r})$, which has to be considered only in one unit cell because of its three-dimensional periodicity, can be written in terms of the electron density of the contributing atoms.

Let $\varrho_j(\mathbf{r})$ be the electron density of the jth atom with the position vector $\mathbf{r}$ referred to an origin at the atomic center (Fig. 3.25). The problem is that the precise electron density distribution of an atom is not precisely known. For chemical reasons, it is evident that ϱ_j depends on both the atom type and its bonding state.

Let us suppose that $\varrho_j(\mathbf{r})$ is known; then we get for the electron density of N atoms with position vectors $\mathbf{r}_j$ in the unit cell

$$\varrho(\mathbf{r}) = \sum_{j=1}^{N} \varrho_j(\mathbf{r} - \mathbf{r}_j)$$

If $f_j(\mathbf{h})$ is the Fourier transform of ϱ_j, we get from the "Shift-Theorem" from 1, equation (1.29),

$$F(\mathbf{h}) = \mathsf{F}(\varrho(\mathbf{r})) = \sum_{j=1}^{N} f_j(\mathbf{h})\, e^{2\pi i \mathbf{h}\mathbf{r}_j}$$

The integral representation for F in (3.5) is now replaced by a series representation, which is more convenient in numerical calculations. The problem remaining is the quantity $f_j(\mathbf{h})$, which must be calculated from the $\varrho_j(\mathbf{r})$.

In crystallography it is customary to make the approximation that the ϱ_j''s are spherical, i.e. the ϱ_j does not depend on $\mathbf{r}$, but only on $r = |\mathbf{r}|$, hence

$$\varrho_j \approx \varrho_j(r) \tag{3.9}$$

This implies that f_j also has spherical symmetry. Since $|\mathbf{h}|$ is proportional to $s = \sin\theta/\lambda$ (Bragg's law, see equation (1.44) in Section 1), we get

$$f_j = f_j(s), \qquad s = \frac{\sin\theta}{\lambda} \tag{3.10}$$

and finally for F,

$$F(\mathbf{h}) = \sum_{j=1}^{N} f_j\, e^{2\pi i \mathbf{h}\mathbf{r}} \tag{3.11}$$

(Usually the argument s for the f_j is omitted.)

Several models have been used to express $\varrho_j(r)$ analytically and to thence derive $f_j(s)$. These calculations are based on the atomic model at rest, so that the f_j corresponds to the scattering power of the stationary atom. The f_j-curves in terms of s obtained by the various calculations (Hartree, Fock and several other authors) are tabulated in the International Tables, Vol. III for all elements and a large number of elemental ions (together with a detailed literature reference list on that field). The f_j's are called "atomic scattering factors". For example, the atomic scattering factor curves for H, C, and O and O^- are drawn in Fig. 3.26. They all start at $s = 0$ with a value equal to the atomic number, and decrease monotonically towards zero for $s > 1.1$.

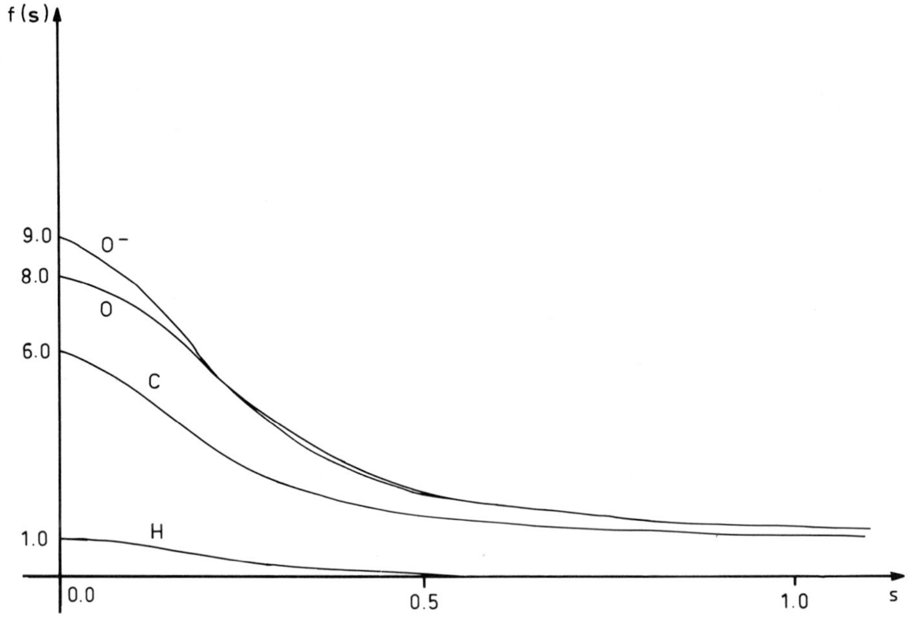

Fig. 3.26. Examples of atomic scattering factor curves.

Analytical representations of scattering factor curves have been published by Cromer & Mann [Cromer, D.T. & Mann, J.B., Acta Cryst. *A24*, 320 (1968)] for several elements. These expressions are advantageous in computer calculations since otherwise the complete table has to be stored and actual values for f(s) must be derived by interpolation.

The approximation for the ϱ_j(r) and the shape of f_j-curves results in two important consequences:

(1) The assumption of a spherical electron density distribution for each atom means that all bonding effects are *a priori* neglected in the molecular model. Thus, if precise results of an X-ray analysis show electron density details in bond directions, for an analysis of these details it is necessary to discard the assumption of a spherical atomic model.

The location of the maximum in the electron density of an atom is usually interpreted as the atomic position. Therefore, although we are concerned with the electron density function, we generally determine the atomic positions rather than the real density distribution.

(2) The magnitude of F [see (3.11)], depends on the magnitude of the f_j's. Since they decrease with increasing s, there is a general trend for the F's to decrease with s. Therefore, reflections with large $\sin\theta/\lambda$-values will usually tend to have weak intensities, which is observed on the film exposures shown in Fig. 2.36 and 2.38, for example.

Furthermore, for s $\gg$ 1.1, the f(s) approaches zero. Reflection intensities for these s-values are generally too weak to be observed. For MoKα radiation with $\lambda \approx 0.71$ Å, for example, we have for $\theta = 45°$ ($\sin\theta \approx 0.71$), $\sin\theta/\lambda \approx 1$. Therefore, reflections with $\theta > 45°$ have intensities which are usually near zero and can only be measured under special conditions. Even below this angle, reflections are very weak from the small f(s)-values and an additional damping from thermal motion effects, so that in general, reflections can only be observed with MoKα radiation to a θ-limit of 25–30°. The situation is different for CuKα radiation with a wavelength $\lambda \approx 1.54$ Å. Even a maximum $\sin\theta = 1$ leads to $\sin\theta/\lambda \approx 0.65$. It follows that for CuKα radiation, all reflections up to the limit given by the Ewald sphere can be observed and should be measured.

(3) Since the atomic scattering factors correspond to the stationary atom model, formula (3.11) for the structure amplitude is also valid only for the molecules *at rest*. In the crystal, the atoms always execute thermal and zero-point vibrations about their restpoints. As was shown by Debye in 1914, the thermal motion of each atom can be taken into account if the atomic scattering factor f for the stationary atom is replaced by the scattering factor f_T for the vibrating atom, of the form

$$f_T = f\, e^{-(B \sin^2\theta)/\lambda^2} \qquad (3.12)$$

The quantity B, denoted as the DEBYE-WALLER factor, is related to the atomic vibration by

$$B = 8\pi^2 U = 8\pi^2 \bar{u}^2 \qquad (3.13)$$

where $\bar{u}$ is the root-mean-square amplitude of the atomic vibration. $\bar{u}$ then has the dimension of a length, and B that of the square of a length. Since λ is usually given in Angstroms, the condition that the exponent in (3.12) have no dimension requires the dimensions of Å for $\bar{u}$ and Å^2 for B.

The description of thermal motion by a single parameter U is based on the assumption of an isotropic vibration of the atom, which means that the atomic motion is equal in all directions in space and that the volume indicating the mean sojourn probability is a sphere. The quantities B or U are then said to be isotropic temperature factors.

It is clear that this simple model can only be an approximate description of the atomic motion. It is not reasonable that the vibration in the bond direction is the same as that normal to the bond. Therefore a more complicated thermal motion than the isotropic assumption will generally take place.

An improved description is obtained by introducing anisotropic temperature factors. This is done as follows:

For each reflection **h** we get, from Bragg's equation,

$$|\mathbf{h}|^2 = \frac{4\sin^2\theta}{\lambda^2}$$

Substituting this in (3.12), we get

$$f_T = f\, e^{-T_{iso}}$$

with

$$T_{iso} = (B|\mathbf{h}|^2)/4 = (Bh^2a^{*2} + Bk^2b^{*2} + Bl^2c^{*2} + 2Bhka^*b^*\cos\gamma^* + 2Bhla^*c^*\cos\beta^* + 2Bklb^*c^*\cos\alpha^*)/4 \quad (3.14)$$

Replacing the isotropic B in each summand by the tensor components B_{ij} $(i, j = 1, 3)$, we get

$$T_{aniso} = (B_{11}h^2a^{*2} + B_{22}k^2b^{*2} + B_{33}l^2c^{*2} + 2B_{12}hka^*b^*\cos\gamma^* + 2B_{13}hla^*c^*\cos\beta^* + 2B_{23}klb^*c^*\cos\alpha^*)/4 \quad (3.15a)$$

or, in terms of

$$U_{ij} = B_{ij}/(8\pi^2) \quad (3.16)$$

$$T_{aniso} = 2\pi^2(U_{11}h^2a^{*2} + U_{22}k^2b^{*2} + U_{33}l^2c^{*2} + 2U_{12}hka^*b^*\cos\gamma^* + 2U_{13}hla^*c^*\cos\beta^* + 2U_{23}klb^*c^*\cos\alpha^*) \quad (3.15b)$$

The probability volume is now an ellipsoid with the U_{ii} expressing the mean square amplitudes of vibration axes and the U_{ij} $(i \neq j)$ representing the ellipsoid orientation. Although a more exact description of thermal motion requires a more complicated model for its correct representation, the ellipsoid model is found in practice to be a good compromise. On one hand it allows the description of an anisotropic behaviour of the vibrating atom, and on the other hand it keeps the number of parameters to an acceptable limit. Note that the transition from isotropic to anisotropic temperature parameters needs the introduction of five more parameters for each atom. More general models for the representation of thermal motion require substantially more parameters.

Another expression for the anisotropic exponent which is frequently used is

$$T_{aniso} = \beta_{11}h^2 + \beta_{22}k^2 + \beta_{33}l^2 + 2\beta_{12}hk + 2\beta_{13}hl + 2\beta_{23}kl \quad (3.15c)$$

A comparison with (3.15b) shows that the β's are defined by

$$\begin{aligned}
\beta_{11} &= 2\pi^2U_{11}a^{*2}\\
\beta_{22} &= 2\pi^2U_{22}b^{*2}\\
\beta_{33} &= 2\pi^2U_{33}c^{*2}\\
\beta_{12} &= 2\pi^2U_{12}a^*b^*\cos\gamma^*\\
\beta_{13} &= 2\pi^2U_{13}a^*c^*\cos\beta^*\\
\beta_{23} &= 2\pi^2U_{23}b^*c^*\cos\alpha^*
\end{aligned} \quad (3.17)$$

(Note that frequently a definition of the U_{ij} is used which includes the cosine terms for the U_{ij} elements with $i \neq j$, i.e. U_{12}, U_{13} and U_{23} are replaced by $U_{12}' = U_{12}\cos\gamma^*$, $U_{13}' = U_{13}\cos\beta^*$, $U_{23}' = U_{23}\cos\alpha^*$.)

With the introduction of temperature factors, we get for the structure factor

$$F(\mathbf{h}) = \sum_{j=1}^{N} f_j \, e^{-T_j} \, e^{2\pi i \mathbf{h} \mathbf{r}_j} \tag{3.18}$$

with T_j either the isotropic expression (3.14) or the anisotropic expression (3.15).

Note that the exponential term of thermal motion causes a further damping of the f(s) curves with increasing s. Therefore the tendency of high-order reflections to have weak intensities is enhanced by the atomic thermal and zero-point motion. Crystals consisting of atoms with low thermal vibration occur mostly with inorganic compounds which give intensity data to high $\sin\theta/\lambda$ values. Crystals of organic compounds usually consist of more strongly vibrating atoms. In consequence, low temperature measurements give better results for crystals in which reflection intensities are not observable at medium $\sin\theta/\lambda$ values.

Transformation into a discrete series representation can also be obtained for $\varrho(\mathbf{r})$. Since reflections are discrete, we can replace the integral by a sum symbol in (3.6). Then dV^* reduces to ΔV^*. Thus,

$$\Delta V^* = V^* = 1/V$$

with V^* the volume of the reciprocal unit cell since F exists only for lattice points of type (3.7). Then we get

$$\varrho(\mathbf{r}) = (1/V) \textstyle\sum_h F(\mathbf{h}) \, e^{-2\pi i \mathbf{h}\mathbf{r}}$$

or, with $\mathbf{h} = h\mathbf{a}^* + k\mathbf{b}^* + l\mathbf{c}^*$,

$$\varrho(\mathbf{r}) = (1/V) \textstyle\sum_h \sum_k \sum_l F(hkl) \, e^{-2\pi i (hx + ky + lz)} \tag{3.19}$$

3.2.2 *Intensity Symmetry, Asymmetric Unit*

With the expression (3.11) for F, we can transform the symmetry properties of the unit cell to those of F and with that to the intensities. The first general symmetry property of F, however, does not depend on any crystal symmetry. From equation (1.30) in 1.2.1 we had obtained for the Fourier transform G of a real function f, the property $G^-(\mathbf{b}) = G^*(\mathbf{b})$. Since $\varrho(\mathbf{r})$ is a real function, this property holds for F, as can be seen easily from (3.11),

$$F(-\mathbf{h}) = F^*(\mathbf{h}) \tag{3.20}$$

It follows that

$$I(\mathbf{h}) \sim F(\mathbf{h})\, F^*(\mathbf{h}) = F(\mathbf{h})\, F(-\mathbf{h})$$
$$I(-\mathbf{h}) \sim F(-\mathbf{h})\, F^*(-\mathbf{h}) = F(-\mathbf{h})\, F(\mathbf{h})$$

hence

$$I(-\mathbf{h}) = I(\mathbf{h}) \tag{3.21}$$

This important property, which expresses the fact that X-ray diffraction is always centrosymmetric, is called "Friedel's law". It follows immediately that the number of independent reflections is reduced to one half of the limiting sphere. For example, the rule of thumb derived in 1.2.3 that the copper limiting sphere includes $9 \times V$ reflections can now be modified to state that the number of independent reflections is at most $4.5 \times V$ in the general asymmetric case.

The triclinic space groups P1 and $P\bar{1}$ have only the identity and the inversion center symmetry elements. This inversion center does not cause further intensity symmetry, so for the triclinic system the half limiting sphere is the so-called "asymmetric unit" of reflections, that is, the number of independent reflections. In practical intensity measurements, this means that only positive values need be considered for one of the indices hkl. If, for instance, l is chosen to be positive, it has to be observed that for $l = 0$, only the half-circle contains all the independent reflections (see Fig. 3.27).

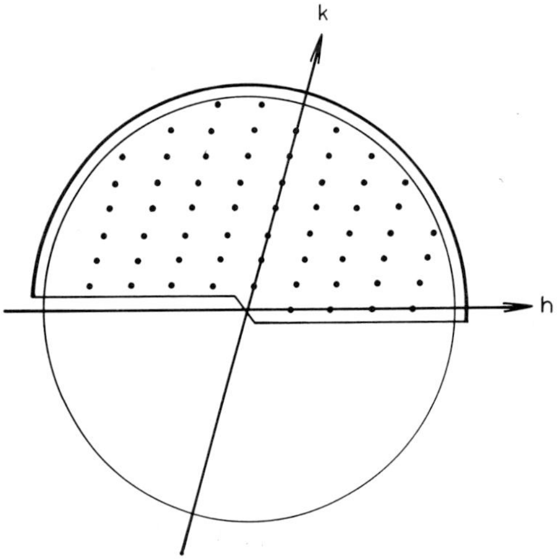

Fig. 3.27. Independent portion of reflections for $l = 0$ in the triclinic system.

Although it has no influence on intensity symmetry, the existence of an inversion center in $P\bar{1}$ has an important consequence on another property of F, i.e. the phase. Since for every atom with its position vector $\mathbf{r} = (x, y, z)$, the centrosymmetric vector $\mathbf{r} = (-x, -y, -z)$ is present, (3.11) reduces to

$$F_{cent}(\mathbf{h}) = \sum_{j=1}^{N/2} f_j \left(e^{2\pi i \mathbf{h}\mathbf{x}_j} + e^{-2\pi i \mathbf{h}\mathbf{x}_j}\right)$$

With Euler's formula

$$2\cos\varphi = e^{i\varphi} + e^{-i\varphi}$$

we get

$$F_{cent}(\mathbf{h}) = \sum_{j=1}^{N/2} f_j \cos 2\pi \mathbf{h}\mathbf{x}_j \tag{3.22}$$

Instead of being a complex number, F is now a real number, i.e. the phase problem, which in the complex representation

$$F = |F|\, e^{i\varphi}$$

includes the problem of determining φ for all possible values from 0 to 2π, now reduces to a "sign problem". For centrosymmetric structures, it has to be determined whether the signs of the F's are "plus" or "minus". The important property of centrosymmetric structures, that the structure factors are real, can also be illustrated graphically (see Fig. 3.28).

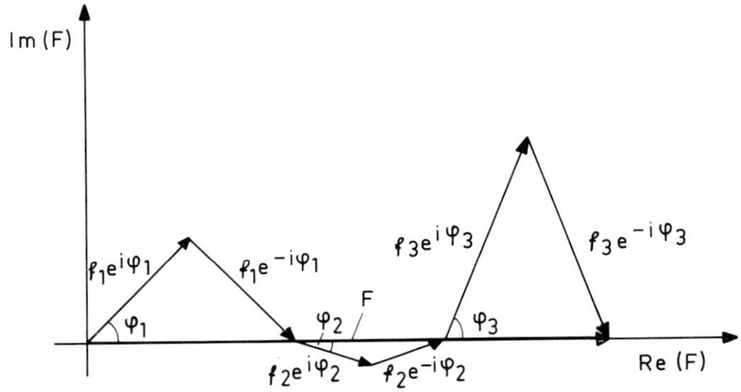

Fig. 3.28. Graphic representation of a centrosymmetric structure factor.

Note that it follows from (3.20) that for centrosymmetric structure factors,

$$F(-\mathbf{h}) = F(\mathbf{h}) \tag{3.23}$$

As the first example of a crystal symmetry element having influence on intensity symmetry, let us consider the mirror plane. Consider a crystal in the monoclinic system with the mirror plane in the x−z plane. Then every atom with its position vector $\mathbf{r} = (x, y, z)$ has its equivalent in $\mathbf{r}' = (x, -y, z)$. It follows for the structure factor

$$F_m(hkl) = \sum_{j=1}^{N/2} f_j \left(e^{2\pi i (hx_j + ky_j + lz_j)} + e^{2\pi i (hx_j + k(-y_j) + lz_j)}\right)$$

and

$$F_m(h-kl) = \sum_j f_j \left(^{2\pi i (hx_j - ky_j + lz_j)} + e^{2\pi i (hx_j - k(-y_j) + lz_j)}\right)$$

hence

$$F_m(hkl) = F_m(h-kl) \tag{3.24}$$

For a glide plane, translational elements are added only to x and z, and do not affect the calculation above. So, as a general rule, any mirror or glide plane in direct space causes a mirror plane in reciprocal space.

A similar rule can be obtained for the n-fold axes, as represented by the following calculation for a 2-fold axis. Assuming again a monoclinic system with the 2-fold axis in the b-direction, the equivalent position vectors are $\mathbf{r} = (x,y,z)$ and $\mathbf{r}' = (-x,y,-z)$. Then for the structure factor we get

$$F_2(hkl) = \sum_{j=1}^{N/2} f_j(e^{2\pi i(hx_j + ky_j + lz_j)} + e^{2\pi i[h(-x_j) + ky_j + l(-z_j)]})$$

We obtain the same result for negative h and l, i.e.,

$$F_2(hkl) = F_2(-hk-l) \tag{3.25}$$

A screw component does not affect this result and since we get analogous properties for the other rotation axes, a general rule can be given for n-fold axes: N-fold axes or screw axes in direct space cause n-fold axes in reciprocal space.

As a representative example for all crystal systems, let us discuss the consequences of these two rules for the space groups in the monoclinic system. Here every space group contains either a 2-fold axis or a mirror plane (or a related symmetry element). Then one of the properties (3.24) or (3.25) is valid, holding also for the intensities. Since Friedel's law is always satisfied, we get as additional conditions:

(a) if $I(hkl) = I(h-kl)$, it follows that $I(hkl) = I(-hk-l)$
because $(-hk-l) = -(h-kl)$,

or

(b) if $I(hkl) = I(-hk-l)$, it follows that $I(hkl) = I(h-kl)$
from Friedel's law.

Since either (a) or (b) is valid, we get for the intensity symmetry in the eleven monoclinic space groups,

$$I(hkl) = I(-h-k-l) = I(h-kl) = I(-hk-l) \tag{3.26}$$

The only octants of the limiting sphere which have different intensities are those which differ in the sign of h or l. Thus the asymmetric unit of reflections in the monoclinic system is one quadrant, for instance that containing reflections of type (hkl) and (−hkl).

The investigation of intensity symmetry in the monoclinic system has shown not only that all space groups of one crystal class have the same intensity symmetry, but that all three classes of this crystal system (2, m, 2/m), have the same intensity symmetry. This is not surprising, when taking Friedel's law into consideration, since these crystal classes differ only by a center of symmetry.

This equivalence of intensity symmetry leads to a further classification among the crystal classes. All crystal classes having the same intensity symmetry, i.e. differing only by a center of symmetry in their symmetry elements, are said to belong to a "Laue group". All 32 crystal classes can be classified by the eleven Laue groups which are listed in Table 3.5, together with their corresponding crystal classes.

Table 3.5. The Eleven Laue Groups.

Crystal System	Laue Group	Included Crystal Classes	Asymmetric Unit of Limiting Sphere
Triclinic	$\bar{1}$	$1, \bar{1}$	one half sphere
Monoclinic	2/m	2; m; 2/m	one quadrant
Orthorhombic	mmm	222; mm2; mmm	one octant
Tetragonal	4/m	4; $\bar{4}$; 4/m	one octant or less, see International Tables, Vol. I for structure factor expressions
	4/mmm	422; 4mm; $\bar{4}$2m; 4/mmm	
Hexagonal	6/m	6; $\bar{6}$; 6/m	
	6/mmm	622; 6mm; $\bar{6}$m2; 6/mmm	
Trigonal	$\bar{3}$	3; $\bar{3}$	
	$\bar{3}$m	32; 3m; $\bar{3}$m	
Cubic	m3	23; m3	
	m3m	432; $\bar{4}$3m; m3m	

Since the X-ray diffraction from a single crystal gives the intensity symmetry and not the crystal symmetry, this equivalence in the intensity symmetry for the various space groups of the same Laue group is a disadvantage for space group determination. Before discussing this problem in further detail, let us conclude the question of the asymmetric reflection unit. In the orthorhombic system, we have only one Laue group, just as in the monoclinic and triclinic case. Similar considerations as above lead to the result that intensities are equal if their absolute values of indices are equal, i.e.,

$$\begin{aligned} I(hkl) &= I(-hkl) = I(h-kl) = I(hk-l) \\ &= I(-h-kl) = I(-hk-l) = I(h-k-l) = I(-h-k-l) \end{aligned} \tag{3.27}$$

The asymmetric unit of reflections is thus the octant hkl with h, k, and l all positive.

For the higher symmetric crystal systems, no general conditions can be given, since the intensity symmetry is described by more than one Laue group. The symmetry relations of F's are given in the International Tables, Vol. I, for every space group.

3.2.3 Systematic Extinctions

In our first inspection of the film exposures, we observed, in addition to symmetry, special groups of reflections in which some were systematically absent. Let us now discuss this phenomenon in detail. We shall see that, without exception, the translational parts of the symmetry elements will cause these systematic absences and that this property will be of significant assistance in space group determination.

As representative for all translational elements, let us prove three propositions:

(1) If a lattice is C-centered, all reflections hkl with $h + k = 2n + 1$ are systematically absent (general extinction rule).

(2) If a lattice contains a glide plane perpendicular to the **b**-direction with glide component **c**/2, all reflections of type h0l with $l = 2n + 1$ are systematically absent (zonal extinction rule).

(3) If a lattice contains a 2-fold screw axis in the **b**-direction, all reflections of type 0k0 with $k = 2n + 1$ are systematically absent (axial extinction rule).

The proof of all three propositions is easily done by calculation of F(**h**), assuming the named symmetry:

(1) For a C-centered lattice, every vector (x, y, z) has its equivalent in $(x + 1/2, y + 1/2, z)$. Then we have

$$F(hkl) = \sum_{j=1}^{N/2} f_j [e^{2\pi i(hx_j + ky_j + lz_j)} + e^{2\pi i(h(x_j + 1/2) + k(y_j + 1/2) + lz_j)}]$$

$$= \sum_{j=1}^{N/2} f_j [e^{2\pi i(hx_j + ky_j + lz_j)} (1 + e^{i\pi(h + k)})]$$

With $h + k$ odd, $e^{i\pi(h+k)} = -1$, hence

$$F(hkl) = 0, \text{ if } h + k = 2n + 1.$$

(2) Assuming the situation as in space group $P2_1/c$, the pair of equivalent vectors concerning the c-glide plane is (x, y, z) and $(x, 1/2 - y, 1/2 + z)$. Then we get

$$F(hkl) = \sum_{j=1}^{N/2} f_j [e^{2\pi i(hx_j + ky_j + lz_j)} + e^{2\pi i(hx_j + k(1/2 - y_j) + l(1/2 + z_j))}]$$

$$= \sum_{j=1}^{N/2} f_j [e^{2\pi i(hx_j + lz_j)} (e^{2\pi iky_j} + e^{2\pi ik(1/2 - y_j)} e^{i\pi l})]$$

For reflections (hkl) with $k = 0$, we get

$$F(h0l) = \sum_{j=1}^{N/2} f_j [e^{2\pi i(hx_j + lz_j)} (1 + e^{i\pi l})]$$

For l odd, $e^{i\pi l} = -1$, hence

$$F(h0l) = 0 \text{ if } l = 2n + 1$$

(3) If a 2-fold screw axis in the direction of **b** is present, we have the equivalent position (x, y, z) and (−x, 1/2 + y, 1/2 − z). It follows that

$$F(hkl) = \sum_{j=1}^{N/2} f_j[e^{2\pi i(hx_j + ky_j + lz_j)} + e^{2\pi i(-hx_j + k(1/2+y_j) + l(1/2-z_j))}]$$

$$= \sum_{j=1}^{N/2} f_j[e^{2\pi i(k/4 + ky_j + l/4)}(e^{2\pi i(hx_j - k/4 + lz_j - l/4)} + e^{2\pi i(-hx_j + k/4 - lz_j + l/4)})]$$

$$= \sum_{j=1}^{N/2} f_j[e^{2\pi i(k/4 + ky_j + l/4)}\, 2\cos 2\pi(hx_j + lz_j - k/4 - l/4)$$

For reflections of type (0k0), we get

$$F(0k0) = \sum_{j=1}^{N/2} f_j[e^{2\pi i(k/4 + ky_j)}\, 2\cos(k\pi/2)]$$

For k odd, $\cos(k\pi/2) = 0$, hence

$$F(0k0) = 0 \text{ if } k = 2n + 1$$

It is evident that calculations with corresponding symmetry elements will have similar results. So we can establish the following rules:

(1) In every non-primitive lattice, *general* systematic extinctions are present. For a C-centered lattice, only reflections hkl with h + k = 2n are present. For A- or B-centered lattices, the conditions are k + l = 2n and h + l = 2n. For an F-centered lattice, only reflections hkl satisfying simultaneously the conditions h + k, k + l, (l + h) = 2n are present (the condition l + h = 2n is redundant, since it follows from the first and second conditions). For an I-centered lattice, only reflections hkl with h + k + l = 2n are present.

(2) Every glide plane causes *zonal* systematic extinctions. If the glide plane is perpendicular to **a**, **b**, or **c**, the extinctions affect reflections of type 0kl, h0l, or hk0. The glide direction is given by that of the non-zero indices of which even parity is required. For a glide plane perpendicular to **b**, we have the possibilities:

Table 3.6. Axial Extinction Conditions for Screw Axes in the c-Direction (reflections present are listed).

Type	00l reflections	Type	00l reflections
2_1	l = 2n	4_2	l = 2n
3_1, 3_2	l = 3n	6_1, 6_5	l = 6n
4_1, 4_3	l = 4n	6_2, 6_4	l = 3n
		6_3	l = 2n

(a) if h0l for $h = 2n$ holds, the glide plane is of type a,

(b) if h0l for $l = 2n$ holds, the glide plane is of type c,

(c) if h0l for $h + l = 2n$ holds, the glide plane is of type n.

(3) Every screw axis causes *axial* systematic extinctions. If the axes coincide with **a**, **b**, or **c**, the reflection series affected are h00, 0k0, or 00l. The extinctions for all possible screw axes in the c-direction are listed in Table 3.6.

3.3 Space Group Determination

3.3.1 *General Rules*

Some, but not all, space groups can uniquely be determined from the recognition of Laue symmetry and systematic extinctions. Let us consider an example where the space group determination from intensity symmetry and systematic extinctions is not unique. If the intensity symmetry corresponds to the monoclinic Laue group 2/m and no systematic extinction can be observed, then we have a primitive lattice with no translational symmetry elements, such as screw axes or glide planes. The space group could be P2 or Pm or P2/m.Recognition of the correct space group without solving the structure is impossible. In some cases, the calculation of unit cell contents can provide additional information. This is done as follows:

The mass per mol of a compound of molecular weight W is

$$m = W \ [g\,mol^{-1}]$$

Then the mass m_n of n molecules is

$$m_n = (nW)/L \ [g]$$

where L is Avogadro's number. If the unit cell has the volume V, the density ϱ can be calculated, if the number n of molecules in the unit cell is known. Since $\varrho = m_n/V$, we get

$$\varrho = (nW)/(VL) [g\,cm^{-3}]$$

This expression for ϱ is usually called "X-ray density" and is designated ϱ_x, since it is derived from X-ray results. Usually V is given in Å^3, i. e. 10^{-24} cm^3. Since $L = 6.023 \times 10^{23}\,mol^{-1}$, ϱ_x is written as

$$\varrho_x = n \frac{W\,10}{V\,10^{-24}\,6.023} 10^{-24} [g\,cm^{-3}]$$

or with $10/6.023 \approx 1.65$

$$\varrho_x = (1.65\,nW)/V [g\,cm^{-3}] \text{ or } [Mg\,m^{-3}] \qquad (3.28)$$

This formula can be used in two ways. The first is the calculation of n from the value of

ϱ measured macroscopically (by flotation methods, for example). Since W is usually known and V is derived from the lattice constants, (3.28) gives n. Since n must be an integer (except for a few special examples), the calculation from (3.28) should give an integral value (with some error caused mainly by the uncertainty of ϱ). With the nearest integral value of n known, ϱ can be calculated very precisely from (3.28), since W and V can be determined very accurately. Thus the X-ray density determination can result in a very precise value.

Applying (3.28) to the structure of KAMTRA, the molecular weight is W = 188.2 g/mol and the volume V will be determined to be 620.1 Å^3 (see Section 4.2.2). The experimentally measured density was found to be 1.95 g cm^{-3}. From (3.28) it follows that n = 3.95, which means n must be equal to 4. With n = 4 we calculate $\varrho_x = 1.97$ g cm^{-3} and get a very precise result for the density. The space group symmetry of KAMTRA (to be determined in the next section) is four-fold. So it is clear that the asymmetric unit of the cell is one molecule.

Returning to the preceding monoclinic example, we can use the knowledge of the unit-cell contents for a decision in favour of one of the space groups as follows: the space groups P2 and Pn have two-fold symmetry, that of space group P2/m is four-fold. If the cell contains four molecules, it is more probable that P2/m is the correct space group. On the other hand, if the cell contains two molecules, the space group P2/m is only possible if the molecule itself contains one of the space group symmetry elements, that is, either a 2-fold axis, a mirror plane, or an inversion center. Then the space group could be P2/m with one half molecule in the asymmetric unit. If the molecule cannot have this symmetry, the space group P2/m can be excluded, except in spacial rare cases where the molecules are disordered.

In some cases, the physical properties of the compound are useful for space group determination.

One major problem is the decision whether a crystal belongs to an acentric or the corresponding centric space group since the Laue symmetry always indicates the cor-

Table 3.7. Noncentrosymmetric Crystal Classes, Polar and Enantiomorphous Classes.

Crystal system	Noncentric (Piezoelectricity)	Polar (Pyroelectricity)	Enantiomorphous
Triclinic	1	1	1
Monoclinic	2; m	2; m	2
Orthorhombic	222; mm2	mm2	222
Tetragonal	4; $\bar{4}$; 422; 4mm; $\bar{4}$2m	4; 4mm	4; 422
Hexagonal	6; $\bar{6}$; 622; 6mm; $\bar{6}$m2	6; 6mm	6; 622
Trigonal	3; 32; 3m	3; 3m	3; 32
Cubic	23; 432$^+$; $\bar{4}$3m	–	23; 432
No.	21	10	11

$^+$ No Piezoelectricity

responding centric class. So from the diffraction experiment a reference to one of the 21 non-centric crystal classes can scarcely be obtained. In this connection the property of certain crystal classes to have polar directions may be useful. A given direction, represented by a vector $\mathbf{r} = [u, v, w]$, is defined as a polar direction if it is not related by one of the crystal class symmetry elements to its corresponding centrosymmetric direction $-\mathbf{r} = [-u, -v, -w]$. From this definition it follows immediately that a polar direction can only occur in one of the 21 non-centrosymmetric classes (Table 3.7).

If for a given crystal class a vector **r** can be chosen in a polar direction having the property that the vector sum of **r** and all symmetry related vectors **r**', **r**'', . . . does not equal zero, the crystal class is said to be a polar class. The ten polar classes are listed in Table 3.7.

Two physical properties are closely connected to crystals belonging to a polar or a noncentric crystal class. If for the non-zero resultant in a polar class an electric dipole moment is produced in terms of a change of temperature this phenomenon is called pyroelectricity. The development of an electrical polarity along a crystal's polar direction from a mechanical compression or tension is defined as piezoelectricity.

Since piezoelectricity requires the existence of a polar direction it is clear that this effect is only observable in one of the noncentric classes. In fact, an exception is the high symmetric cubic class 432 where piezoelectricity is no longer possible so that actually 20 crystal classes remain (see Table 3.7). Since pyroelectricity is restricted to one of the ten polar crystal classes, the presence of one of these two physical properties may be used in the course of crystal class decision. However, it should be pointed out that nothing can be concluded from the absence of these effects because they may be too weak to be observable.

The last column of Table 3.7 shows those eleven of the twenty-one noncentrosymmetrical classes which are enantiomorphous, i. e. they contain no symmetry element which produces the reflected image of a motif.

If, for example, a structure consists of one enantiomer of an optically active compound, only one of the 64 space groups belonging to these crystal classes is possible. If we had such a compound in our monoclinic example, we could immediately exclude Pm and P2/m, resulting in a unique solution, in this case P2. In general, even if additional informations from the physical behaviour of a compound are used a unique determination of the space group is not obtained, although there are 50 out of the 230 space groups for which the determination is unambiguous. Two of these are $P2_1/c$ and $P2_12_12_1$, which occur very frequently with organic molecules.

In the monoclinic system, one axial extinction and one zonal extinction determines the unique space group $P2_1/c$. The orthorhombic space group $P2_12_12_1$ is also uniquely determined. If the intensities show the symmetry of the orthorhombic system and three perpendicular axial extinctions are present, the unique space group is $P2_12_12_1$.

Space group determination is made in the following sequence:

(1) Determine the Laue class from the intensity symmetry.

(2) Select the unit-cell vectors to be consistent with the observed intensity symmetry.
(3) Check whether these unit-cell vectors are equal in length and determine the angles between them.
(4) Determine the systematic extinctions, if any.
(5) Calculate the cell volume and the unit-cell contents.
(6) Check whether the results of (1) to (5) are in agreement with the properties of one (or more than one) of the 230 space groups.

3.3.2 *Space Group of KAMTRA*

With our knowledge of crystal and related intensity symmetry, we can now analyze the film exposures of KAMTRA in more detail. On the zero layer Weissenberg exposure (Fig. 2.36a), for which the characteristic "festoon"-character is represented in Fig. 3.29, the lines marked A and B have special properties already discussed in 2.3.2. The Δx-distance of those two lines on the film is exactly 4.5 cm, corresponding to an angular increment of 90^0 (since $g = 0.5\,\mathrm{mm}/^0$). These two lines therefore constitute a rectangular lattice plane, as shown in Fig. 3.30, which shows the plot of the undistorted lattice layer from the *upper* half of the Weissenberg exposure. This is obtained by application of formula (2.13) of 2.1.3.
Because of the symmetry and extinctions, the directions of A and B are chosen as the reciprocal unit-cell vectors. If **a*** and **b*** are defined as indicated in Fig. 3.30, the vector

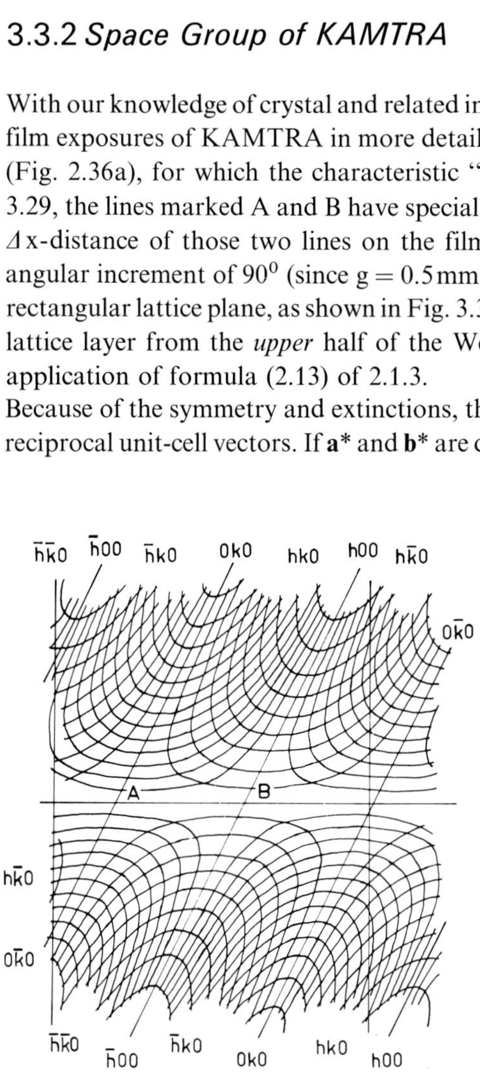

Fig. 3.29. Schematic representation of the zero-layer Weissenberg exposure of KAMTRA (see Fig. 2.36a). Vertical lines indicate film borders.

of the rotation axis must be **c** to satisfy the layer line condition (2.3) of 2.1.1. With $\mathbf{h} = h\mathbf{a}^* + k\mathbf{b}^* + l\mathbf{c}^*$ and $\mathbf{r} = 0\mathbf{a} + 0\mathbf{b} + 1\mathbf{c}$, we get $\mathbf{hr} = h0 + k0 + l1 = n$ ($n = 0, 1, 2, \ldots$). Then the zero layer contains all hk0-reflections, the first layer all hk1-reflections, etc.

With these definitions and the undistorted representation of Fig. 3.30, the reflections can be indexed, i.e., they can be identified by their integer coordinates. The result of indexing the reflections for the zero layer is shown for several reflections in Fig. 3.30. Note the relation between reflections on the upper and lower half of a Weissenberg exposure (compare Fig. 3.29 and 3.30). As Fig. 3.29 shows, one complete lattice layer would be recorded on the upper half as well as on the lower half of the film, if the translation were large enough. In practice, the length of the film cylinder (which usually does not exceed 100-120 mm, corresponding to an angular increment of 200 to 240°

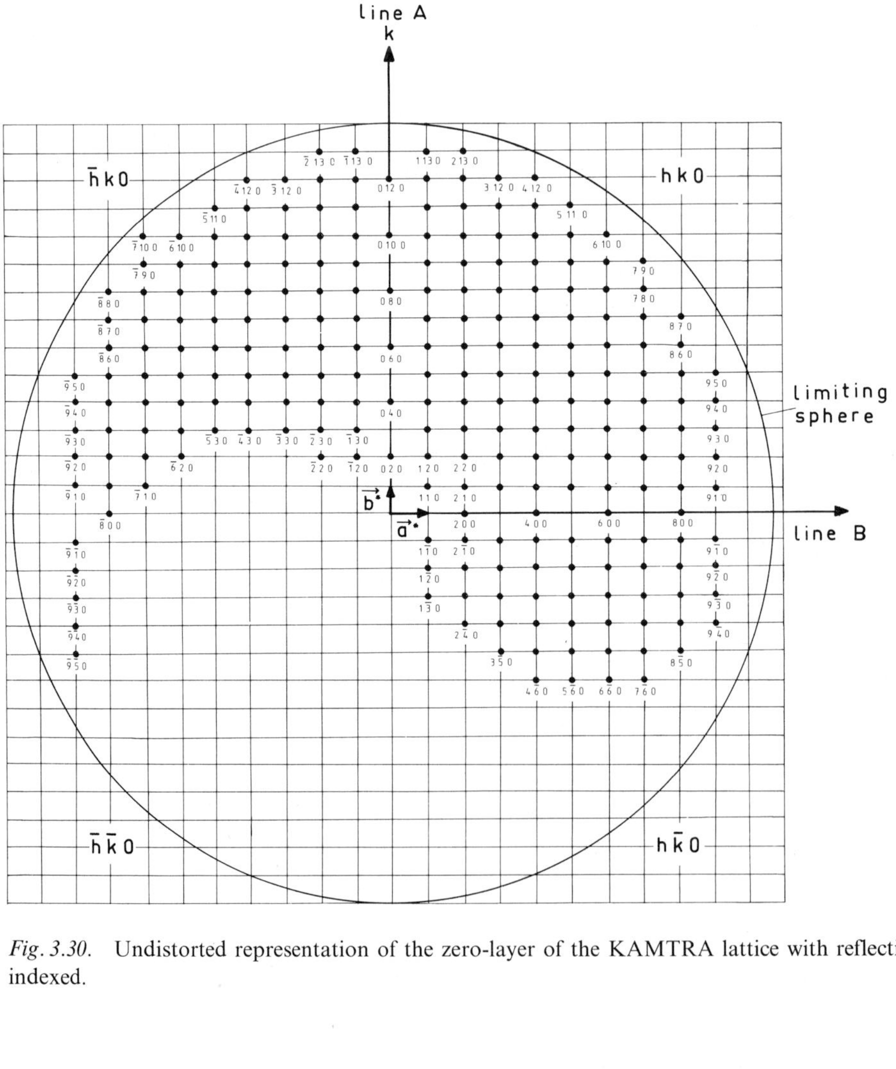

Fig. 3.30. Undistorted representation of the zero-layer of the KAMTRA lattice with reflections indexed.

for film rotation with $g = 0.5\,mm/^0$), restricts the range covered to two or two and one-half quadrants on each half of the film. Usually this is sufficient, since two quadrants of a reciprocal lattice layer include all the symmetry independent reflections in a layer in most cases.

The magnitude of **c** is 7.70 Å, obtained as 1/D* from the rotation photograph. The magnitudes of **a*** and **b*** can be derived from the θ-values of reflections on the lines B and A, which have the indices h00 and 0k0 respectively. Application of Bragg's law in the form of equation (2.7a) to axial reflections of the type h00 and 0k0 gives the magnitude of the corresponding reciprocal lattice constants.

The precision of a* depends only on the precision in measurement of θ, which is, in practice, always less than that of λ. The error $\Delta\theta$ of θ can therefore be assumed to be independent of θ. Since $\Delta \sin\theta$ decreases when θ approaches 90^0, the precision of a lattice constant determination increases with high angles of θ. Precise measurements of lattice constants should therefore be made by using high-order reflections.

At this stage of space group and preliminary lattice constant determination, medium precision is sufficient. Choosing the reflections no. 5 on line A and no. 19 on B (see Fig. 2.36b and 3.30), with the indices (0 10 0) and (8 0 0), we find $y_5 = 46.0\,mm$ and $y_{19} = 51.8\,mm$, corresponding to the angles $\theta_5 = 46.0^0$ and $\theta_{19} = 51.8^0$. Then we have $a^* = 0.127\,Å^{-1}$ and $b^* = 0.0933\,Å^{-1}$, and we know that γ^*, the angle between a* and b*, is 90^0.

With the indexing of reflections from Fig. 3.30, the intensity symmetry can be formulated as

$$I(hk0) = I(\bar{h}k0)$$
$$I(hk0) = I(h\bar{k}0)$$
and
$$I(hk0) = I(\bar{h}\bar{k}0)$$

Since the upper level Weissenberg exposures show the same intensity symmetry along A' and B' and along A" and B", the intensity symmetry can be formulated generally as

$$I(hkl) = I(\bar{h}kl) = I(h\bar{k}l) = I(\bar{h}\bar{k}l)$$

Intensity symmetry of that type is found in the orthorhombic and higher symmetry systems. Additional information on the crystal system is obtained from the inspection of the angles between the axes.

c* is the vector from the origin to the intersection point of lattice lines A' and B' in the first layer. If **c*** forms right angles with **a*** and **b***, the point of intersection of the rotation axis coincides with the top of **c*** at point C (Fig. 3.31a). It then follows that the lattice lines A' and B' intersect the rotation axis also in C and in consequence are present as *lines* on the Weissenberg photographs (see equation 2.9). If one of the angles differs from 90^0 (in Fig. 3.31b this is illustrated for the angle β^*), the point C at the top of **c*** is not identical with the intersection point D, and the lattice line B' no longer intersects the rotation axis. Its image on the Weissenberg film is no longer a line. If both angeles between **c*** and the two other unit cell vectors are different from 90^0, neither A' nor B' are straight lines of the film. The question whether **c*** forms one or two non-

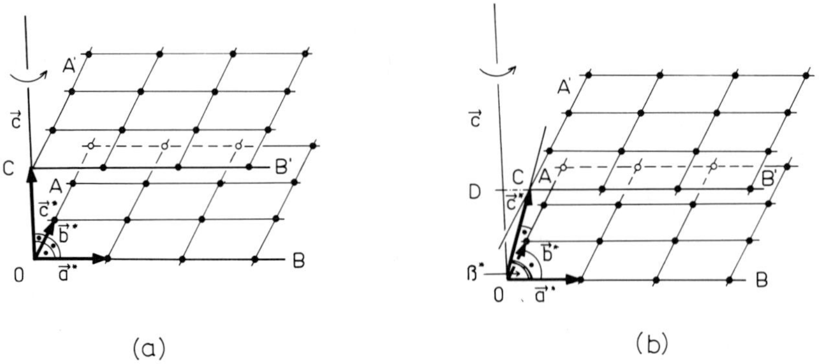

Fig. 3.31. Illustration of a shift of the intersection point: (a) orthogonal system; (b) non-orthogonal system.

right angles with the vectors in the zero layer can easily be answered by inspecting the Weissenberg exposure of first layer where the absence of straight lines for A' or B' (or both) indicates non-right angles. If only one angle differs from 90^0, the magnitude of displacement $p = \overline{DC}$ can be used for an estimation of the angle. Since $\overline{OD} = D^*$, we get (see Fig. 3.31b)

$$\cot \beta^* = p/D^* \tag{3.29}$$

However, the determination of p is not very accurate. Furthermore, since $p \ll D^*$, the accuracy of β^* derived from this shift is low, especially if the alignment of the crystal is not very precise and the rotation axis differs slightly from the crystal c-axis.

Inspecting the first layer of KAMTRA (Fig. 2.36c), we find that A' and B' are straight lines, and conclude that all the angles are right angles. We then have all the direct lattice constants, since a and b are just the reciprocals of a* and b* (Table 3.4):

$$a = 1/a^* = 7.87\,\text{Å} \qquad b = 1/b^* = 10.72\,\text{Å} \qquad c = 7.70\,\text{Å}$$
$$\alpha = \beta = \gamma = 90^0$$

From these lattice constants and the intensity symmetry, it is found that KAMTRA belongs to the orthorhombic system. The only systematic extinctions are on the lines A and B in the zero layer, indicating the axial extinctions

$$h00 \text{ for } h = 2n + 1 \qquad 0k0 \text{ for } k = 2n + 1$$

Extinctions of that type are caused by 2_1-screw axes. We do not know whether an axial extinction of type 00l for $l = 2n + 1$ is present since the 00l reflections are obscured by the primary beam stop when the crystal is rotated about the c-axis. If we have only the two screw axes along the x- and y-direction, the space group is $P2_1 2_1 2$; if the third screw axis is also present, the space group is $P2_1 2_1 2_1$. From the extinctions listed in the International Tables for the orthorhombic system, we see that no other space groups are possible. Both space groups require four asymmetric units. From the density and

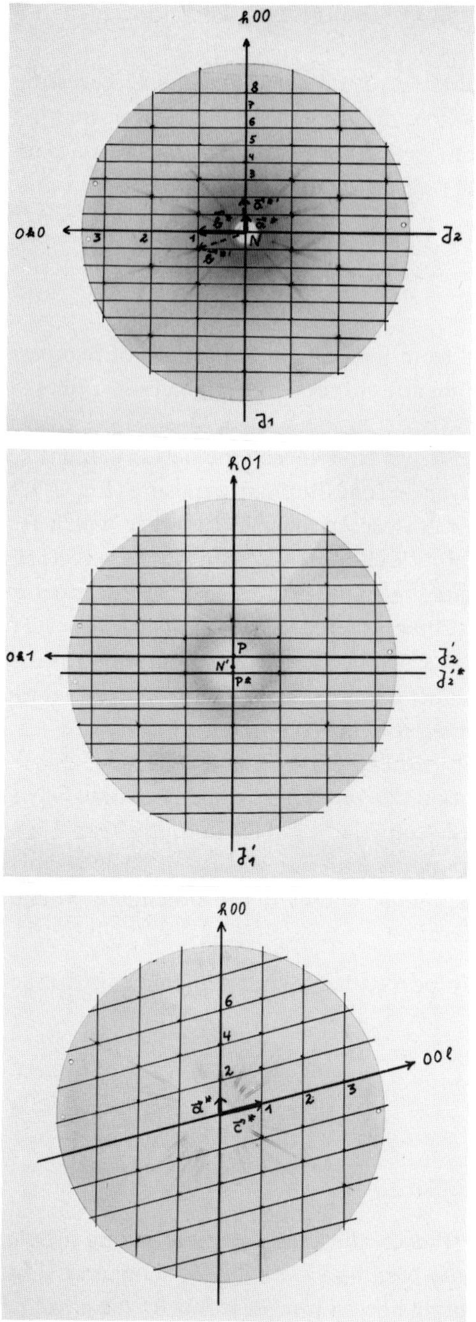

(a)

(b)

(c)

Fig. 3.32. Determination of the space group for NITROS:
(a) from the zero-layer de Jong-Bouman exposure;
(b) from the first-layer de Jong-Bouman exposure;
(c) from the zero-layer precession photograph.

cell volume, we found that the unit cell contains four molecules. Therefore, each molecule represents the asymmetric unit.

A decision on the third screw axis can be obtained by taking an additional precession photograph, but it is not necessary to do so, because we get the desired information from intensity measurements. If every second reflection of the 001 series has an unobservable intensity, we have a 2_1-axis in the z-direction, otherwise only a 2-fold axis.

3.3.3 *Space Group of NITROS*

The space group of NITROS is determined from the undistorted images of two perpendicular reciprocal lattice planes on the de Jong-Bouman and precession photographs (see 2.3.3).

Following the procedure suggested in 3.3.1, we first check whether any intensity symmetry can be recognized. On the zero layer de Jong-Bouman exposure (Fig. 3.32) we find that intensity symmetry is present with respect to the lines J_1 and J_2, which are perpendicular to each other (see also Fig. 2.39). We choose two reciprocal base vectors, say **a*** and **b*** in these directions, as illustrated in Fig. 3.32a, to ensure integer coordinate for all reflections. However, with this choice, there is a reflection only for every second lattice point. We see from Fig. 3.32a that reflections are present only for the lattice points for which $h + k$ is odd. Some reflections are systematically absent, but we cannot decide whether they are a zonal extinction of the type hk0 for $h + k = 2n + 1$, caused by a n-glide plane, or a general extinction of type hkl for $h + k = 2n + 1$ due to a C-centered lattice. Note that the extinctions in this zero layer would be avoided if we decided to choose other basis vectors (dotted vectors, **a***' and **b***' in Fig. 3.32a). Then we could assign a reflection to every lattice point and vice versa, but basis vectors would not lie on the symmetry lines. With **a*** and **b*** chosen in the direction of J_1 and J_2, we have derived the following properties:

(1) The zero layer contains all reflections of type hk0, hence the rotation axis is parallel to the direct basis vector **c**.
(2) The angle between **a*** and **b*** is a right angle, hence $\gamma^* = 90^0$.
(3) The following symmetry properties hold:

$$I(hk0) = I(\bar{h}k0) = I(h\bar{k}0)$$

(4) Reflections hk0 are present only if $h + k = 2n$.

In the de Jong-Bouman exposure of the first layer, the line J_1' corresponding to J_1 is easily recognized, since it passes through the origin and is a line of symmetry (Fig. 3.32b). However, the choice of a corresponding line J_2' is more difficult. We have the two lines J_2' and J_2'*, neither of which pass through the origin N' of the film, which is the intersection of the rotation axis. From the last section we know that this "origin shift" is caused by the third reciprocal lattice vector making an oblique angle with one (or both) of the two in the plane of the zero layer. The question now arises whether the line J_2' or the line J_2'* is the line of 0k1 reflections. That is, whether the point P, which is the intersection of J_1' and J_2', or the point P*, which is the intersection of J_1' and J_2'*, is the 001-reflection.

From a mathematical standpoint, both selections are equivalent. In one case, the reciprocal lattice vector **c*** would point from N to P, in the other case to P*. However, from a crystallographic point of view, only P is a good choice, making J_2' the line of 0k1 reflections. This is shown by inspection of the systematically absent reflections. We see that for both choices systematic extinctions are present. We have a general extinction in this lattice, due to centering, rather than a zonal extinction due to a glide plane.

In the zero layer we found that hk0 reflections with h + k odd were systematically absent. Since we already know that we have at least one oblique angle, we can have at most a monoclinic system. So we have to observe the convention that only a C-centered lattice is observed, causing reflections hkl with $h + k = 2n + 1$ to be systematically absent. Therefore provision has to be made that this is valid also for the hk1 reflections. As can be seen easily on Fig. 3.32b, this is only true for the choice of P as the 001 reflection. So J_1' is the line of h01 reflections, J_2' is the line of the 0k1 reflection, and P is the top of **c***.

From the position of P on the h01 line, it follows that **c*** is perpendicular to **b***, hence $\alpha^* = 90^0$. The magnitude of the oblique angle β^*, between **c*** and **a***, can be derived approximately from the shift p = N'P of the intersecting point. We find p = 3.9 mm, corresponding to a distance 0.0337 $Å^{-1}$ in the reciprocal lattice. The spacing between the zero and first layers was calculated from the cone axis exposure (see 2.3.3) to be 0.141 $Å^{-1}$. So we get, from (3.29), $\beta^* \approx 76.5^0$. (Observe the limited accuracy of that method. Let us assume that p has an accuracy of, say, $\pm$ 0.3 mm. With p = 4.2 mm, we get $\beta^* = 75.6^0$, with p = 3.6 mm, β^* is 77.6^0. So the precision of β^* is at most a degree).

Returning to the intensity symmetry of the first layer exposure, we note that we have symmetry only along the line J_1' (Fig. 3.32b and Fig. 2.39c). So we have

$$I(hk1) = I(h\bar{k}1)$$

but

$$I(hk1) \neq I(\bar{h}k1)$$

Together with the results of the zero layer, we can state generally that

$$I(hkl) = I(h\bar{k}l)$$

but

$$I(hkl) \neq I(\bar{h}kl)$$

This is the symmetry of a monoclinic lattice (see 3.26), and all the results derived from these two films indicate a monoclinic space group.

From the general extinctions, we know that we have a C-centered lattice. Since no further extinctions (zonal or axial) are visible, the number of possible space groups reduces to three. These are C2, Cm, or C2/m. For a further decision, we need the lattice constants, which can now be derived. The direct lattice constant c is the reciprocal of the spacing derived from the cone axis exposure, hence $c = 1/0.141 = 7.09\,Å$. a^* and b^* can be measured directly on the zero layer exposure. Using the distance between 800 and -800 reflections for the estimation of a^*, we find $16 \times (a^*) = 107.0\,mm$. From (2.16) we get ($d_F = 75\,mm$) $a^* = 0.0578\,Å^{-1}$. Similarly (using the 020 and 0-20 reflections) we get $4 \times (b^*) = 73.6\,mm$ and $b^* = 0.159\,Å^{-1}$. With β^* being known, all direct lattice constants can be calculated. With $b = 1/b^*$ and $a = 1/(a^* \sin \beta^*)$, we get $b = 6.28\,Å$, $a = 17.79\,Å$ and finally $\beta = 180^0 - \beta^* = 103.5^0$.

All these lattice dimensions and space group information have been derived from the two de Jong-Bouman exposures. So it is superfluous, in principle, to take further films. However, it is a good practice to confirm the results obtained from one set of exposures by a photograph taken from another lattice plane.

Let us look at the precession photograph of a zero layer (Fig. 3.32c) perpendicular to the layers shown by the de Jong-Bouman exposures. The spacing obtained from the corresponding cone axis exposure was $D = 6.16\,Å$, which is almost the same value as b. It follows that the precession photograph shows the h0l reflections,which is supported by the fact that the h00 series, known from the zero layer de Jong-Bouman exposure, reappears on this film.

Note that is is very advantageous for this precession exposure to have the monoclinic angle in the plane of the film, so we can check the value obtained from the shift of the intersection point. We find $\beta^* = 76.0$, in good agreement with the previous value, within the limits of accuracy.

When indexing the reflections of this precession photograph, it must be remembered that from general extinctions, only reflections with h even are present. Then, calculating the reciprocal lattice constants a^* and c^*, we find (using 003 and 00 – 3), $6 \times (c^*) = 105.5$ mm, $16 \times (a^*) = 107.3$ mm, hence $c^* = 0.145\,Å^{-1}$, $a^* = 0.0580\,Å^{-1}$, in good agreement with the results derived above.

Since the precession photograph confirms all results derived from the de Jong-Bouman exposures, we can be certain that the space group and lattice constant determinations for NITROS are correct. These are

$$
\begin{aligned}
a &= 17.79\,Å \\
b &= 6.28\,Å \\
c &= 7.09\,Å \\
\beta &= 103.5^0,
\end{aligned}
$$

monoclinic space group C2, Cm or C2/c. The cell volume is then $V = abc \sin \beta$ (Table 3.4) $= 770\,Å^3$.

Now we can calculate the number of molecules per unit cell, n, using (3.28). The density of NITROS was determined by Steudel & Rose (unpublished results) to be $\varrho = 1.96\,g\,cm^{-3}$. With a molecular weight (formula $NH_4[S_4N_5O]$) of $W = 232.33\,g\,mol^{-1}$, we get $n \approx 3.94$, hence the nearest integer is $n = 4$. That means we have four formula units $NH_4[S_4N_5O]$in the cell. The number of symmetry operations

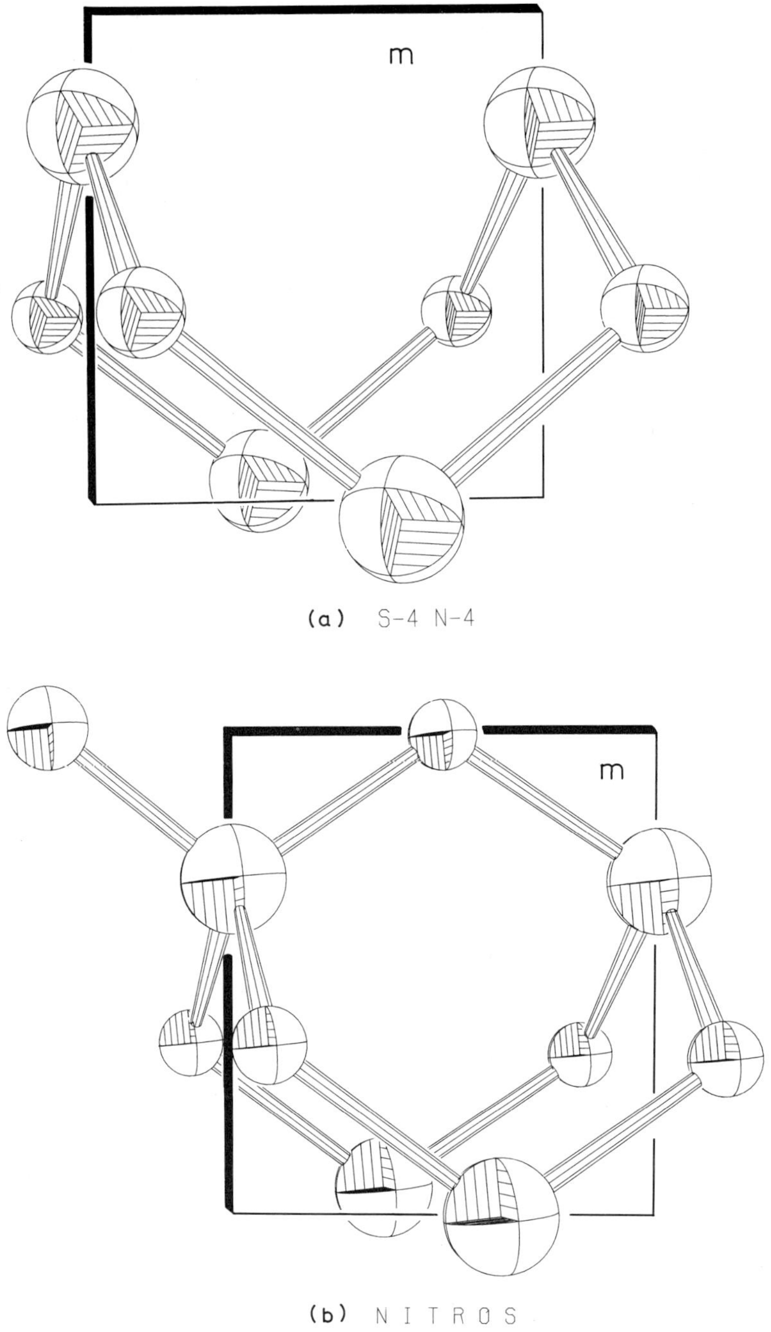

Fig. 3.33. (a) Mirror symmetry of the S_4N_4-molecule;
(b) Assumed mirror symmetry for the $S_4N_5O^-$-anion (sulfur atoms are representend by large spheres, nitrogen and oxygen by smaller spheres).

is four in the space groups C2 and Cm; it is eight in C2/m. So it seems more probable that either C2 or Cm is the correct space group. However, from IR and mass spectroscopy, it was proposed that the $[S_4N_5O]^-$ anion had a cage-like form similar to the known structure of S_4N_4 [Sharma, B. D. and Donohue, J., Acta Cryst. 16, 891 (1963)]. Since a mirror symmetry was observed within the S_4N_4 molecule (Fig. 3.33) it is possible that this symmetry element could be present for the $[S_4N_5O]^-$ anion.

Since a mirror plane is a symmetry element of C2/m and mirror symmetry is also consistent with the shape of the NH_4^+ cation, it is possible that we have the centric space group C2/m with one half cation and one half anion as the asymmetric unit. The final decision must come from the complete structure determination.

3.3.4 *Space Group of SUCROS*

We can now derive the space group of SUCROS. Inspecting the Weissenberg films of the zero, first, and second layer, which were represented in Fig. 2.38, it is observed that the lines A and B reappear as straight lines A' and B' andA" and B" respectively on the upper layers. From what we have learned about the shift of the intersection point, we know that A' and B' as well as A" and B" intersect the rotation axis and that therefore the reciprocal vector from the origin O to the intersection of A' and B' forms a right angle with both lines A and B (Fig. 3.34). Two right angles are already a strong indication of a monoclinic system, since the third angle between A and B differs from 90°. This can be seen immediately from the Δx-distance of these two lines (Fig. 2.38a),

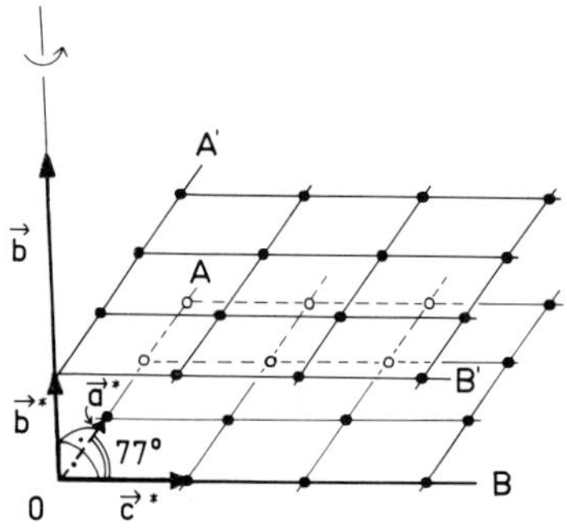

Fig. 3.34. Choice of basis vectors for SUCROS.

which is 38.5 mm, corresponding to an angle of 77^0 ($g = 0.5\,mm/^0$). Hence let us denote the reciprocal vectors in the A- and B-direction as **a*** and **c*** and that from O to the intersection of A' and B' as **b*** (Fig. 3.34). Then we have a system of reciprocal basis vectors corresponding to the conventions of a monoclinic system. However, the question of whether we really have a monoclinic lattice must be decided by the intensity symmetry. First we can estimate the numerical values of the lattice constants. From our selection of **a*** and **c***, we have the h 0 l reflections in the zero layer (h 1 l in the first layer, h 2 l in the second layer). The rotation axis is then **b** (see layer line condition 2.3). From the results of the rotation photograph we obtain $b = D = 8.79\,Å$, a* and c* can be taken from reflections of the h 0 0 line (line A) and 0 0 l (line B) series. Using $\theta_{800} = 35.3^0$ and $\theta_{009} = 65.5^0$, we get $a^* = 0.0937\,Å^{-1}$, $c^* = 0.1312\,Å^{-1}$. Since we know the monoclinic angle β^* (77⁰), we can calculate all direct lattice constants:

$$a = 10.95\,Å$$
$$b = 8.79\,Å$$
$$c = 7.83\,Å$$
$$\alpha = \gamma = 90^0, \beta = 103^0$$

Now let us see whether an intensity symmetry supporting the assumption of a monoclinic system is present. This can best be determined from the de Jong-Bouman exposures (Fig. 3.35a and b). Since the rotation axis was the same as for the Weissenberg photographs, they represent the same reciprocal lattice layers. Thus we have the h 0 l reflections on the zero layer, the h 1 l reflections on the first layer exposure. On both exposures it can clearly be seen that the intensity distribution shows the symmetry of a two-fold axis. That is, we have

$$I(h0l) = I(\bar{h}0\bar{l})$$
$$I(h1l) = I(\bar{h}1\bar{l})$$

Since we can assume that this holds for all k, we have

$$I(hkl) = I(\bar{h}k\bar{l})$$

which is the symmetry for a monoclinic crystal.

Having proven that SUCROS belongs to the monoclinic crystal system, we must now find the space group. A precise inspection of all Weissenberg and de Jong-Bouman exposures shows no systematic extinctions present. However, the 0 k 0 reflections are not visible on either the Weissenberg or the de Jong-Bouman exposures.

Fortunately, we have made some additional precession photographs of SUCROS. The axis in the direction of the primary beam had a length of 7.84 (derived as 1/D from the cone axis exposure, see 2.3.3) indicating this axis to be **c**. Then we have the h k 0 reflections on the zero layer and the h k 1 reflections on the first layer precession photograph. The lines P_A and P_C (Fig. 3.35c) indicate the h 0 0 and 0 k 0 series. Since the angle between **a*** and **b*** is a right angle, the precession photographs now show a *rectangular* lattice. The most important result, is the absence of every second reflection on the 0 k 0 series, indicating an axial extinction. It follows that we have a 2_1-screw axis in the b-direction. Since no further extinctions are present, we have as possible space groups for SUCROS, $P2_1$ or $P2_1/m$.

From the chemical properties of SUCROS we can get a unique solution of the space group problem. It is well-known that SUCROS is optically active. It is therefore impossible for the molecule and its mirror image to be present in the crystal, as would be

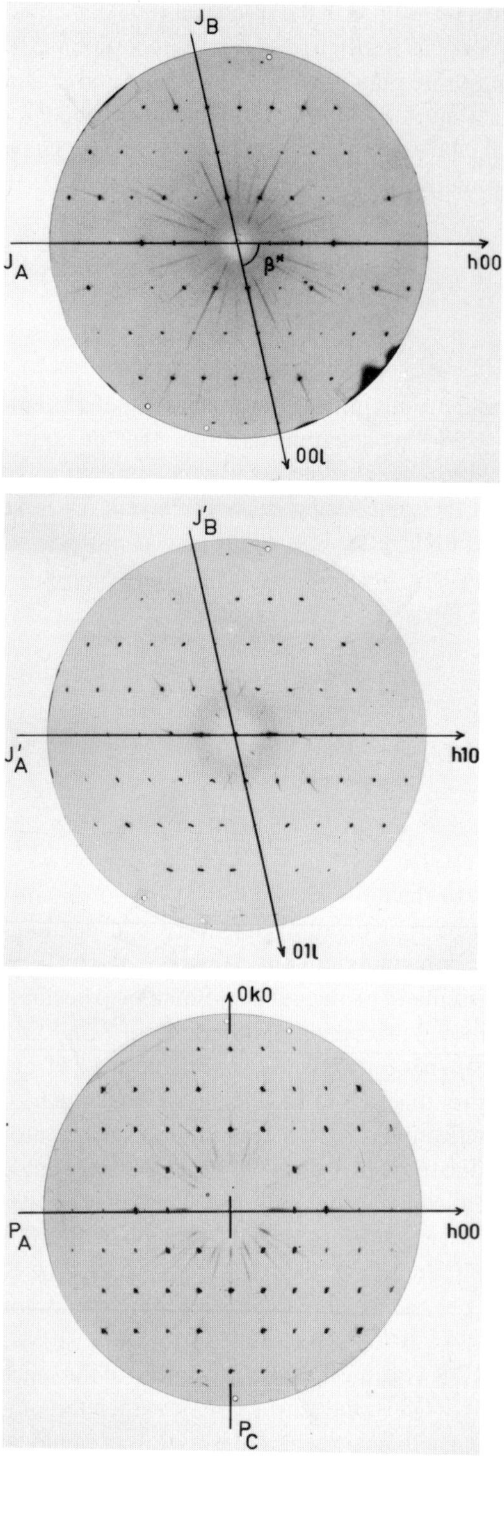

(a)

(b)

(c)

Fig. 3.35. Space group determination for SUCROS:
(a) from the zero-layer de Jong-Bouman exposure;
(b) from the first-layer de Jong-Bouman exposure;
(c) from the zero-layer precession photograph.

the case in the space group $P2_1/m$, because of the existence of a mirror plane as a symmetry element. We can therefore conclude that the correct space group for SUCROS is $P2_1$.

Finally, let us calculate the unit cell contents. SUCROS, formula $C_{12}H_{22}O_{11}$, has a molecular weight of $W = 342.30\,g\,mol^{-1}$. With a preliminary cell volume $V = abc\sin\beta = 734\,Å^3$, a density of $\varrho = 1.574\,g\,cm^{-3}$ [d'Ans-Lax, 'Taschenbuch für Chemiker und Physiker', (1964), Vol. II, 3rd ed., p. 776, Berlin: Springer], we get from (3.28)

$$n = \frac{1.574 \times 734}{1.65 \times 342.3} = 2.05$$

hence n is equal to to 2, i. e., we have two molecules per unit cell, and, since the symmetry in $P2_1$ is two-fold, there is one molecule in the asymmetric unit.

4 Diffractometer Measurements

4.1 Main Characteristics of a Four-Circle Diffractometer

In the last part of Section 3 we found that significant information about the crystal structure can be obtained from the preliminary information provided by films. The intensity symmetry and the presence or absence of extinctions, the lattice constants, space group (although not unambiguously in all cases), and the size of the asymmetric unit can usually be determined. These preliminary results give all (or nearly all) the necessary information about the crystal lattice geometry. The structure within one unit cell, or more precisely, the positions of the atoms in the asymmetric unit, are impossible to obtain from this type of data. The determination of these atomic positions requires the measurement of the reflection intensities and the solution of the phase problem in order to calculate the Fourier transform as described in 3.2.1, formula 3.19. The number of reflections that can be measured for a given structure is determined by the volume of the limiting sphere, which in turn depends on the unit cell volume and the wavelength used. Although the intensity symmetry may reduce the number of symmetry-independent reflections to one-half or one-quarter (or even more), several hundreds, up to some thousands of reflections have to be measured for structures of medium size. This work is done today by means of an automatic four-circle diffractometer. The great advantages of this instrument are:

(1) It runs automatically by computer control and thus allows a rapid and convenient collection of data with only minor supervision.

(2) The accuracy of the results is significantly better than that from film measurements.

(3) The results can usually be stored on a device such as a magnetic tape or disc, for further processing on a computer.

(4) Since the reflections can be positioned very precisely, the lattice constants can be determined with high accuracy.

Therefore all final X-ray measurements, both for the lattice constants and for the reflection intensities are made on a diffractometer. Although we cannot describe the features of a diffractometer in all details (such information is given in the equipment instruction manual), some remarks on the main characteristics will be made.

A four-circle diffractometer consists of four parts (see Fig. 4.1):

(1) the X-ray source (high voltage generator and X-ray tube);

(2) the diffractometer itself, usually constructed, but not necessarily, as a 'Eulerian cradle';

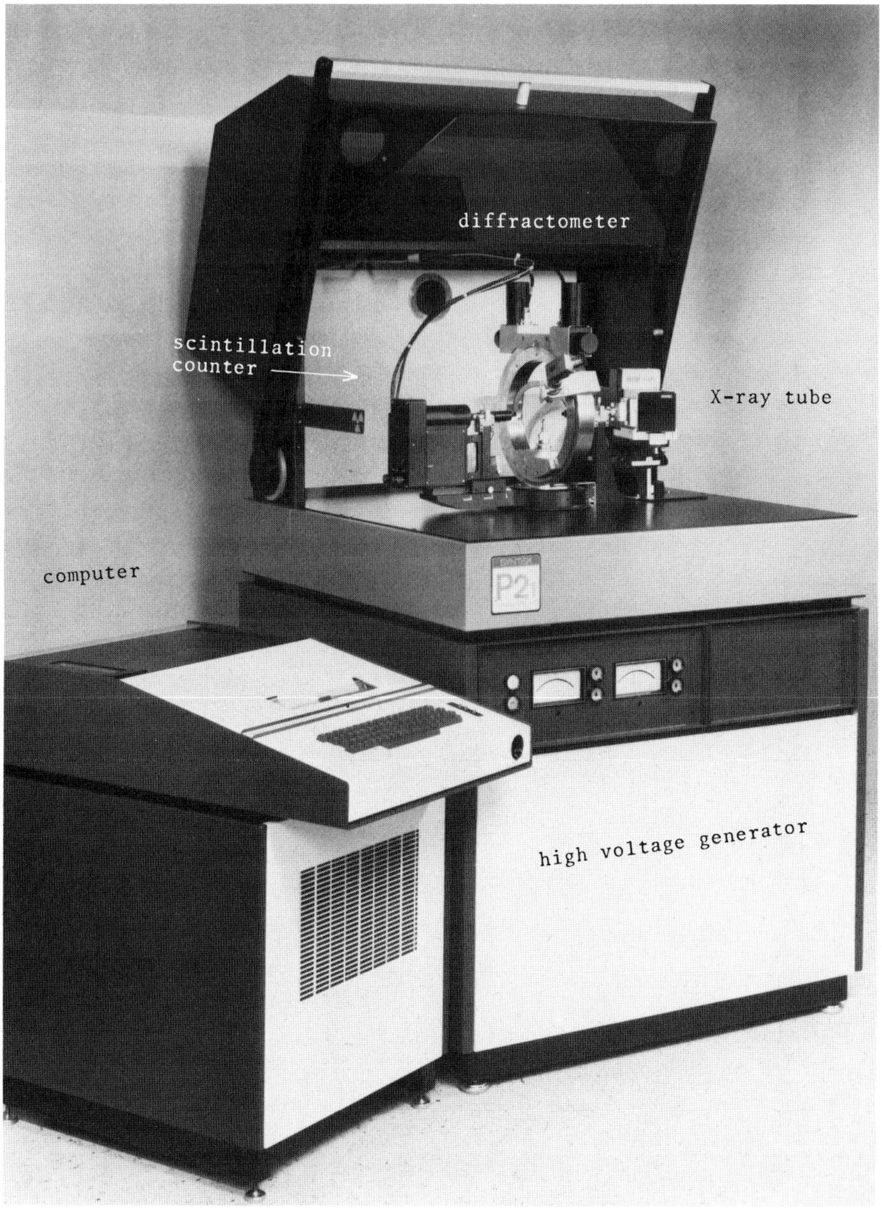

Fig. 4.1. A four-circle diffractometer (Syntex $P2_1$, Syntex Analytical Instruments, Inc.).

(3) the detector (scintillation counter) and its counting chain;
(4) the computer, for control of the diffractometer and storage of results from the detector.

4.1.1 *Eulerian Cradle Geometry*

The central part of a four-circle instrument is the diffractometer with its Eulerian cradle geometry which will be discussed in detail.Of the four-circle diffractometers available commercially, one type, the Nonius CAD-4 diffactometer, has a so-called 'Kappa-axis geometry' instead of a Eulerian cradle. Since this is the only exception, we will restrict our consideration to the most frequently used, the Eulerian geometry.

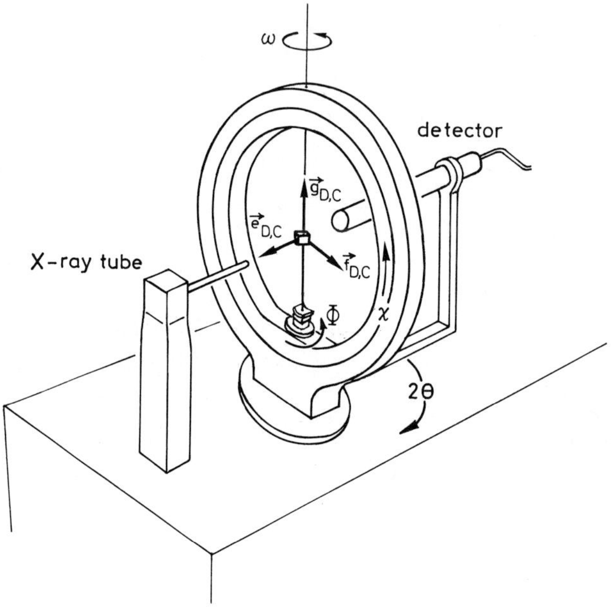

Fig. 4.2. Eulerian cradle geometry.

The diffractometer consists of four circles (see Fig. 4.2) allowing the crystal to be brought into various orientations. Two circles, denoted by ϕ and χ, are used to adjust the crystal orientation relative to the diffractometer coordinate system. A third circle, ω, permits the orientation of the crystal lattice planes at a given angle ω to the direction of the primary beam, and finally, on the fourth circle, 2θ, the detector can be moved to lie at an angle 2θ to the primary beam.

A schematic representation of the four circle geometry and the coordinate system used is given in Fig. 4.3a. Several conventions concerning the choice of the diffractometer coordinate system and the different circles are in use. Here we follow the definitions given in the International Tables [Vol. IV, pp. 276, (1974)] for that problem.

The primary beam direction, the crystal and the detector collimator are fixed in a horizontal plane. This plane is called the diffraction plane since every reciprocal lattice vector **h** has to be brought into this plane so that the Ewald diffraction condition can be realized. The axis perpendicular to the diffraction plane passing through the crystal is called the main axis of the apparatus. The detector rotates around this axis and the ω circle just so and we have thus the main axis as the ω and the 2θ axis.

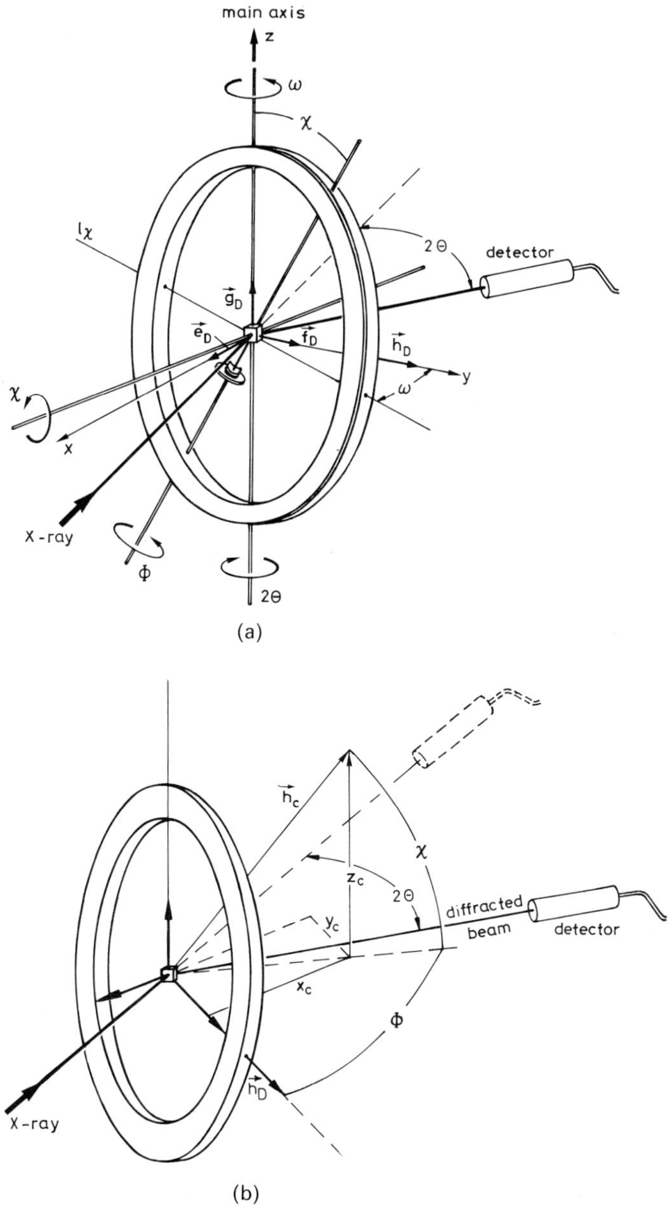

Fig. 4.3. (a) Schematical representation of the four circles and the diffraction coordinate system; (b) diffractometer in bisecting position.

Another axis, the χ axis, is defined as passing through the crystal in the diffraction plane. The plane of the χ circle is then situated perpendicular to the diffraction plane. The ϕ axis is the spindle axis of the goniometer head. The sense of rotation is indicated by the curved arrows in Fig. 4.3a.

The magnitudes of the four angles are defined as follows. 2θ is the angle between the direction of incident and diffracted beam, i. e. the detector has to be rotated by 2θ from its zero position to registrate the diffracted beam. The diffraction vector and the plane of the χ circle define the angle ω. χ is the angle between the ϕ axis and the diffractometer main axis, an absolute origin of ϕ is usually not defined.

A diffraction coordinate system with orthonormal unit vectors $\mathbf{e}_D, \mathbf{f}_D, \mathbf{g}_D$ is chosen as indicated Fig. 4.3a. $\mathbf{g}_D$ has the direction of the main axis. Then $\mathbf{e}_D$ and $\mathbf{f}_D$ must be in the diffraction plane. To facilitate the further calculation it is convenient to choose $\mathbf{f}_D$ parallel to the diffraction vector $\mathbf{h}$ (for positive 2θ). Then $\mathbf{e}_D$ points towards the X-ray source for $2\theta = 0$.

A second orthonormal coordinate system is used which is fixed with respect to the goniometer head and is denoted the crystal-fixed system. Its basis vectors $\mathbf{e}_C, \mathbf{f}_C, \mathbf{g}_C$ are defined to be coincident with the diffraction system if all diffractometer angles are zero (see Fig. 4.2).

If at least one angle differs from zero, the two systems are related by a transformation matrix as was shown in 1. 1.5, formula 1.17). Assuming to have a rotation about ω, then this rotation takes place in the x-y plane and we can use formula 1.3 from section 1. 1.2.

Setting

$$E_D = \begin{pmatrix} \mathbf{e}_D \\ \mathbf{f}_D \\ \mathbf{g}_D \end{pmatrix}, \qquad E_C = \begin{pmatrix} \mathbf{e}_C \\ \mathbf{f}_C \\ \mathbf{g}_C \end{pmatrix}$$

we have

$$E_C = R(\omega)\, E_D$$

with $R(\omega) = (r_{ik})$ a 3×3 matrix. For $\mathbf{e}_C$ we have then

$$\mathbf{e}_C = r_{11}\,\mathbf{e}_D + r_{12}\,\mathbf{f}_D + r_{13}\,\mathbf{g}_D$$

with $r_{13} = 0$ and (with formula 1.3)

$$\begin{pmatrix} r_{11} \\ r_{12} \end{pmatrix} = \begin{pmatrix} \cos\omega & -\sin\omega \\ \sin\omega & \cos\omega \end{pmatrix} \begin{pmatrix} 1 \\ 0 \end{pmatrix} = \begin{pmatrix} \cos\omega \\ \sin\omega \end{pmatrix}$$

Similary for $\mathbf{f}_C$

$$\mathbf{f}_C = r_{21}\,\mathbf{e}_D + r_{22}\,\mathbf{f}_D + r_{23}\,\mathbf{g}_D$$

with $r_{23} = 0$ and

$$\begin{pmatrix} r_{21} \\ r_{22} \end{pmatrix} = \begin{pmatrix} \cos\omega & -\sin\omega \\ \sin\omega & \cos\omega \end{pmatrix} \begin{pmatrix} 0 \\ 1 \end{pmatrix} = \begin{pmatrix} -\sin\omega \\ \cos\omega \end{pmatrix}$$

(Note that at this point the sense of rotation plays an important role. The sense for the ω-rotation was chosen clockwise as was done in Fig. 1.1 in 1.1.2. So formula 1.3 can be applied without alterations.)

Since $\mathbf{g}_C = \mathbf{g}_D$, we get finally

$$E_C = \begin{pmatrix} \cos\omega & \sin\omega & 0 \\ -\sin\omega & \cos\omega & 0 \\ 0 & 0 & 1 \end{pmatrix} E_D$$

In the same way we get the matrices for rotation about ϕ and χ

$$R(\phi) = \begin{pmatrix} \cos\phi & \sin\phi & 0 \\ -\sin\phi & \cos\phi & 0 \\ 0 & 0 & 1 \end{pmatrix}$$

$$R(\chi) = \begin{pmatrix} 1 & 0 & 0 \\ 0 & \cos\chi & \sin\chi \\ 0 & -\sin\chi & \cos\chi \end{pmatrix}$$

In the general case of a rotation about all three angles the transformation is described by the product of the three matrices

$$E_C = R(\phi)\, R(\chi)\, R(\omega)\, E_D = S\, E_D$$

$$S = R(\phi)\, R(\chi)\, R(\omega) =$$

$$\begin{pmatrix} \cos\phi\cos\omega - \sin\phi\sin\omega\cos\chi & \cos\phi\sin\omega + \sin\phi\cos\omega\cos\chi & \sin\phi\sin\chi \\ -\sin\phi\cos\omega - \cos\phi\sin\omega\cos\chi & -\sin\phi\sin\omega + \cos\phi\cos\omega\cos\chi & \cos\phi\sin\chi \\ \sin\chi\sin\omega & -\sin\chi\cos\omega & \cos\chi \end{pmatrix}$$

A lattice vector $\mathbf{h}$ of reciprocal space is represented in term of the reciprocal lattice constants by

$$\mathbf{h} = \begin{pmatrix} h \\ k \\ l \end{pmatrix} = h\mathbf{a}^* + k\mathbf{b}^* + l\mathbf{c}^*$$

This vector may have the representations in the E_C and E_D systems

$$\mathbf{h}_C = \begin{pmatrix} x_C \\ y_C \\ z_C \end{pmatrix}, \qquad \mathbf{h}_D = \begin{pmatrix} x_D \\ y_D \\ z_D \end{pmatrix}$$

Then if follows from (1.18), that

$$\mathbf{h}_C = (S^{-1})'\, \mathbf{h}_D$$

Since for each R matrix $(R^{-1})' = R$ holds, this is valid also for S and we get

$$\mathbf{h}_C = S\, \mathbf{h}_D$$

If a lattice vector **h** is in diffraction position it follows from the special choice of the E_D system (with $D^* = |\mathbf{h}|$)

$$\mathbf{h}_D = \begin{pmatrix} 0 \\ D^* \\ 0 \end{pmatrix}$$

Thus it follows

$$\mathbf{h}_C = S\,\mathbf{h}_D = D^* \begin{pmatrix} \cos\phi \sin\omega + \sin\phi \cos\chi \cos\omega \\ -\sin\phi \sin\omega + \cos\phi \cos\chi \cos\omega \\ -\sin\chi \cos\omega \end{pmatrix} \qquad (4.1)$$

If the components of $\mathbf{h}_C$ are known, (4.1) can be used for a calculation of the setting angles. However, one of the angles is redundant so that in terms of one angle, which can be varied, different diffraction positions can be calculated.

Frequently people prefer a "symmetric" diffraction position, which is defined by setting $\omega = 0$ causing the diffraction vector to be in the plane of the χ-circle (Fig. 4.3b). In other words, the plane of χ-circle bisects the angle between the incident and the diffracted beam. Therefore this diffraction position is also called "bisecting position".

With this provision (4.1) reduces to

$$x_C = D^* \sin\phi \cos\chi$$
$$y_C = D^* \cos\phi \cos\chi$$
$$z_C = -D^* \sin\chi$$

It follows

$$\sin\chi = -z_C/D^* \qquad (4.2a)$$

$$\cos\phi = \frac{y_C}{\sqrt{x_C^2 + y_C^2}} \qquad (4.2b)$$

2θ is obtained from Bragg's equation and all setting angles can either be calculated by solving (4.1) or in the symmetric case from (4.2). From the special choice of the E_D system it appears that the X-ray tube has to be moved for each reflection to be positioned at an angle θ against $\mathbf{e}_D$. Since the construction of a diffractometer requires the X-ray tube to remain stationary, the diffractometer is turned about the ω-axis by the angle θ towards the primary beam direction.

Provision has to be made to obtain the components of $\mathbf{h}_C$. For that purpose, the orientation of the crystal reciprocal axes relative to the E_C system has to be determined in a previous experimental procedure. If the orientation matrix obtained is U, we have

$$A^* = U\,E_C \qquad (4.3a)$$

with

$$A^* = \begin{pmatrix} \mathbf{a}^* \\ \mathbf{b}^* \\ \mathbf{c}^* \end{pmatrix}$$

and, using (1.18a)

$$\begin{pmatrix} x_C \\ y_C \\ z_C \end{pmatrix} = U' \begin{pmatrix} h \\ k \\ l \end{pmatrix} \tag{4.3b}$$

From the orientation matrix, the lattice constants are obtained. Using (4.3a) we get

$$A^* A^{*\prime} = (U E_C)(U E_C)' = U E_C E_C' U' = U U',$$

since $E_C E_C'$ is the unit matrix.

On the other hand

$$A^* A^{*\prime} = \begin{pmatrix} a^{*2} & a^* b^* \cos\gamma^* & a^* c^* \cos\beta^* \\ a^* b^* \cos\gamma^* & b^{*2} & a^* c^* \cos\alpha^* \\ a^* c^* \cos\beta^* & b^* c^* \cos\alpha^* & c^{*2} \end{pmatrix} \tag{4.4}$$

So we can summarize. A four circle diffractometer is operated in such a way, that at first the orientation matrix is determined, from which the reciprocal lattice constants and the components of $\mathbf{h}_C$ are obtained.

Then for every lattice vector **h**, the four angles, χ, ϕ, ω and 2θ are calculated from (4.1) or (4.2) and **h** is brought into a position to satisfy the Ewald condition (see Fig. 4.3b). Usually it is not necessary to make these calculations separately from the instrument, since they are done by the software supplied with the diffractometer.

4.1.2 *X-Ray Source, Detector and Controlware*

Cu- and Mo-target X-ray tubes are usually supplied for crystal structure analysis. These two target materials are sufficient for most normal problems. For monochromatic radiation, either a β-filter or a graphite crystal monochromator is used. The decision whether to use a filter or a monochromator is not obvious. Nearly all diffractometers are supplied with a monochromator, since it was considered the best way to obtain monochromatic radiation. However, this opinion is contested, and laboratory experience is that filtered radiation frequently gives a more accurate set of intensity data. Two problems arise with the use of monochromators. The first is a greater attenuation of the intensity than for filters, and the second is the profile of the monochromatic beam. If the monochromator crystal is not of excellent quality, and if it is not adjusted very precisely, it can give a non-uniform beam profile which can be a source of serious errors in the intensity measurements. Even with a well-aligned monochromator, the plateau of constant intensity of the monochromated beam is very small. This implies that only small crystals can be used, causing a further decrease of the diffracted intensity. Filtered radiation is less attenuated and the filter involves no alignment problems other than that with the tube and the collimator. The plateau (see Fig. 4.4) is larger than that of a monochromated beam, but has the disadvantage of a

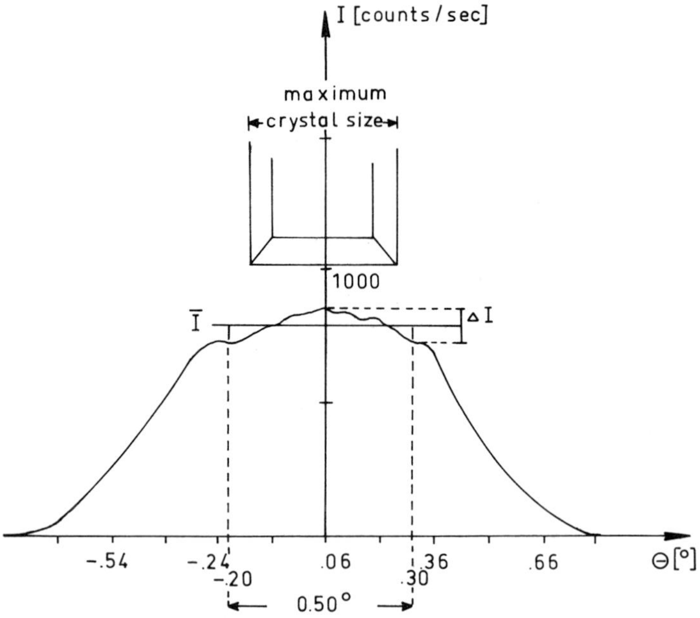

Fig. 4.4. Experimentally taken primary beam profile. Within a $\Delta\Theta$ range of 0.5 (corresponding to a diameter of 1 mm) the intensity is approximately constant with $\Delta I/I$ less than $10^0/_0$. The maximum crystal size must not exceed the primary beam limits.

less monochromatic radiation, resulting in a more diffuse reflection profile and therefore a deterioration of reflection to background ratio.

The detector usually consists of a scintillation counter with pulse height analyzer. Care has to be taken in avoiding an overload of the counter. Depending on the counter's so-called "dead-time", only a limited number of counts/second can be recorded. Reasonable counting rates are some ten thousand counts/sec, the actual rate depending on the counter type used. If, for strong reflections, more counts enter the detector window, the amount exceeding maximum rate is lost, causing errors in the intensity values. In this case, there is provision for the insertion of attenuation filters. The attentuation factors can be obtained by measuring a reflection of medium intensity with and without filters.

The precision of the counter measurements is a problem which can be treated by statistical methods. The theory of counter statistics is complicated, and we consider only the experimental aspect. In statistics, the number, N, of counts from a counter measurement is a random variable, the distribution of which can be described by a distribution function. To obtain this function experimentally we plotted the result of 1024 independent measurements of one counting rate in Fig. 4.5. The distribution maximum occurs near $N_o = 5000$ counts/sec (exactly at $N_o = 4972$ c/s). A theoretical calculation of a so-called "Poisson distribution" defined by

$$p(N) = \frac{N_o^N}{N!} e^{-N_o} \tag{4.5}$$

with that parameter N_o, is plotted as the full line in Fig. 4.5. It shows a close agreement with the experimental distribution. This confirms that the distribution of counting rates can be derived theoretically by a Poisson function.

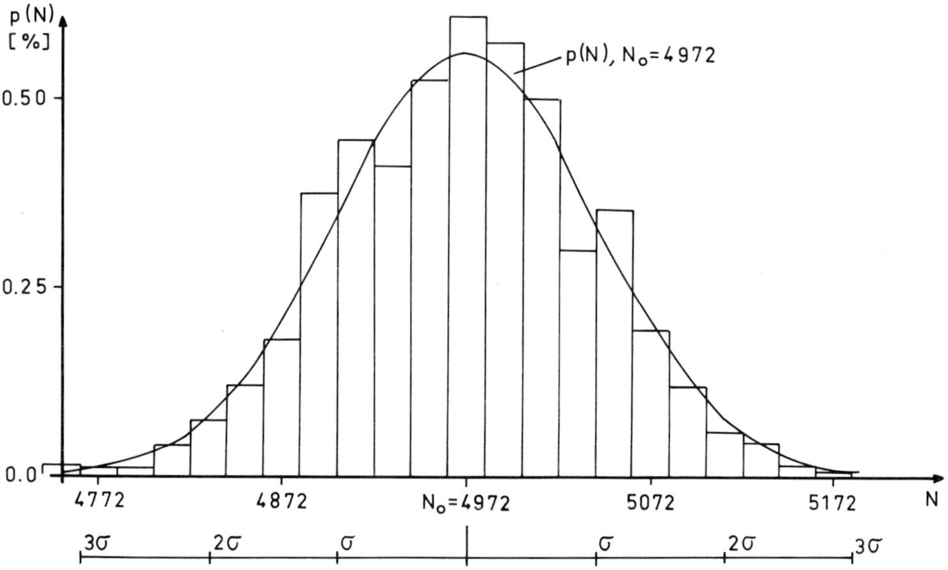

Fig. 4.5. Experimentally obtained distribution of a counting rate plotted together with a theoretical Poisson distribution.

For the experimentor, the two most important parameters of an arbitrary distribution f(x) are the mean value

$$\bar{x} = \int_{-\infty}^{+\infty} x f(x) dx \tag{4.6a}$$

and the standard deviation, σ, which is defined as the root-mean-square deviation of a set of observations from the mean value

$$\sigma^2 = \overline{(x - \bar{x})^2} \tag{4.7a}$$

or, by using (4.6),

$$\sigma^2 = \int_{-\infty}^{+\infty} (x - \bar{x})^2 f(x) dx \tag{4.7b}$$

If the number of observations is finite, as in an actual experiment, $\bar{x}$ and σ are approximated by x_m and s, given by

$$x_m = (\sum_{i=1}^{n} w_i x_i)/n \tag{4.6b}$$

$$s^2 = [\sum_{i=1}^{n} w_i (x_i - x_m)^2]/(n-1) \tag{4.7c}$$

if n observations x_i ($i = 1, \ldots, n$) have been made. The w_i, which replace the probability element $f(x)\,dx$, can be regarded as weighting parameters, taking the accuracy of each observation x_i into account. The quantity $(n-1)$, which denotes the degree of freedom of the experiment, is generally that number of observations that have to be made in addition to those necessary for the unique determination of the desired parameters. It can be shown that for $n \to \infty$, x_m and s converge towards $\bar{x}$ and σ, so it is customary to use the symbols $\bar{x}$ and σ, even if they are calculated from (4.6b) or (4.7c).

In the actual case of a counting rate, the distribution variable is restricted to discrete positive integer values, so that the integrals in (4.6a) and (4.7b) can be replaced by sums. Then, from the Poisson distribution, we get (the summation can be started with $N = 1$, since for $N = 0$, the sum is equal to zero):

$$\bar{N} = \sum_{N=1}^{\infty} \frac{N_o^N}{N!} e^{-N_o} N = N_o e^{-N_o} \sum_{N=1}^{\infty} \frac{N_o^{N-1}}{(N-1)!}$$

$$(\text{set } N' = N-1) = N_o e^{-N_o} \sum_{N'=0}^{\infty} \frac{N_o^{N'}}{N'!} = N_o e^{-N_o} e^{N_o}$$

hence,

$$\bar{N} = N_o \tag{4.8}$$

Equation (4.8) implies that the mean value is given by the parameter N_o of the Poisson distribution. For the standard deviation, σ, we get, from (4.7a) and (4.5)

$$\sigma^2 = \overline{(N - N_o)^2} = \overline{\Delta^2}$$

with

$$\begin{aligned} \Delta^2 &= N^2 - 2NN_o + N_o^2 \\ &= N(N-1) + N - 2NN_o + N_o^2 \\ &= N(N-1) + N(1 - 2N_o) + N_o^2 \\ &= A + B + C \end{aligned}$$

with

$$\begin{aligned} A &= N(N-1) \\ B &= N(1 - 2N_o) \\ C &= N_o^2 \end{aligned}$$

Then for σ^2, we get

$$\sigma^2 = \sum_{N=0}^{\infty} (A + B + C)\, p(N)$$

The calculation is best done for A, B and C separately. For C we get

$$\sigma_C^2 = N_o^2 \sum_{N=0}^{\infty} p(N) = N_o^2$$

because p(N) is a normalized distribution. For A we obtain

$$\sigma_A^2 = \sum_{N=2}^{\infty} N(N-1) \frac{N_o^N}{N!} e^{-N_o}$$

$$= e^{-N_o} N_o^2 \sum_{N=2}^{\infty} \frac{N_o^{N-2}}{(N-2)!}$$

$$= e^{-N_o} N_o^2 e^{N_o} = N_o^2$$

and finally, for B,

$$\sigma_B^2 = \sum_{N=0}^{\infty} N(1-2N_o)\, p(N)$$

$$= (1-2N_o) \sum_{N=0}^{\infty} N p(N) = (1-2N_o) N_o$$

hence

$$\sigma^2 = \sigma_A^2 + \sigma_B^2 + \sigma_C^2 = N_o^2 + N_o - 2N_o^2$$

$$\sigma^2 = N_o$$

or

$$\sigma = \sqrt{N_o} \qquad (4.9)$$

Usually a counting rate N will be different from N_o but not too different, so that $N \approx N_o$ and $\sqrt{N} = \sqrt{N_o}$. So (4.9) can be interpreted as follows:

If a counting rate N is measured, a good estimation of the standard deviation for the mean value N_o is $\sqrt{N}$. Then $\sqrt{N} \approx \sigma$ is the estimated standard deviation (e.s.d.) of N_o.

The probability $w(x_1, x_2)$ for an observation x to fall into an intervall $[x_1, x_2]$ is given by the integral over the distribution function

$$w(x_1, x_2) = \int_{x_1}^{x_2} f(x)\, dx \qquad (4.10)$$

If f(x) is a Poisson distribution, the numerical calculation of (4.10) is not so easy. Fortunately, for large N_o, i.e. $\sqrt{N_o} \gg 1$, a Poisson distribution can be approximated by a Gaussian distribution

$$g(N) = \frac{1}{\sqrt{2\pi N_o}} e^{-\frac{(N-N_o)^2}{2N_o}} \qquad (4.11)$$

having N_o as the mean value and $\sqrt{N_o}$ as the standard deviation, as before.

For a Gaussian distribution, (4.10) has been calculated for various intervals. It follows that

$$w(N_o \pm \sigma) = 68.3\%$$
$$w(N_o \pm 2\sigma) = 95.4\%$$
and $$w(N_o \pm 3\sigma) = 99.7\%$$

Interpreting the results for counting rates, we can state that, if we observe a value N, say $N = 5000$, we have $\sigma \approx 70$ and a 68.3% probability exists that N_o lies between 5000 $\pm$ 70, a 95.4% probability exists that N_o lies between 5000 $\pm$ 140, and a nearly 100% (99.7%) probability that N_o lies between 5000 $\pm$ 210. The knowledge of the standard deviation of a counting rate will be of great importance when the problem of estimating the standard deviations of reflection intensities must be treated.

The controls of the various diffractometers differ and only a general description can be made. Usually, a medium-size computer operates the diffractometer and the communication with the user is via the keyboard of the computer. Software packages allow more or less automatic operation. Software routines are usually available for

(1) Determination of the crystal orientation with respect to the Eulerian cradle,
(2) Automatic or semi-automatic lattice constant determination and refinement,
(3) Automatic operation and supervision of intensity measurement,
(4) Storage of results.

With some diffractometers, an additional program package for structure determination and refinement is available. However, since refinement programs run best on high-speed computers with large storage, it seems more profitable to use the diffractometer and its computer for measurement purposes only, and to do the structure determination and refinement calculations on a larger computer, if available.

4.2 Single Crystal Measurements

4.2.1 *Choice of Experimental Conditions*

The final X-ray measurements consist of the precise determination of lattice constants and the collection of reflection intensities. *All* such measurements should be made as precisely as possible. Lower than optimum accuracy in the experimental data may result in difficulties with the structure determination or the refinement. In some cases, poorly measured data can result in complete failure of a structure determination. In this sense, single crystal analysis is a method which provides either very extensive and precise results or nothing at all (which happens if the phase problem cannot be solved). Single crystal analyses seldom provide only a partial description of a structure.

To avoid common sources of errors the experimental conditions for diffractometer measurements should be chosen very carefully. The most important question is the choice of a single crystal of good quality and of optimal size. The crystal quality is good if the specimen mounted on the goniometer head consists of only *one* single crystal with no fractions of other crystals attached to it. Powder residues from the crystallization and preparation process should be avoided as much as possible. If possible, the natural crystal faces should be recognizable, especially if absorption and extinction corrections are to be made. Otherwise the crystal may be ground to a spherical or cylindrical shape. The best way to check crystal quality is by examination of a film photograph of the specimen. Fig. 4.6a–d shows Weissenberg photographs of crystals with defects which are severe enough that they are unsuitable for intensity measurement. The crystal shown in Fig. 4.6a consists of several fragments attached to each other. An overlap of neighbouring reflections can be recognized, mainly in the high order reflections, indicating that individual reflection intensities can no longer be taken.

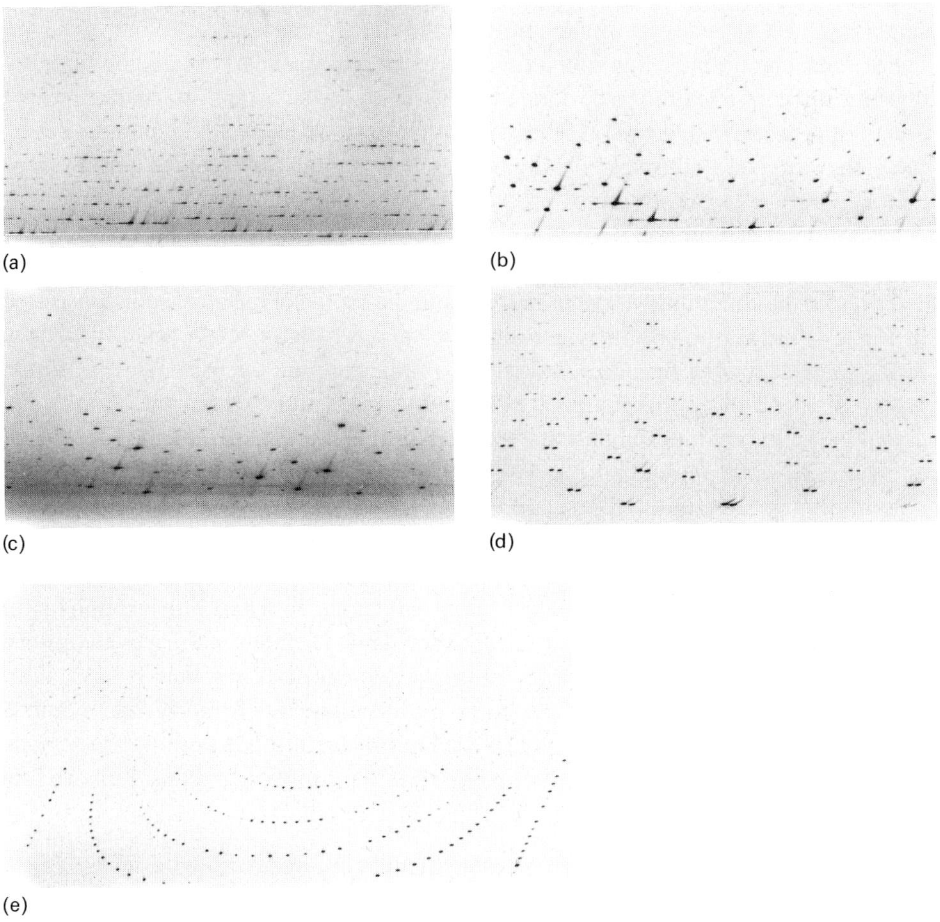

Fig. 4.6. (a–d) Weissenberg exposures (upper half) of crystals of minor quality for various reasons; (e) exposure of a good single crystal.

The photograph in Fig. 4.6b shows additional reflections due to some randomly distributed crystalline satellites. If one of these satellite reflections happens to coincide with a main reflection, that intensity measurement will be incorrect. The photograph in Fig. 4.6c is taken from a crystal the surface of which is covered with powder. The powder pattern, which is visible as straight lines, influences the background intensities and thus the reflection intensities also. Fig. 4.6d shows the Weissenberg photograph of a twinned crystal which was used for intensity measurement in our laboratory and resulted in a good structure determination. In this case, the two individual crystals were attached with their orientation relative to the crystallographic b-axis turned by a few degrees. Since the intensity spots of all reflections are separated without overlap, the intensities of one individual crystal could be measured with sufficient accuracy. This crystal was used for intensity measurement because no untwinned specimen was available. Generally, the use of twinned crystals for intensity measurement should be avoided whenever possible. Finally, in Fig. 4.6e, the diffraction pattern of an excellent single crystal is shown with sharp undistorted reflection spots.

For some compounds, it is very difficult or even impossible to obtain good quality crystals. If the information to be derived from the crystal structure determination is of great importance, then a crystal of lesser quality must be used. Sometimes, however, the preparation of a derivative having better single crystals will provide a more satisfactory solution to the structural problem with less complications.

The optimal crystal size depends on two parameters; (a) the size of the primary beam, and (b) the linear absorption coefficient of the compound.

To check the uniformity of the primary beam, the profile of the focal spot, illustrated in Fig. 4.4, should be checked carefully after a tube alignment or when using the diffractometer for the first time. For filtered radiation, the intensity plateau should have a diameter of approximately 0.8–1.0 mm, so that to be within these limits, the crystals should have no dimensions larger than 0.6–0.8 mm. From the absorption equation (2.38), it follows that the pathlength of the X-ray beam through the crystal determines the absorption. There are two ways to reduce absorption errors for radiation of a given wavelength. The first is to reduce the crystal size and the second is to choose a crystal which is as spherical as possible. Then the pathlength depends only on θ, and does not significantly differ for various reflections, so that the absorption factor acts more or less as a constant factor. It was shown in 2.2.3 that $x = 3/\mu$ was an optimal choice for the crystal diameter. Since, for organic compounds, μ generally has a value less than 10 cm^{-1} for $CuK\alpha$ radiation, a crystal diameter of more than 1 mm is possible, so that the primary beam size is the limiting factor in determining the crystal size. Summarizing our conclusions with regard to the choice of crystals, we have four general rules:

(1) Choose a single crystal of best possible quality,

(2) Choose the crystal size as large as is consistent with $3\mu^{-1}$ and the size of the primary beam,

(3) The crystal shape should be as spherical as possible,

(4) Choose a crystal with visible natural faces.

To realize conditions (2) and (3), often larger crystals have to be cut. From our experience, a razor blade is an excellent tool for organic compounds. Since crystal cutting damages the crystal faces and sometimes destroys the crystal, it should be minimized. The resulting powder and crystalline fragments should be removed from the cut face.

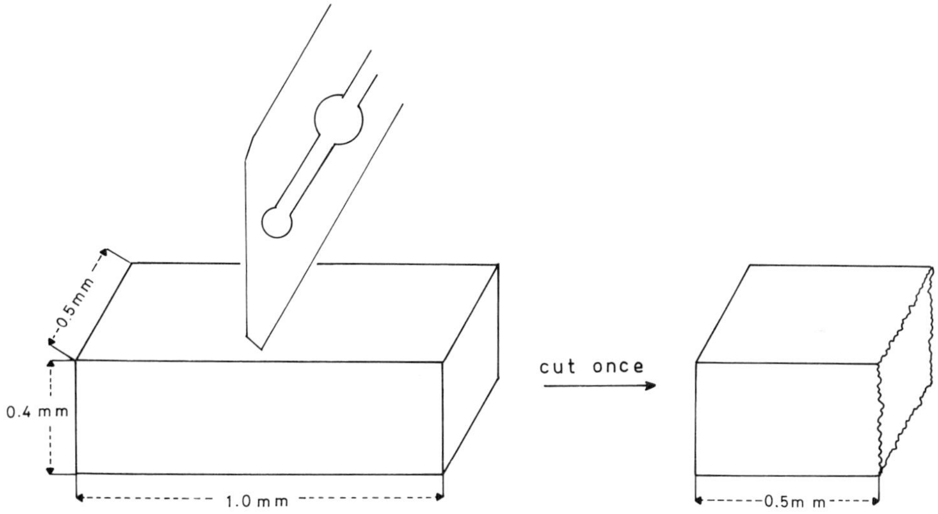

Fig. 4.7. Cutting a crystal.

The next question is the choice of the most appropriate radiation to use. As mentioned above, it is customary in single crystal analysis to use X-ray tubes with copper or molybdenum targets. More than 99% of all modern crystal structure investigations are made with these two radiations. Only for very few special problems is CrKα radiation used. When deciding between molybdenum and copper radiation, the following properties should be considered:

(1) If filters are used (Ni for CuKα, Zr for MoKα radiation), more white radiation can be eliminated for CuKα radiation. It follows that the intensity to background ratio is generally more favorable for copper radiation. This is especially true for crystals with a large number of weak intensities.

(2) The optimum operating voltage for CuKα radiation is 35–40 KV, whereas it is 55 KV for MoKα radiation. For some high voltage generators, it is difficult to stabilize the higher voltage. With some modern transformers, this higher voltage can be the source of breakdown in the electronic circuitry of the diffractometer.

(3) The linear absorption coefficient is always much smaller for MoKα radiation.

(4) The resolution of the diffraction spectra can never exceed $\lambda/2$. Therefore the resolution of a structure determination cannot exceed 0.75 Å if all reflections of the

copper sphere are taken. Since no atoms approach closer than this separation, this is not a significant problem. However, in principle, resolution can be improved by using Mo-radiation, but as pointed out in the discussion of form factors (3.2.1), this can only be realized under special conditions. Conversely, with very large unit cells, as in proteins, intensities cannot be separated with Mo-radiation.

Summarizing, it is recommended that for organic compounds with the smaller absorption coefficients, CuKα radiation is used, while for inorganic compounds, MoKα radiation will generally be more convenient.

For KAMTRA the linear absorption coefficients (see 2.2.2, Table 2.1) were found to be 72.8 cm^{-1} for CuKα and 8.2 cm^{-1} for MoKα radiation. Because of the large μ for CuKα radiation, MoKα radiation is chosen. The reflection-to-background ratio is still good, since the film exposures of KAMTRA show that many high-order reflections are observed.

The crystal used for film exposures was too large for diffractometer measurements, since one dimension was 0.8 mm. We therefore chose another specimen of $0.3 \times 0.4 \times 1.0$ mm which was cut perpendicular to the needle direction to obtain a specimen with dimensions $0.3 \times 0.4 \times 0.4$ mm. (It might be a problem to use different crystal specimen for film and diffractometer experiments. When the amount of crystals is small or there is doubt about their quality, it is wise to use the same crystal for the photographic examination as for the intensity measurements.)

NITROS has a very large μ (106.5 cm^{-1}) for CuKα radiation due to the high absorption coefficient of sulfur. For MoKα radiation, $\mu = 11.8$ cm^{-1}, which is more advantageous. A crystal of size $0.2 \times 0.3 \times 0.4$ mm was used.

For SUCROS, $\mu_{Cu} = 12.5$ cm^{-1}, and this organic compound has an excellent diffraction power, as can be seen from the Weissenberg exposures (Fig. 2.38). An almost cube-shaped crystal of $0.6 \times 0.6 \times 0.6$ mm provided an excellent specimen for intensity measurements.

4.2.2 *Precise Determination of Lattice Constants*

Usually the orientation of the reciprocal crystal axes relative to the crystal-fixed system has to be determined as a first step. As was show in 4.1.1 the lattice constants can then be obtained from the result of the orientation matrix calculation [see formula (4.4)]. There exist two further methods for the precise determination of lattice constants which do not utilize the results of the orientation matrix. The first is based on the precise location of axial reflections. The second is based on the precise location of non-axial reflections. From the 2θ reflection angles, the lattice constants can be derived from Bragg's law.

The quantity $D^* = |\mathbf{h}|$ fully written reads (with $\mathbf{h} = h\mathbf{a}^* + k\mathbf{b}^* + l\mathbf{c}^*$)

$$D^{*2} = \mathbf{hh} = h^2 a^{*2} + k^2 b^{*2} + l^2 c^{*2} + 2hka^* b^* \cos\gamma^* + 2hla^* c^* \cos\beta^* + 2klb^* c^* \cos\alpha^*$$

If the Bragg angle is known for six independent non-axial reflections, the six quantities a*, b*, c*, α^*, β^*, γ^* can be derived from Bragg's law. Usually more than six reflections are determined. Their Bragg angles are then used in a "least-squares" procedure to obtain refined values for the lattice constants (the principles of "least-squares" will be discussed in Section 6).

If the 2θ-values of axial reflections are measured, then the quantities a*, b*, c* can be derived immediately from Bragg's law as already shown in 2.1.2 (equation 2.7a). When using the method of measuring only axial reflections, the crystal must be assumed to be in a special position relative to the diffractometer axes, such that the reciprocal axes involving a non-right angle are situated in a special plane of the diffractometer, say the horizontal plane. As illustrated in Fig. 4.8, the angle β^* of a monoclinic system can be measured directly if the monoclinic **b**-axis is oriented in the direction of the main axis. Then **a*** and **c*** are required to be in the perpendicular plane $\chi = 0$. For $\chi = 0$, the ω- and ϕ-axis are identical. If ϕ_1 is then the ϕ-value of a 0 0 l reflection, ϕ_2 that of a h 0 0 reflection, β^* is given by $\phi_2 - \phi_1$. Since it is not always easy to obtain such a special crystal orientation (for a triclinic crystal, this must be done three times), the method of axial reflection measurement is more suitable for the orthogonal systems, where it gives high precision, if axial reflections with high θ-values, e.g. θ-values above 50°, or better, 60°, (for CuKα radiation) are measured. For our structure examples we have applied both methods for lattice constant determination.

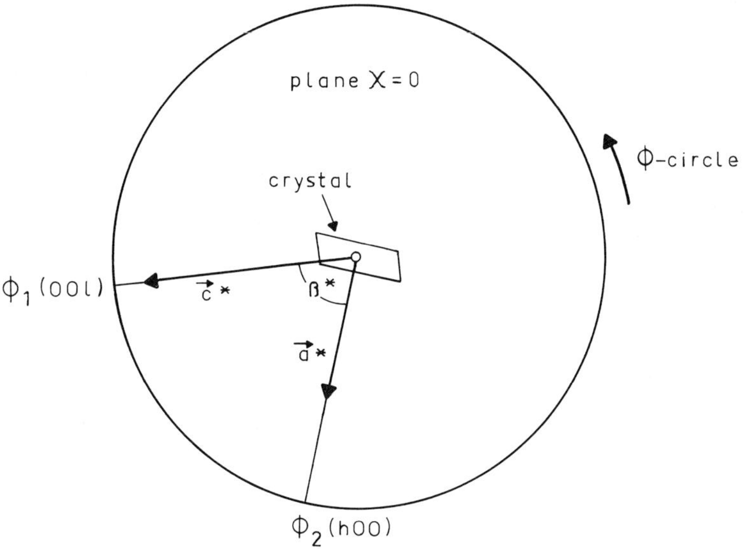

Fig. 4.8. Measuring reciprocal angles on the ϕ-circle.

For KAMTRA, the 2θ-angles for 20 high-order non-axial reflections were determined precisely by an ω-2θ scan over the reflection profile. For all reflections, 2θ was greater than 30° and for 13 reflections it was greater than 45° (using MoKα radiation).

From these 20 reflections, a least-squares lattice parameter refinement was calculated. Programs for this are usually supplied with the software of each modern diffractometer, and moreover, in the X-ray program system, two sub-routines (LATCON and PARAM) are available for that purpose. Our results for the lattice constants of KAMTRA, calculated with PARAM, are a = 7.775(2)Å, b = 10.625(3)Å, c = 7.615(2)Å, with all angles 90° since we have an orthorhombic system. The quantities given in parentheses are the estimated standard deviations, which derive directly from the least-squares calculations (the details of the least-squares calculations are discussed in 6.1). The precision of the lattice constants obtained here is in accordance with present-day standards, which cover a range of some thousands of Å for a length and a few hundreds of degrees for an angle. Note that this is a relatively high precision, better than 0.1%. From a comparison with the preliminary lattice constants derived from film data (a = 7.87, b = 10.72, c = 7.70 Å) it can be seen that the accuracy was improved by at least one order of magnitude. Nevertheless, such high level of precision for the lattice constants is necessary for the correct positioning of reflections for the intensity measurements (as can be seen from (4.1) to (4.4), all diffractometer angles depend on the lattice constants) and for the accurate determination of molecular geometry.

The lattice constants of NITROS were determined by the method of measuring high-order axial reflections. Since only the lengths of the reciprocal lattice vectors can be obtained in this way, some provision for obtaining the monoclinic angle β^* was necessary. For that purpose, the crystal was adjusted on the diffractometer with the b-axis parallel to the main axis. Then **a*** and **c*** are in the equatorial plane and a ϕ-scan over one reflection on each axis permits the determination of β^*, found to be 75.17°. Since the ϕ-value on each axis has an accuracy of, say 0.01 to 0.015°, a crude estimate of the standard deviation may be 0.02°, so we have for the reciprocal monoclinic angle of NITROS, $\beta^* = 75.17(2)°$. The lattice vector lengths were obtained by measuring two high-order reflections on each axis. Since all of these reflections had 2θ angles above 40°, the $K\alpha_1$ and $K\alpha_2$-maximum were clearly separated and two maxima could be obtained for each reflection. Thus a total of four observations was present for each lattice constant. A summary of results for these measurements is given in Table 4.1.

The mean values and standard deviations of the reciprocal lattice constants were obtained using equations (4.6b) and (4.7c). A short comment on the question of using the proper weights w_i should be made here. It can be assumed that the error of 2θ or θ is independent of the magnitude of θ. However, D* derives from $\sin\theta$ by Bragg's law, and a constant error of θ causes a decreasing error of $\sin\theta$ with increasing θ. This is the reason for preferring high-order reflections for a lattice constant determination. Since $d\sin\theta = \cos\theta\, d\theta$, the error $\Delta\sin\theta$ is proportional to $\cos\theta$ for constant $\Delta\theta$. So a good choice is to set the weights w_i proportional to $1/\cos\theta$ when calculating lattice constants from axial reflections using (4.6b) and (4.7c).

In our case the θ-values for NITROS do not differ much; the 0 6 0 and 0 10 0 reflections on the b*-line have θ-values of approximately 20 and 35°. The cosines of these angles are 0.94 and 0.92 and their ratio is ≈ 0.9, which justifies the use of unit

Table 4.1. Results of Lattice Constant Measurements for NITROS.

Reflection	MoKα	θ [°]	D* [Å^{-1}]
18 0 0	1	21.705	0.05794
18 0 0	2	21.848	0.05795
20 0 0	1	24.275	0.05796
20 0 0	2	24.440	0.05798
0 6 0	1	19.975	0.16051
0 6 0	2	20.097	0.16052
0 8 0*	–	–	–
0 10 0	1	34.666	0.16039
0 10 0	2	34.915	0.16043
0 0 9	1	27.732	0.14580
0 0 9	2	27.908	0.14578
0 0 10	1	31.144	0.14584
0 0 10	2	31.352	0.14584

* The weak 0 8 0 reflection was not observable in a reasonable scanning time. The wavelengths used were $\lambda(\text{MoK}\alpha_1) = 0.70926$ Å and $\lambda(\text{MoK}\alpha_2) = 0.71354$ Å.

weights. Generally however, if a large θ-range is covered, different weights, calculated as described above, should be used.

The numerical calculation for a* results in mean values of $a^* = 0.05796$ Å^{-1} and $s = 2.5 \times 10^{-5}$ Å^{-1}. With $a = (a^* \sin\beta^*)^{-1}$, we get a = 17.848 Å. The e.s.d. of a can be calculated with the law of propagation of errors, which is in this case

$$\left(\frac{\sigma(a)}{a}\right)^2 = \left(\frac{\sigma(a^*)}{a^*}\right)^2 + \left(\frac{\sigma(\sin\beta^*)}{\sin\beta^*}\right)^2$$
$$= (4.3 \times 10^{-4})^2 + (9.2 \times 10^{-5})^2 = 9.2 \times 10^{-8}$$

It follows that $\sigma(a) = 0.005$ [$\sigma(\sin\beta^*)$ is obtained in the following way: setting $\sigma(\beta^*) = 0.02°$, then $\sigma(\beta^*)$ in radians is 3.5×10^{-4}. With $\sigma(\sin\beta^*) = \cos\beta^* \, \sigma(\beta^*)$, we get $\sigma(\sin\beta^*) = 8.9 \times 10^{-5}$].

Lattice constants b and c with their e.s.d. 's are obtained in the same way, with the exception that β* does not contribute to b. The final numerical results are a = 17.848(5) Å, b = 6.232(2) Å, c = 7.095(2) Å, β = 104.83(2)°.

For SUCROS, the work was done as for NITROS, giving the results: a = 10.867(4) Å, b = 8.705(3) Å, c = 7.764(3) Å, β = 102.92(2)°.

4.2.3 *Intensity Measurement*

With precisely known orientation matrix and lattice constants, the reflections can be positioned exactly and their intensities can be measured. The procedure for intensity data collection is done in different ways on the various types of diffractometers, but the principles are the same. Although in theory single crystal reflections are treated as points, this is only true for perfect crystals of zero dimensions irradiated from an ideally monocromatic point X-ray focus. In reality, reflections have finite dimensions, depending on the crystal size and quality and on the finite dimensions and wavelength dispersion of the X-ray source.

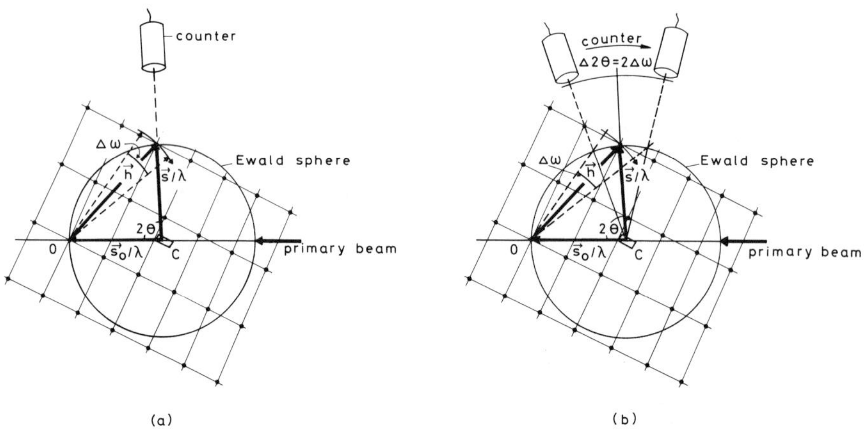

Fig. 4.9. Schematical representation of an ω-scan (a); (b) ω-2θ-scan.

For this reason, as the crystal is rotated by an angular increment $\Delta\omega$, during the intensity measurement, the lattice vector **h** has a finite size as it passes through the reflection sphere. Scanning procedures can be used in two different ways: the "ω-scan" method and the "ω-2θ-scan" method. The principles of both methods are shown in Fig. 4.9a and b. In the ω-scan mode, the detector is held stationary at the theoretical 2θ angle of the actual reflection and the crystal and thus the corresponding reciprocal lattice vector **h** is rotated by an angular increment $\Delta\omega$. In the ω-2θ mode, both crystal and detector are moved. The crystal is rotated by $\Delta\omega$ as for the ω-scan, while the detector is rotated in the 2θ-circle by an angular velocity which is twice that of the crystal rotation, so that $\Delta 2\theta$ is equal to $2\Delta\omega$.

Interpreting these two scan modes in terms of the reciprocal lattice, we can say that in the ω-scan mode, rotation takes place perpendicular to the lattice vector **h**, while in the ω-2θ-scan mode, the scan is made in the direction of **h** (see Fig. 4.10a). The main disadvantage of the ω-scan technique occurs with poorly monochromatized radiation, when the reflection intensities on a lattice row include a linear trail of additional intensity from other wavelengths which are extended in the direction of the reflection (see Fig. 4.10b). This results in an unsymmetrical background measurement, which in the ω-scan mode cannot be easily recognized. In the ω-2θ-scan mode, this background asymmetry can be measured and taken into consideration when net intensity is calculated. More details of ω-scan and ω-2θ-scan methods are discussed by Alexander and Smith in Acta Cryst. *15*, 983–1004 (1962). It is common practice to use the ω-2θ-scan whenever *filtered* radiation is taken as the X-ray source. But if graphite monochromated radiation and a small crystal are used, there is no significant difference between the two methods. Since the scan range for the ω-scan is usually smaller and this

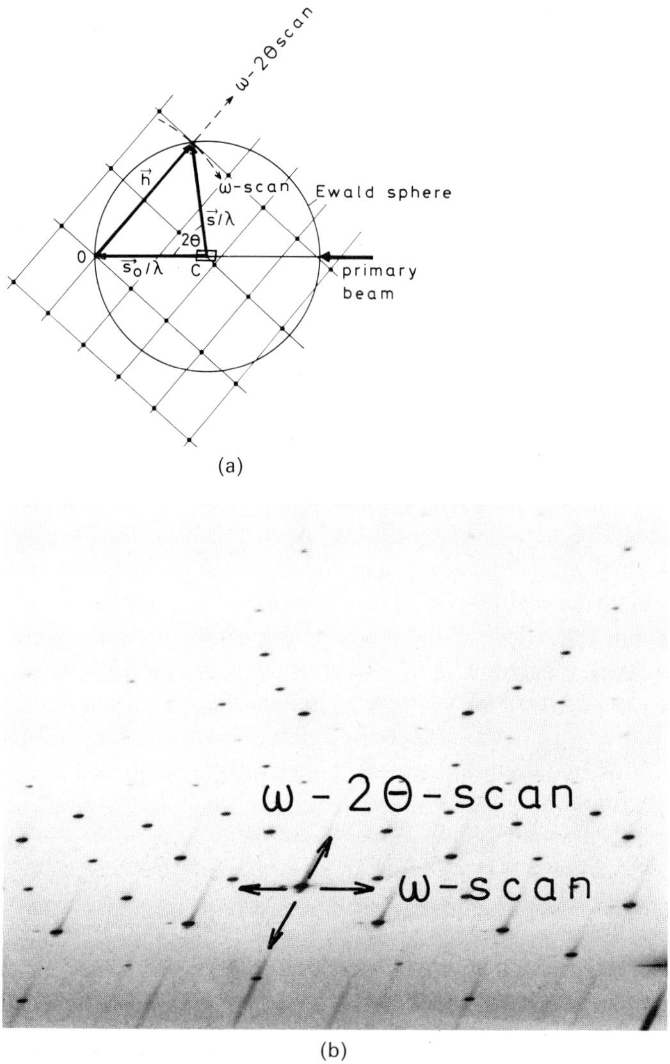

Fig. 4.10. Scan directions in the ω-scan and ω-2θ-scan mode: (a) represented schematically; (b) on a Weissenberg lattice row.

measurement procedure is faster, it is frequently used when a large number of intensities have to be measured, as with protein crystals. Preliminary ω-scans should also be used to confirm that the crystal is indeed satisfactory for the intensity measurements. In low temperature work, it is sometimes found that a crystal which appears to be satisfactory by film examination at room temperatures has crack at low temperature and is not suitable for intensity measurements.

A single reflection has an intensity distribution in terms of θ shown in Fig. 4.11. In Fig. 4.11a, a reflection profile at a low angle is taken. Fig. 4.11b shows a reflection at medium θ and in Fig. 4.11c, the profile of a reflection with large θ is plotted showing the splitting due to the $K\alpha_1$-$K\alpha_2$-doublet. The net intensity of a reflection is defined as the integral over the intensity profile minus the background intensity:

$$I_{net}(h,k,l) = \int_{\Delta\omega} I(\omega)\,d\omega - I_{Background} \qquad (4.12)$$

To get I_{net} from (4.12), every intensity measurement is made in such a way that a scan is made over the reflection profile with the counter integrating the complete intensity. The background is then measured on the left and on the right edge of the reflection profile using the same counting time as was used for the reflection scan. The net intensity is then calculated from the rate of scan minus the mean value of the two backgrounds.

In practice, some preparation is necessary before data collection can be started. The first is to estimate the interval $\Delta\omega$ which defines the flanks of a reflection. The determination of the proper value of $\Delta\omega$ is not a trivial problem. For a given diffractometer alignment, it depends on the crystal size and quality, and moreover, it increases with θ, as can be seen from Fig. 4.11. The first question is: Where does the reflection end and the background start? This question must be answered as precisely as possible, since if $\Delta\omega$ is chosen too small (Fig. 4.12a), some of the flanks of the reflection are included in the background, resulting in too small a net intensity. If $\Delta\omega$ is chosen too large (Fig. 4.12b), the ratio of the reflection to background intensity becomes disadvantageous, especially for weak reflections. The correct value of $\Delta\omega$ can be obtained experimentally from the measurement of some reflection profiles from which the flanks can be estimated. If these profiles are taken for different θ-values, the dependence $\Delta\omega$ versus θ can be found and used in the measurement of the integrated intensities. It has been found that a good analytical approximation of $\Delta\omega(\theta)$ is given by

$$\Delta\omega(\theta) = a + b\tan\theta \qquad (4.13)$$

If the experimentally measured $\Delta\omega$-values are plotted versus $\tan\theta$ (see Fig. 4.13), a least-squares line should be obtained from which the parameters a and b can be estimated. Typical values for a and b are in a range of 0.70–1.00° for a and 0.40–0.50° for b, so that for $\theta = 45°$ (with $\tan\theta = 1$), reflections have a width of 1.0–1.5°.

Further considerations determine the minimum and maximum scan speeds. All diffractometer controls provide for a preliminary estimation of intensity, either by a rapid scan over the reflection or by a point measurement in the theoretical maximum.

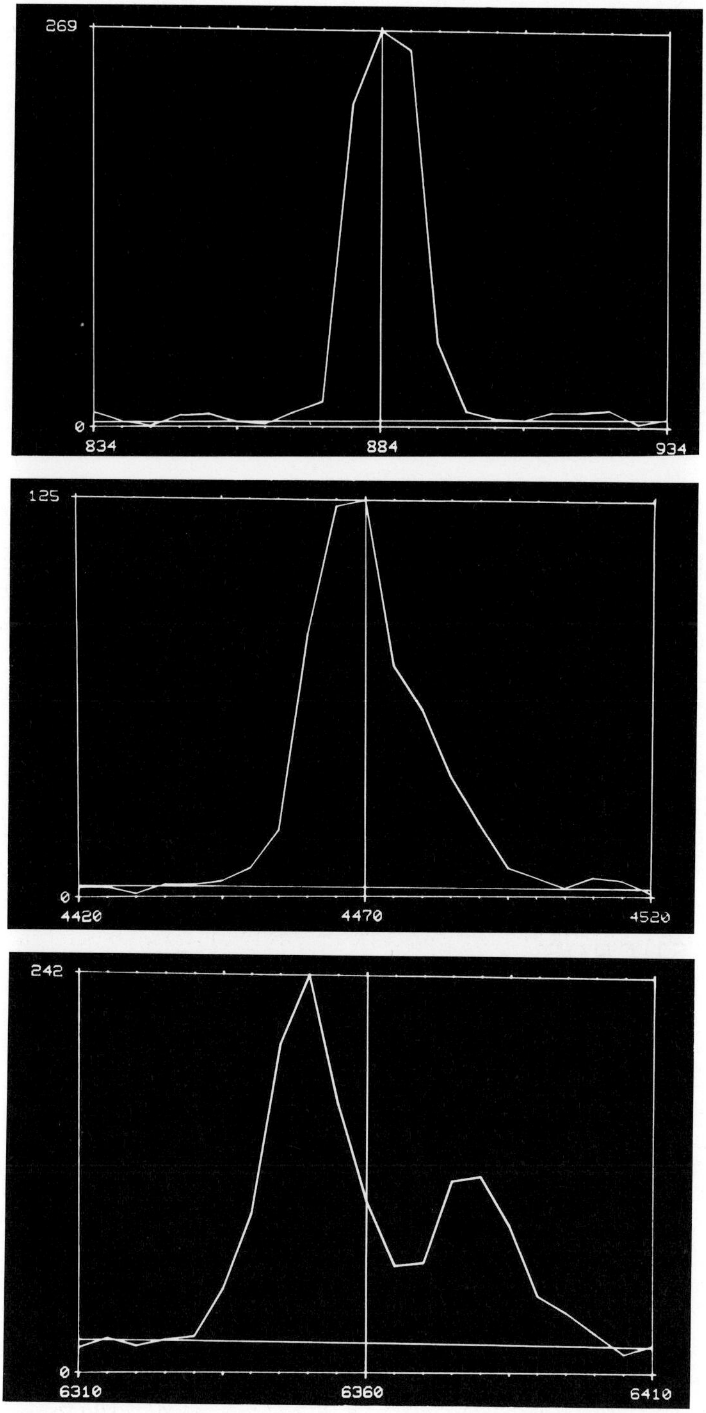

Fig. 4.11. Reflection profiles at (a) low, (b) medium, and (c) high 2θ-angle.

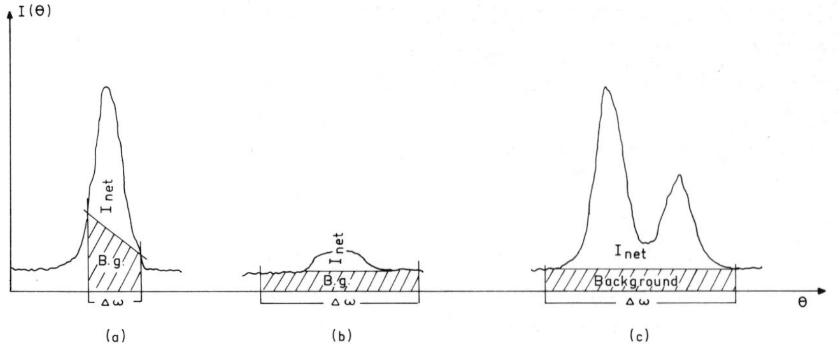

Fig. 4.12. Choice of a proper $\Delta\omega$:
(a) chosen too small; (b) chosen too large; (c) properly chosen $\Delta\omega$.

From this preliminary measurement of every reflection, the individual scan speed is calculated for each reflection so as to ensure the collection at a sufficient counting rate for a significant measurement of weak reflections. Nevertheless, the minimum scan speed has to be chosen by the user. It should be small enough to avoid bad counting statistics for weak reflections, and it must be fast enough to allow the data collection to be completed in a reasonable time. The maximum scan speed usually cannot be preset by the user, since it depends on the diffractometer type.

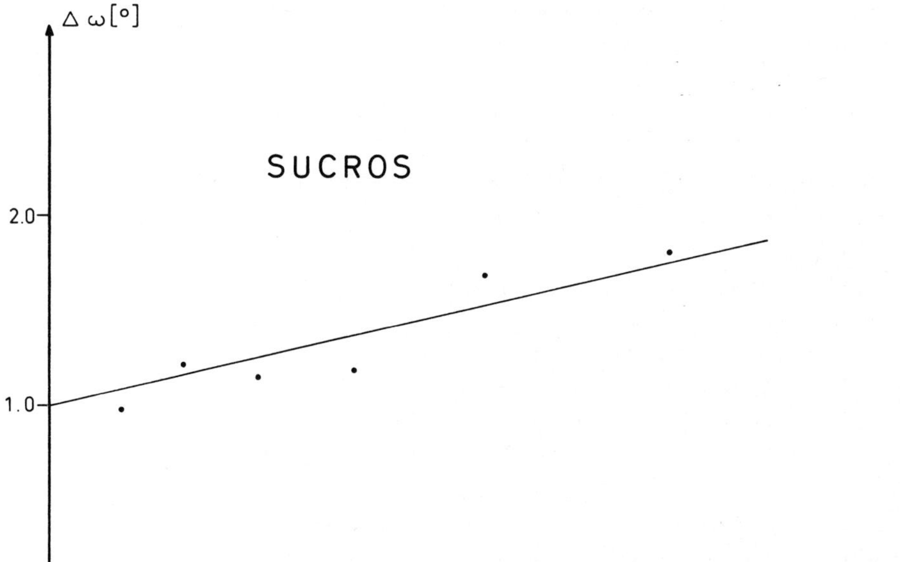

Fig. 4.13. Estimation of the proper scan range from plotting $\Delta\omega$ versus $\tan\theta$.

The preliminary estimation of intensity enables the controls to insert attenuating filters if the counting rate exceeds the capacity of the detector. If this option is provided, the user has to determine the attenuating factors of the filters before starting the intensity measurements.

The last factor to be considered is the set of reflections for which the intensities will actually be measured. This is usually the asymmetric unit of reflections, however, if the highest possible accuracy is desired, all available reflections, i.e. also the symmetry related should be measured. This should only be done in cases where it is intended to make accurate absorption and extinction corrections. From averaging the corrected intensities usually a very precise data set is obtained. In addition to the limitations due to the limiting sphere and symmetry requirements, further limitations arise from the 2θ-range of the diffractometer. Due to the construction of the Eulerian cradle, 2θ cannot generally exceed 150°. For low 2θ-values, say 8–10°, the primary beam intensity may penetrate into the detector window and affect the measurement. So 2θ is generally restricted to a range from about 8 to 150°, with the actual limits differing with the various types of instruments.

The full 2θ-range is generally used when CuKα radiation ($\lambda = 1.5418$ Å) is selected. With $2\theta = 150°$, $\sin\theta/\lambda$ has a maximum value of 0.63 Å^{-1}. For MoKα radiation ($\lambda = 0.7107$ Å) this $\sin\theta/\lambda$ value is already obtained by $2\theta = 53.2°$ so that the amount of data for the copper sphere up to $2\theta = 150°$ corresponds to a molybdenum sphere up to 53.2°. This should be taken into consideration when MoKα radiation is used. It should be noted, however, that due to the rapid decrease of form factors with $\sin\theta/\lambda$ (see 3.2.1), intensity measurement for larger $\sin\theta/\lambda$ is not very useful with structures with only first row atoms. If the film exposures show no reflections for high 2θ-values, say 120°, a lower 2θ-limit than that given by the diffractometer should be chosen, since it only wastes time when unobservable reflections are measured.

Finally, systematically absent hkl reflections should be excluded from the data set to be measured, since they are recorded as very weak in the preliminary measurements and are therefore measured with the lowest speed in the data collection. However, if there is any doubt whether an axial or zonal extinction is present, measurements should be made to check the systematic extinctions.

Summarizing the necessary preparations for intensity data collection, we recommend the following check list.

(1) Determine the crystal orientation and lattice constants precisely.

(2) Choose the scan mode, either ω-scan or ω-2θ-scan.

(3) Ensure that the scanning range $\Delta\omega$ is properly selected.

(4) Choose a reasonable minimum scan speed and determine, if necessary, the filter factors, which are the reciprocals of the attenuation factors (i.e. usually >1).

(5) Check that at least one symmetry independent set of reflections is selected for intensity measurement, choose a suitable 2θ-range and remove the systematically absent reflections from the data collection.

(6) Check that the optimum voltage and current are chosen on the X-ray generator.

(7) Check the adjustment of the pulse height discriminator.

With these preparations, the intensity data collection should run automatically. Although the measurement routines differ for the various types of diffractometers, the following information will be recorded for each reflection:

(1) I(r), the integrated intensity from the scan over $\Delta\omega$.

(2) The time t needed for the scan over $\Delta\omega$, where t is derived from the scan speed (estimated from preliminary intensity measurement) s [deg/sec] and $\Delta\omega$ by

$$t = \Delta\omega / s.$$

(3) $I(b_r)$ and $I(b_l)$, the right and left background intensity with t/2 chosen as the counting time for *each* background.

(4) The filter factor, f.

(On some diffractometers, provision is made to carry out step scans and record the complete reflection profile in the intensity measurement process. This type of data collection is extremely time and storage consuming and should only be considered for very high precision work.)

Then the net intensity for a given reflection with the indices hkl is given by

$$I(hkl) = \{I(r) - [I(b_r) + I(b_l)]\} \frac{f}{t} \tag{4.14}$$

with the dimension of counts/sec. The standard deviation $\sigma(I)$ is then (see 4.9)

$$\sigma(I) = \sqrt{I(r) + I(b_r) + I(b_l)}\ \frac{f}{t} \tag{4.15}$$

Weak reflections where I is less than or comparable to $\sigma(I)$ are referred to as "unobserved" reflections, since their intensity is "less than" the estimated errors. A reflection (hkl) is said to be unobserved if

$$I(hkl) < n\sigma(I) \tag{4.16}$$

where it is common practice to chose $n = 2$ (some researchers use $n = 1$ or $n = 3$ or some slightly different values). The number of "less thans" depends on the crystal size, the power of incident radiation, and the specific properties of the individual crystal structure. A first judge of the quality of a set of intensity data can be made from the number of "less thans", which should not be greater than 10 to 30% of the total data set. If this amount lies between 30 and 50%, only a fair data set is obtained, and if the unobserved reflections exceed 50%, a different crystal or better experimental conditions should be sought in order to measure more observed reflections. With more than 50% unobserved reflections, it is unlikely that the phase problem, and thus the structure analysis, can be solved. In general, it is a waste of time to collect a data set of such poor quality.

During the data collection procedure, the repeated measurement of at least one reference reflection is recommended. If the reference reflection intensities are constant, it indicates a satisfactory stability of the experimental conditions. If the reference reflection intensities change significantly, i.e. greater than a few percent, possible sources of errors, such as crystal disorientation, radiation damage to the crystal, or instability of the primary radiation source, etc., have to be determined.

For some compounds, the crystal cannot be kept stable for the complete period of data collection under normal conditions. Sometimes the inclusion of the crystal in a glass capillary or measurement at a lower temperature may solve the problem. If the crystal is stable for several hours or even a few days, measurement should be continued until the reference reflection indicates an intensity decrease to 70–80% of the initial value. Then a new specimen should be mounted to continue the data collection. In this way, several crystals may be used to obtain a data set for one compound. This is particularly true for protein crystals. The decrease in intensity during the measurement of *one* specimen may be corrected. If I_0 is the intensity of the reference reflection at the beginning, and $I(t)$ the intensity of the actual reference reflection, a scaling can be derived from the ratio $I_0/I(t)$. The scaling of intensities for different crystals may also be done from the reference reflection intensities.

The number of reflections collected within 24 hours depends on several conditions. The diffractometers in use in the late sixties gave 200 to 300 reflections in 24 hours. Today, with increasing power of the X-ray sources and higher speeds of the diffractometer motion, 400–500 reflections can be measured in 24 hours. Only when the crystals deteriorate very quickly, as with proteins, should reflections be measured more quickly, since high-speed data collection is always accompanied by loss of accuracy.

The number of reflections for medium-sized structures lies between 1000 and 3000, so that the time necessary for data collection is usually less than one week.

We shall now explain some of the details of the intensity measurements on KAMTRA and the two other test structures. For all compounds, we used filtered radiation (MoKα for KAMTRA and NITROS, CuKα for SUCROS), so the data collection was done in the ω-2θ mode. The scanning range $\Delta\omega$ was determined experimentally. As shown for SUCROS in Fig. 4.13, we took some reflection profiles at different angles θ and plotted $\Delta\omega$ versus $\tan\theta$. The approximation by a line gives values for constants a and b, by the application of (4.13), of $a = 1.00$, $b = 0.45$. We get the scan range for the two other compounds in the same way. As the minimum scan speed we choose 1°/min; the maximum preset speed on our Siemens diffractometer is 10°/min.

An independent set of reflections for KAMTRA is one octant of the limiting sphere, since we have an orthorhombic crystal system. We collected intensity data only for reflections with h, k, and l all ≥ 0, and utilizing a θ-range corresponding to the copper sphere, 2θ was chosen less than 55°. For the Siemens diffractometer, it is not recommended that reflections are measured below $2\theta = 6°$, since they are strongly affected by the primary beam. We already know of the axial extinctions h00 for $h = 2n + 1$ and 0k0 for $k = 2n + 1$, so these reflections need not be measured. We do not know whether an axial extinction is also present on the third axis, so all reflections of type 00l are measured.

Since our X-ray tube has only a power of 750 W, we use a voltage of 50 kV and restrict the current to 15 mA. With these conditions, we start measurement and obtain about 300 reflections per 24 hours. The unit cell volume is 630 $Å^3$, so we can expect 600–650 reflections to be measured within 2 or 3 days. In practice, 620 reflections were measured, of which 20 were recorded as "less thans", using the factor $n = 2$ in criteria (4.16). At this stage, the main result was that all 00l reflections with $l = 2n + 1$ were found to be unobserved. Thus we know that a third screw axis is present, and that the space group for KAMTRA is $P2_12_12_1$.

For SUCROS, which is measured with CuKα radiation (35 kV, 20 mA), the conditions are the same, except the 2θ-range is set to vary from 8 to 140° and the independent set of reflections differs. Since we have a monoclinic space group, an independent quadrant must be taken. We measure the two octants hkl and $\bar{h}$kl (h, k, and l all ≥ 0). However, in this data set, the hk0 reflections are contained twice, since they are symmetry related to the $\bar{h}$k0 reflections. Therefore care must be taken that either these symmetry-dependent reflections are measured only once, or that all symmetry dependent reflections are removed from the data set later. Furthermore, we know of an axial extinction 0k0 for $k = 2n + 1$, therefore these reflections are not measured. The cell volume is 716 $Å^3$. Since SUCROS is monoclinic, the number of reflections is approximately twice that of KAMTRA. We get 1297 reflections including 10 less-thans.

For NITROS, we considered using MoKα radiation and a 2θ range with $6° < 2\theta < 60°$ was chosen. From monoclinic symmetry the same quadrant as for SUCROS represents the asymmetric unit. Since we have a general hkl extinction, the reflections hkl with $h + k = 2n + 1$ need not be measured. With these conditions, we get a total of 1207 reflections, 82 of which are less thans.

With the intensity data for all three structures completed, no further experiments are necessary. All further work will be done by means of computer calculations.

5 Solution of the Phase Problem

The task of structure determination and refinement requires extensive numerical calculations for which a fast computer with sufficient storage is necesssary. The main reason for the difficulty of single crystal analysis up to the fifties was the impossibility of doing the numerical calculations by hand in a reasonable period of time. Today, most research groups have the services of a computer center, or their own computer, and it is a question of obtaining suitable programs for the X-ray analysis calculation. It is now not necessary to write one's own programs for most of the numerical problems, since excellent programs already exist, and except for very special problems, they are easily accessible. One program system used worldwide is the "XRAY" system, which is distributed by the University of Maryland, USA. (The last version is the XRAY76 system, edited by J.M. Stewart, Technical Report TR-446 of the computer center, University of Maryland, College Park, Maryland.) This program system includes subprograms for nearly all calculations necessary in the course of an X-ray (or neutron) diffraction analysis. It is up-dated every few years and can be acquired for a nominal charge.

A large number of other programs or program systems exists and anyone wishing to start research on single crystal analysis should discover how to make those programs operational on his particular computer.

5.1 Preparation of the Intensity Data

When a set of reflection intensity data is recorded for a given structure, the list of data should be printed and checked carefully. Examine the listing for any of the following errors and remove them from the data set.

(1) If the reference reflections are not constant within certain limits, the reason must be determined.

(2) Redundant symmetry-related reflections may be contained in the data set, and must be deleted, or the average of the equivalent reflections must be calculated. Comparison of symmetry-equivalent reflections gives a good internal check on the accuracy of the data.

(3) Occasionally, reflections are recorded with impossible intensity values, due to a temporary diffractometer or electronic failure. These should be recognized and eliminated.

(4) Sometimes data transfer from one mass storage to another (e.g. from the diffractometer's magnetic tape to the computer's disc) is the source of errors which should be checked.

A careful check of the experimental intensity data is strongly recommended, since it may avoid long and expensive calculations which are unsuccessful due to a few serious intensity errors caused by mis-measurement or incorrectly transferred data. We once had the problem that a reflection with the intensity 0004520 was transferred as 9004520 due to a change of the leading "0" to a "9". This one bad reflection caused a failure in the structure determination, until the error was recognized and eliminated.

5.1.1 *Data Reduction*

The procedure for converting the measured intensity data to a form suitable for further calculation is called "data reduction". In 1.2.2, we had learned that the intensities are proportional to F F*, i.e.

$$I(\mathbf{h}) \sim F(\mathbf{h})\, F^*(\mathbf{h})$$

This proportionality can be expressed by

$$I(\mathbf{h}) = t^2 K F(\mathbf{h})\, F^*(\mathbf{h}) \tag{5.1}$$

where t is a scale factor and K is a correction factor which includes all the necessary corrections to derive $|F(\mathbf{h})|$ from $I(\mathbf{h})$. K depends on the experimental conditions. For Eulerian cradle diffractometers, it is sufficient in most cases to reduce the corrections to two terms. One term is the "Lorentz and polarization correction". This correction (usually abbreviated Lp-correction) consists of a Lorentz correction L and a polarization correction p. p takes into account the partial polarization of the unpolarized primary beam, which occurs in the diffraction process. p does not depend on the crystallographic apparatus, but only on the diffraction angle 2θ, and can be expressed analytically by

$$p = \frac{1 + \cos^2 2\theta}{2} \tag{5.2}$$

The finite penetration of the Ewald sphere by the reciprocal lattice 'spot' occurs with different velocities for different experimental conditions and for different 2θ-values. Since the intensity is inversely proportional to the "time of penetration", it must be considered for every reflection. This correction factor is denoted L, the Lorentz factor. As mentioned above, L depends on the geometry of the diffraction experiment and on the angle 2θ. For the Eulerian cradle, we have the geometrical situation illustrated in Fig. 5.1 (it is the same as that of the zero-layer Weissenberg technique). The lattice vector **h** passes through the surface of the Ewald sphere when the crystal is rotated by an angular velocity ω about O. The velocity of penetration is given by the component of the velocity **v** vector in the direction of $-\mathbf{s}$. This component v_s is the scalar product $-\mathbf{vs} = v1\cos\theta$. Since v is equal to $\omega|\mathbf{h}| = \omega 2\sin\theta/\lambda$, we get

$$v_s = \frac{\omega}{\lambda}\, 2\sin\theta\cos\theta$$

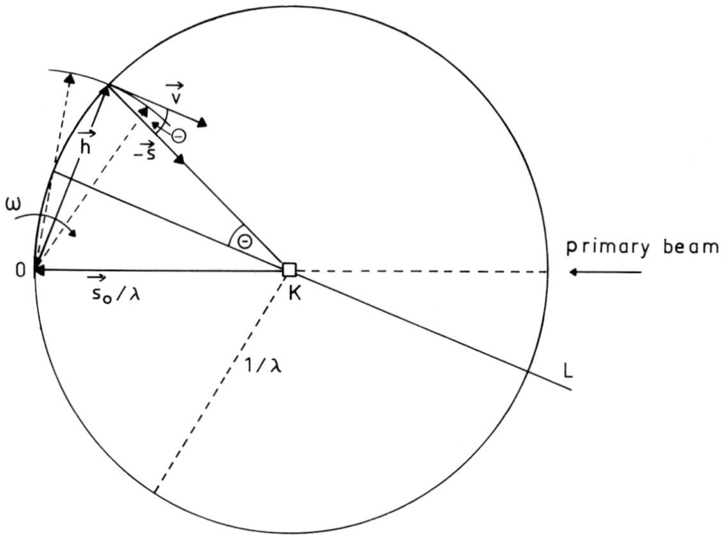

Fig. 5.1. Geometric representation of the derivation of the Lorentz factor.

Since all constant proportionality factors may be included in the scale factor, we may set L equal to the reciprocal of the variable part of the expression above, hence we get

$$L = \frac{1}{2\sin\theta\cos\theta} = \frac{1}{\sin 2\theta} \tag{5.3}$$

A common expression for the Lp-factor is

$$Lp = \frac{1 + \cos^2 2\theta}{2\sin 2\theta} \tag{5.4a}$$

(5.4a) is valid for Eulerian cradle geometry and filtered radiation. If a monochromator is used, a further polarization occurs and p must be replaced by a more complicated expression. For a crystal monochromator with the monochromator 2θ angle denoted ψ the Lp-factor expression reads

$$Lp = \frac{(1 + \cos^2 2\theta)\cos\psi}{\sin 2\theta\,(1 + \cos\psi)} \tag{5.4b}$$

A second correction is that of absorption, although it is not necessary for all structures, especially organic compounds. The attenuation by absorption effects depends on the linear absorption coefficient, μ, and the pathlength, x, of radiation through the crystal. As was pointed out in 2.2.2, it can be expressed by $I = I_o A$ (see equation (2.38)) with

$$A = e^{-\mu x} \tag{5.5}$$

where A is the absorption factor. A general rule on whether an absorption correction should be applied cannot be given. Nevertheless, with some reservations, it can be said that

(1) for $0 < \mu < 25$ (cm^{-1}), an absorption correction is not usually necessary unless a crystal of very unfavorable shape is used, since if applied, the absorption corrections will not significantly improve the accuracy of the results.

(2) for $25 < \mu < 50$ (cm^{-1}), an absorption correction should generally be made. It can only be omitted if the crystal is very small and has an isotropic shape so that μx is small and approximately constant. Then A acts as an additional scale factor and may be included in t. But in general, an absorption correction will significantly improve the results.

(3) for $50 < \mu$ (cm^{-1}), an absorption correction should always be applied. Furthermore, if $\mu \gg 100\ cm^{-1}$, it should even be considered whether a radiation source with smaller wavelength can be used to reduce μ.

For a crystal of arbitrary shape bathed in X-rays, the absorption factor may be expressed by

$$A = \frac{1}{V} \int_V e^{-\mu(p+q)}\, dV \qquad (5.6)$$

dV is a volume element of a specimen having a crystal volume V, and p and q are the lengths of the paths of the incident and reflected beams for the volume dV (Fig. 5.2).

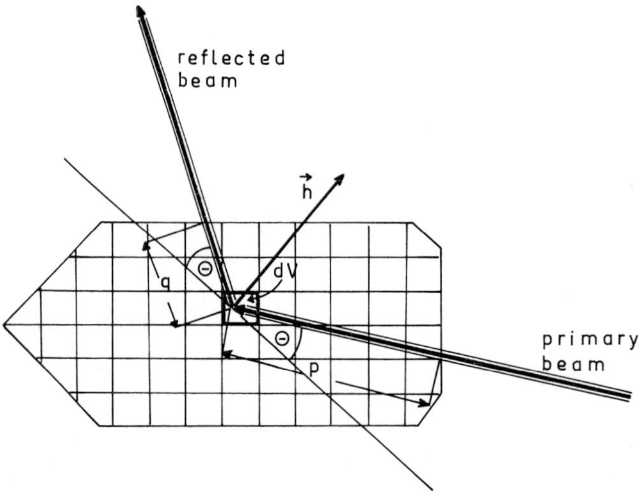

Fig. 5.2. Incident and reflected ray corresponding to the crystal volume element dV.

Several methods for the application of absorption corrections are available (in more or less sophisticated computer programs), all based on the calculation of pathlengths through the crystal, which vary from reflection to reflection. The determination of these pathlengths is not a simple problem. For regularly shaped crystals (e.g. spherical and cylindrical), (5.6) can be calculated analytically. The results are tabulated in the International Tables, Vol. II, pp. 291 (1967), and may be used for crystals of that shape.

It is common practice, especially for compounds with large μ, to grind a crystal in order to obtain spherical crystals. However, it should be noted that this procedure does not work for soft crystals, and there is always the danger that the crystal will be damaged, so the procedure should only be used if sufficient crystals are available.

If the crystal shape cannot be approximated by a sphere or a cylinder, the calculation of (5.6) must be done numerically for the individual crystal. For this calculation, the precise shape is used as input from a measurement of the crystal faces. For that purpose, it is advisable to use crystals with well-developed natural faces.

The Lorentz and polarization and the absorption corrections are in most cases the only ones necessary for data reduction. We can therefore modify (5.1) by

$$I(\mathbf{h}) = t^2 \, LpA \, |F(\mathbf{h})|^2 \tag{5.7}$$

The quantity which is impossible to obtain is the scale factor, which can only be calculated if the structure is already known. It is therefore customary to define a so-called "$F_{rel}(\mathbf{h})$", which differs from $|F(\mathbf{h})|$ only by the scale factor;

$$F_{rel}(\mathbf{h}) = t\,|F(\mathbf{h})| \tag{5.8}$$

Then, from (5.7), we get

$$F_{rel}(\mathbf{h}) = +\sqrt{\frac{I(h)}{LpA}} \tag{5.9}$$

To get $\sigma(F_{rel})$, we express I as a function of F_{rel} and calculate the first derivative:

$$I = LpA\,F_{rel}^{\,2}$$

$$\frac{dI}{dF_{rel}} = 2\,LpA\,F_{rel}$$

Setting $\sigma(F_{rel}) \approx dF_{rel}$ and $\sigma(I) \approx dI$, we get

$$\sigma(F_{rel}) = \frac{\sigma(I)}{2\,LpA\,F_{rel}} \tag{5.10}$$

If no absorption correction is made, $A = 1$ is always used.

The calculation of F_{rel}, together with the calculation of $\sigma(F_{rel})$, is done from the experimental data in the data reduction for every reflection.

5.1.2. Normalization

A calculation immediately following the data reduction is the so-called "normalization". In this process, quantities are derived which play an important role in phase determination. These are known as "normalized structure factors" or "E-values", since they are usually represented bx the capital letter E(**h**).

Although we shall discuss the use of E-values later in connection with "Direct Methods", we describe their definition and calculation at this point, since it is customary to do the normalization either together with or immediately after data reduction, in order to derive further information which can be obtained by this course of normalization procedure.

Before giving the definition of E-values, let us explain their physical background. The structure factor F(**h**), as derived in 3.2.1, [formula (3.18)], is based on a model of N atoms with different electron densities of finite dimensions, with their thermal motion taken into account by the temperature factor expression. For some theoretical considerations, it is advantageous to consider an idealized model of point-shaped atoms of unique electron density with no thermal motion. For this model, the entire electron density, ϱ_n can be expressed as

$$\varrho_n(\mathbf{r}) = \sum_{j=1}^{N} \Delta(\mathbf{r} - \mathbf{r}_j) \tag{5.11}$$

where $\Delta(\mathbf{r})$ is the Dirac's delta function. Using the shift theorem [1.2.1, formula (1.29)], we get for the structure factor G(**h**) of that model

$$G(\mathbf{h}) = \sum_{j=1}^{N} g(\mathbf{h})\, e^{2\pi i \mathbf{h}\mathbf{r}_j}$$

where g(**h**) is the Fourier transform of $\Delta(\mathbf{r})$. Since it can be shown that the Fourier transform of a delta function is equal to 1, we get

$$G(\mathbf{h}) = \sum_{j=1}^{N} e^{2\pi i \mathbf{h}\mathbf{r}_j} \tag{5.12}$$

The "quasi-normalized" structure factor E′(**h**) is defined [Karle, J. and Hauptman, H, Acta Cryst. *9*, 635 (1956)] by the following equation

$$E'(\mathbf{h})^2 = \frac{|F_e(\mathbf{h})|^2}{\sum_{j=1}^{N} f_j^2} \tag{5.13}$$

with

$$F_e(\mathbf{h}) = \sum_{j=1}^{N} f_j\, e^{2\pi i \mathbf{h}\mathbf{r}_j}$$

being the structure factor of a model with thermal motion excluded. If all atoms in the unit cell are alike for all j, then $f_j = f$, and (5.13) reduces to

$$E'(\mathbf{h})^2 = \frac{f^2 \left| \sum_{j=1}^{N} e^{2\pi i \mathbf{h}\mathbf{r}_j} \right|^2}{N f^2}$$

It follows that

$$|E'(\mathbf{h})| = \frac{1}{\sqrt{N}} |G(\mathbf{h})|$$

From this result it follows that, for a structure of one type of atoms, the absolute values of E′ and G differ by a constant factor. We can therfore interpret the E′-values as structure factors of an idealized model with atoms having the same unique electron density and no thermal motion.

The normalized structure factors, or E-values, are defined by

$$E(\mathbf{h})^2 = \frac{E'(\mathbf{h})^2}{\varepsilon} = \frac{|F_e(\mathbf{h})|^2}{\left(\sum_{j=1}^{N} f_j^2\right) \cdot \varepsilon} \tag{5.14}$$

with ε being related to some symmetry properties of the space group. Note that neither (5.13) nor (5.14) defines a phase for the E′- and E-values; these formulae give only the absolute value of these quantities. It is customary to assign the phase of F to the E-value of the same reflection, so that a phase determination of E′s would determine that for the F′s, and vice versa.

The advantage of using E-values instead of G-values is that the E's can be calculated (at least approximately) from the experimental data, while the G's can only be derived from the known structure. The quantity ε is a small positive integer which depends on the space group symmetry. It is usually equal to 1 for general reflections, but may have larger values for some reflection classes, depending on the symmetry. So for instance, in space group $P2_12_12_1$, $\varepsilon = 2$ for the series h00, 0k0, 00l, while for all other reflections, ε is equal to 1. The correct ε-values for each reflection class in every space group are listed in the International Tables, (1967), Vol. II, pp. 355–356 and may be taken from there if the data reduction or normalization program used does not contain an ε-calculation routine.

Since the quasi-normalized and the normalized structure factors differ only by that factor ε, the physical interpretation of E-values is in principle the same as that of E′-values. The main difficulty with the calculation of E-values from the experimentally derived F_{rel}'s arises from the scale and temperature factors, which are originally unknown. It follows from (5.8) that

$$F_{rel}(\mathbf{h}) = t \left| \sum_{j=1}^{N} f_j \exp[2\pi \mathbf{h}\mathbf{r}_j] \exp[-(B_j \sin^2\theta)/\lambda^2] \right|$$

Assuming, as a first approximation, that the B_j are equal for all atoms (i.e. the individual isotropic B_j's are replaced by an "overall isotropic" B), we get

$$F_{rel}(\mathbf{h}) = t \exp[-(B \sin^2\theta)/\lambda^2] |F_e(\mathbf{h})|$$

or

$$|F_e(\mathbf{h})|^2 = \frac{F_{rel}(\mathbf{h})^2}{t^2 \exp[-(2B \sin^2\theta)/\lambda^2]} \tag{5.15}$$

$|F_e|^2$ is needed for the calculation of E-values and F_{rel} is obtained from the experimental data. The unknown quantities are t and B. They can be estimated following a method proposed by Wilson [Wilson, A.J.C., Nature *150*, 151 (1942)]. Introducing the variable

$$s = \sin\theta/\lambda$$

and taking the average $\langle |F_e(\mathbf{h})|^2 \rangle$ and $\langle F_{rel}(\mathbf{h})^2 \rangle$ for a given s, we get

$$t^2 \exp[-2Bs^2] = \frac{\langle F_{rel}(\mathbf{h})^2 \rangle}{\langle |F_e(\mathbf{h})|^2 \rangle}$$

The average $\langle |F_e(\mathbf{h})|^2 \rangle$ can be calculated as follows:

$$|F_e(\mathbf{h})|^2 = F_e(\mathbf{h})\, F_e(\mathbf{h})^* = \Big(\sum_{j=1}^{N} f_j\, e^{2\pi i \mathbf{h}\mathbf{r}_j}\Big)\Big(\sum_{j=1}^{N} f_j\, e^{-2\pi i \mathbf{h}\mathbf{r}_j}\Big)$$

This can be separated into

$$|F_e(\mathbf{h})|^2 = \sum_{j=1}^{N} f_j^2 + \sum_{n=1}^{N} \sum_{m=1}^{N} f_n f_m\, e^{2\pi i \mathbf{h}(\mathbf{r}_n - \mathbf{r}_m)}$$

(with $n \neq m$ in the double sum).

If the average is taken, a summation over all **h** is necessary so that an equal number of positive and negative terms are produced in the second sum, which will therefore tend to zero. It follows that

$$\langle |F_e(\mathbf{h})|^2 \rangle = \sum_{j=1}^{N} f_j^2 \tag{5.16}$$

Then we get

$$t^2 \exp[-2B s^2] = \frac{\langle F_{rel}(\mathbf{h})^2 \rangle}{\sum_{j=1}^{N} f_j^2} = K(s).$$

Taking the natural logarithm on both sides, we get

$$\ln t^2 - 2B s^2 = \ln[K(s)] \tag{5.17}$$

(5.17) is the equation of a straight line if $\ln[K(s)]$ is plotted versus s^2, having as slope $-2B$ and the intersection at $s^2 = 0$ of $\ln t^2$. In practice, this plot is made in the following way. A number of $\sin\theta/\lambda$ intervals is chosen. For all reflections in each interval, the average $F_{rel}{}^2$ is calculated together with the sum in the denominator of K(s), using the medium atomic form factors f_j of this interval. In this way, $\ln[K(s)]$ can be obtained and plotted versus s^2 (Wilson plot). By approximating that plot by a straight line, its slope $-2B$ gives the "overall temperature factor", B, and the intersection at $s^2 = 0$ is $\ln t^2$, from which the scale factor t can be derived.

With B and t known, the $|F_e(\mathbf{h})|^2$ values can be calculated from (5.15) and then the E-values from (5.14). Each computer program for this calculation requires as input, the unit cell and space group information, the F_{rel}'s (usually the output of the data reduction program), the atomic form factors, and finally, the unit cell contents, in order to calculate the f_j^2 sums. The last calculation is difficult to realize if the chemical identity of the compound being investigated is not completely known. In practice, however, it is sufficient to input the approximate unit cell contents. It has been shown that even considerable deviations from the true unit cell contents do not significantly affect the E-value calculation. So, for example, if there are four molecules in a unit cell, each having the formula say, $C_{18}H_{35}O_5$, it is quite sufficient at this stage to approximate the cell contents by $4 \times (C_{20}H_{40})$.

It follows from (5.14) and (5.16) that for the average $\langle E^2 \rangle$

$$\langle E^2 \rangle = 1 \tag{5.18}$$

Thus the magnitude of E-values is independent of the actual structure and does not differ too much from 1. Usually the highest E-values of a structure are less than 3 or 4. It is a good check on severe errors in intensity data to examine whether extremely large E-values are present. If a reflection is found having an E-value of 8 or so, it is quite certain that there is an error in the intensity of this reflection.

Similarly from the results of (5.16) and (5.18), further averages for E-values can be calculated. These averages, as well as the distribution of E-values, do not depend on the actual structure, but only on whether or not the structure is centrosymmetric. Based on calculations made by Wilson [Acta Cryst. *2*, 318 (1949)], it can, for instance, be shown that the average $\langle E \rangle$ has the following values:

$$\begin{aligned} &\langle E \rangle = \sqrt{2/\pi} = 0.798, \text{ for centrosymmetric structures} \\ &\langle E \rangle = (1/2)\sqrt{\pi} = 0.886, \text{ for acentric structures.} \end{aligned} \tag{5.19}$$

Table 5.1a gives some theoretical values depending on the presence of an inversion center, which can be used for the decision whether or not the structure being investigated is centrosymmetric. Every computer program concerned with the normalization of intensity data will usually calculate these values and this enables the user to compare them with the theoretical values. Although the experimental values will never agree completely with the theory (due to experimental errors and to the fact that a finite number of reflections is considered, whereas the theoretical averages are based on an infinite number of reflections), it is usually possible to decide whether they indicate the centric or acentric values. Also, if these values are very far from either the centric or the acentric values, this is a definite indication that something is wrong with the experimental data set.

The following is an example: Once we found for a structure, $\langle E \rangle = 0.35$ and $\langle E^2 - 1 \rangle = 1.70$, which is in agreement with neither of the theoretical values. As we discovered later, we had failed to take the square root of the intensities for calculating F_{rel}, so that the normalization was calculated with F_{rel}^4 instead of F_{rel}^2. It is therefore recommended that the results of normalization are inspected very carefully for two reasons. First, a comparison with the theoretical values given in Table 5.1a will allow a

Table 5.1a. Some Theoretical Values Relating to the Distribution of Normalized Structure Factors.

	Centrosymmetric	Acentric
$\langle E^2 \rangle$	1.0	1.0
$\langle E \rangle$	0.798	0.886
$\langle E^2 - 1 \rangle$	0.968	0.736
Amount with $\|E\| > 1$	31.7%	36.8%
Amount with $\|E\| > 2$	4.6%	1.8%
Amount with $\|E\| > 2.5$	1.2%	0.2%
Amount with $\|E\| > 3$	0.3%	0.01%

Table 5.1b. Actual Values Found for the Example Structures.

	KAMTRA	NITROS	SUCROS
$\langle E \rangle$	0.880	0.806	0.870
$\langle E^2 - 1 \rangle$	0.748	0.950	0.792
Amount with $\|E\| > 1.0$	38.1%	31.5%	33.5%
Amount with $\|E\| > 2.0$	1.9%	5.1%	2.9%
Amount with $\|E\| > 3.0$	0.1%	0.5%	0.1%
Conclusion	Acentric	Centric	Acentric

decision between a centric or an acentric structure and will assist in or verify the space group selection. Second, abnormally large E-values or a large deviation from either type of theoretical values will indicate serious errors in experimental data or the data reduction procedures.

If the results of normalization obtained from computer programs which differ in their calculation procedure are compared, differences of approximately 10% in B and t and about 10–20% in E-values are observed. From this it is realized that these calculations of E-values are not very accurate.

The results of normalization for our test structures are shown in Fig. 5.3.

Significant deviations from a Wilson plot straight line can be seen especially for SUCROS. Such deviations, in the form of an S, are frequently observed and are not the result of experimental errors. From the approximation of the Wilson plots by least squares lines, the slope and intercept are obtained. The interceptions at $s^2 = 0$ corresponding to $\ln t^2$ are 3.15 (KAMTRA), 2.0 (NITROS) and 4.20 (SUCROS), respectively. So we get as first approximations for the scale factors, $t = 4.83$ (KAMTRA), $t = 2.75$ (NITROS), and $t = 8.17$ (SUCROS). Frequently the quantity $c = 1/t$ is used. This is the F_{rel} scale factor and it should be noted that in several computer programs the input of a scale factor refers to that of F_{rel}.

The slopes $(-2B)$ are -2.0, -4.8, -7.8 Å^2, so we get estimates for the overall B's of $B = 1.0$ (KAMTRA), $B = 2.4$ (NITROS), $B = 3.9$ (SUCROS) Å^2.

The actual mean values for our example structures (Table 5.1b) confirm that KAMTRA and SUCROS crystallize in acentric space groups. For NITROS, as dis-

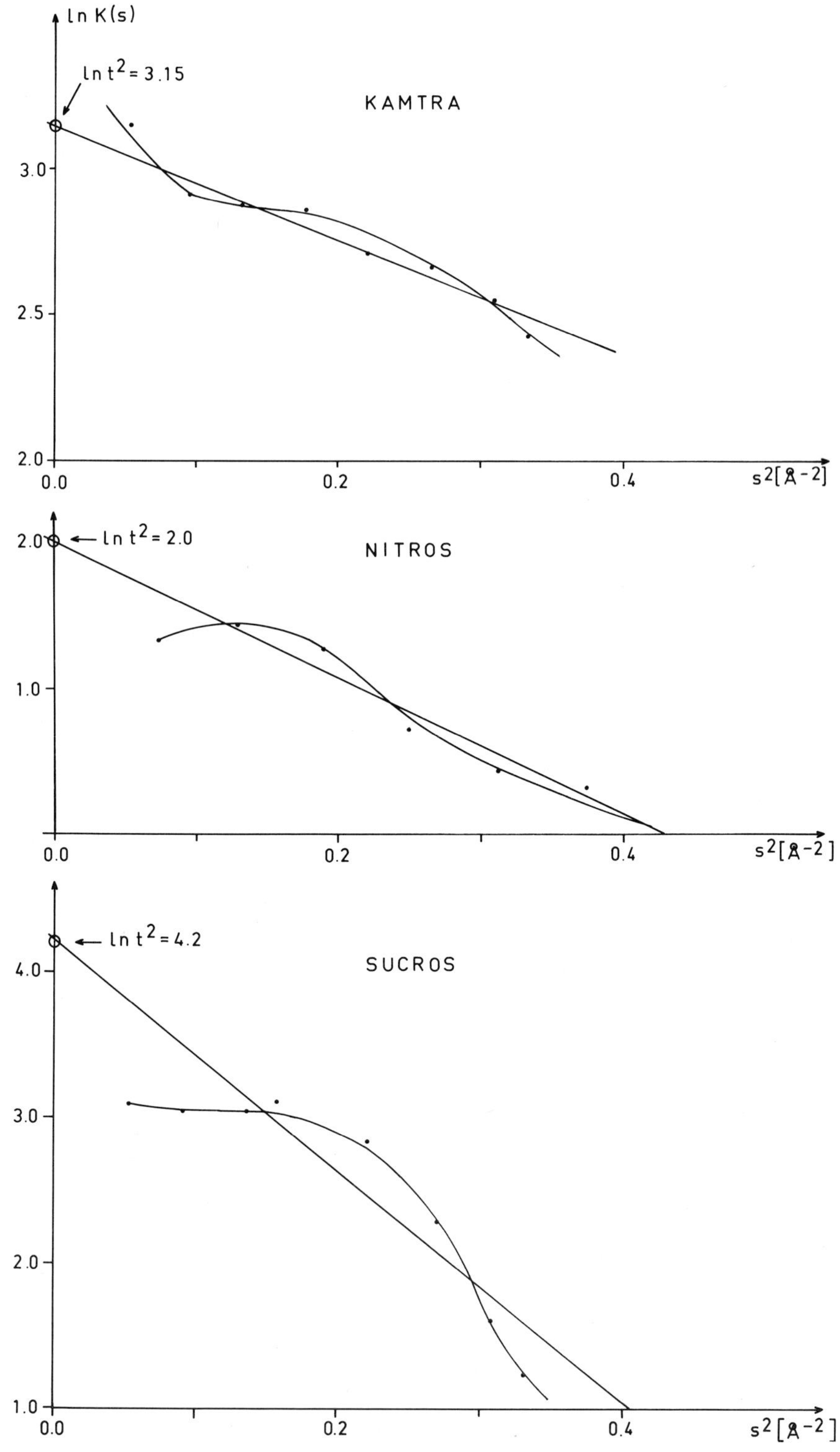

Fig. 5.3. Wilson plots: (a) KAMTRA; (b) NITROS; (c) SUCROS.

cussed in 3.3.3, it was undecided whether the correct space group was C2, Cm or C2/m. The results given in Table 5.1b for NITROS clearly indicate a centric structure, so that C2/m is the correct space group.

5.2 Fourier Methods

5.2.1 *Interpretation of the Patterson Function*

As has been pointed out in 1.2.2, the phase problem is the central problem in single crystal analysis. It results from the impossibility of calculating the Fourier transform of F to obtain ϱ because of the missing phases of F's. To avoid this difficulty, the F's may be replaced by FF*, for which the Fourier transform is possible. The result is the Patterson function

$$P(\mathbf{u}) = \int_{V^*} F(\mathbf{h})\, F^*(\mathbf{h})\, e^{-2\pi i(\mathbf{hu})}\, dV^* \qquad (1.39)$$

which we have already shown to be the convolution square of ϱ

$$P(\mathbf{u}) = \int_{V} \varrho(\mathbf{r})\, \varrho(\mathbf{r}-\mathbf{u})\, dV \qquad (1.40)$$

From the property that $P(\mathbf{u})$ is the convolution square of ϱ, useful geometrical interpretations can be derived which permit the solution of the phase problem in structures that have some special properties. Let us suppose that the structure is represented by point-shaped atoms with a weight given by their atomic numbers, as indicated in the two-dimensional model of benzene in Fig. 5.4. The integral (1.50) then degenerates to a sum (because we have discrete points)

$$P(\mathbf{u}) \sim \Sigma_{\text{atoms}}\, \varrho(\mathbf{r})\, \varrho(\mathbf{r}-\mathbf{u})$$

This can be rationalized as follows. To obtain $P(\mathbf{u})$, shift the structure ϱ by the vector $\mathbf{u}$ (dotted representation in Fig. 5.4a, b, c) to get $\varrho(\mathbf{r}-\mathbf{u})$, calculate for every $\mathbf{r}$ the product $\varrho(\mathbf{r})\, \varrho(\mathbf{r}-\mathbf{u})$, and sum over all $\mathbf{r}$. It follows that $P(\mathbf{u})$ differs from zero only if in a product $\varrho(\mathbf{r})\, \varrho(\mathbf{r}-\mathbf{u})$, *both* factors are different from zero. This happens only if $\mathbf{u}$ is the difference vector of any pair of position vectors of two atoms of the structure (Fig. 5.4a), since only under this condition can one atom of the dotted structure coincide with an atom of the original model. The weight of $P(\mathbf{u})$ is then proportional to the product of weights of contributing atoms. If $\mathbf{u}$ does not satisfy this condition, we have $P(\mathbf{u}) = 0$ (Fig. 5.4b).

It is, of course, possible that more than one pair of position vectors coincide for a special $\mathbf{u}$ (Fig. 5.4c). In this case, $\mathbf{u}$ is said to have a multiplicity larger than one and $P(\mathbf{u})$ has a weight proportional to the sum of the product weights of contributing atoms.

From this interpretation of the convolution square, it can be said that the Patterson function represents all difference vectors of a given structure with a weight propor-

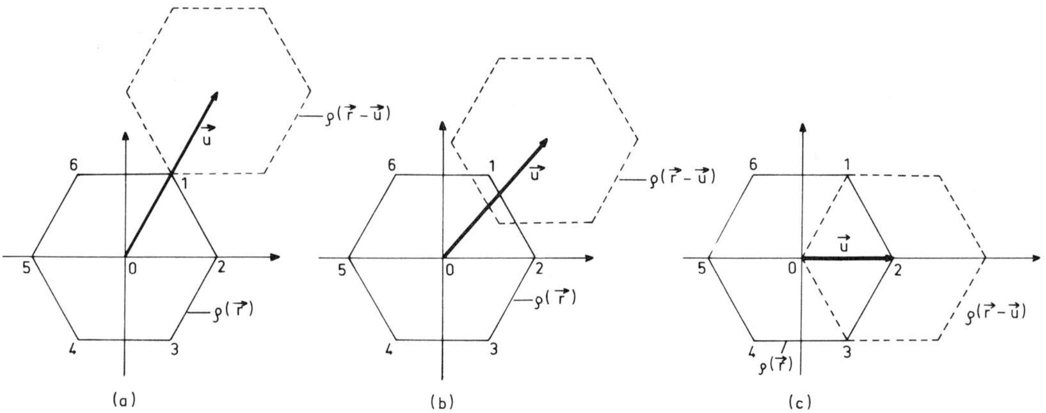

Fig. 5.4. Interpretation of the Patterson function as a convolution square demonstrated by the benzene molecule:
(a) **u** is equal to the difference vector between two atomic positions [C(1)–C(4)], causing $P(\mathbf{u}) \neq 0$;
(b) **u** does not coincide with one atomic difference vector, thus $P(\mathbf{u}) = 0$;
(c) **u** is identical with two atomic difference vectors [C(6)–C(1) and C(4)–C(3)], $P(\mathbf{u}) \sim 2 \times 6^2$.

tional to the frequency of occurrence times the weight of contributing atoms. In section 1.1.3, we discussed this problem when, as an exercise, we solved the problem of drawing all difference vectors in the benzene ring. This was nothing other than the representation of the Patterson function of a benzene ring. For a real structure which has, instead of atomic points, continuous electron density with maxima at the atomic centers, there is no fundamental difference. The Patterson function of a real electron density has in principle the same properties as for the point-shaped model. Only the discrete Patterson points have to be replaced by more or less broad maxima and a decrease to exactly zero can no longer be expected outside the difference vector positions.

With this geometrical interpretation of the Patterson function, the problem of structure determination can be expressed as the following problem: how to find the structure from a given distribution of all difference vectors between each pair of atomic position vectors. This problem cannot be solved in this general formulation. However, the situation becomes quite favorable if additional provisions are made. As we shall show, there will be a good chance for a solution if the structure has a very simple geometry or contains one or two atoms having an atomic number significantly larger than that of most of the other atoms.

5.2.2 *Heavy Atom Method, Principle of Difference Electron Density*

Suppose that we have the structure of benzene as in Fig. 5.4, but with one hydrogen replaced by a substituent of high atomic number, e.g. bromine (Fig. 5.5). The strategy of solving this structure would depend in detail on the space group symmetry; nevertheless, some general remarks can be made. If, for instance, the space group contains at least a center of symmetry, the Patterson function will not only show the difference vectors within one molecule (intramolecular difference vectors) but also those between each pair of atoms in different molecules (intermolecular difference vectors). The weights of difference vectors can be

$1 \times 1 = 1$ from H-H vectors,
$1 \times 6 = 6$ from H-C vectors,
$1 \times 35 = 35$ from H-Br vectors,
$6 \times 6 = 36$ from C-C vectors,
$6 \times 35 = 210$ from C-Br vectors,
$35 \times 35 = 1225$ from Br-Br vectors.

The Br-Br vector will have the highest maximum in the Patterson function (except for the origin maximum). This vector, however, combines the position vector $\mathbf{r}(\mathrm{Br})$ with its centrosymmetric $-\mathbf{r}(\mathrm{Br})$, so that for $\mathbf{u}(\mathrm{Br\text{-}Br})$

$$\mathbf{u}(\mathrm{Br\text{-}Br}) = \mathbf{r}(\mathrm{Br}) - [-\mathbf{r}(\mathrm{Br})] = 2\mathbf{r}(\mathrm{Br})$$

Therefore, the vector having the highest maximum is likely to be twice the position vector of the bromine atom. Hence the location of this atom in the unit cell is determined.

This derivation was only possible because only one "heavy atom" was present in the

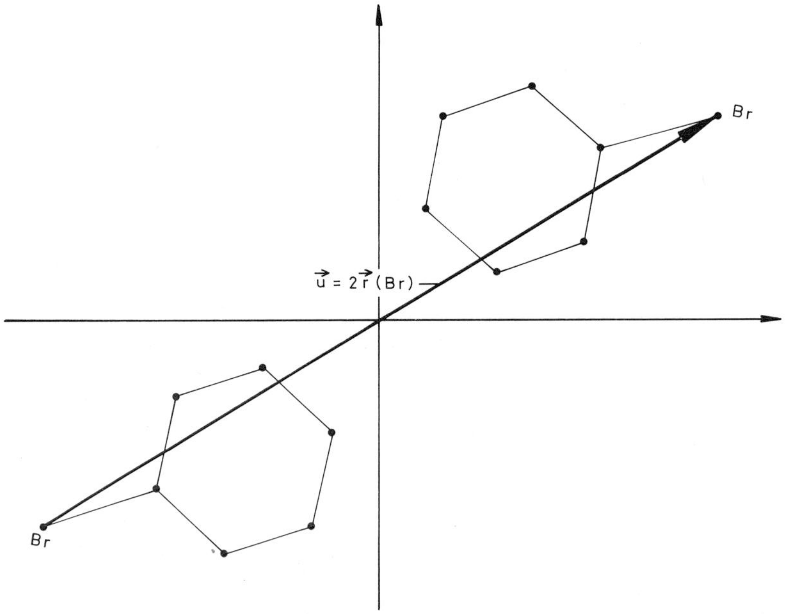

Fig. 5.5. Obtaining the heavy atom position from a centrosymmetric heavy atom structure.

structure, resulting in only one significant high maximum of the Patterson function. With the heavy atom position known, it becomes possible, in favorable cases, to complete the whole structure determination.

This is done using the principle of difference electron density calculation (difference Fourier synthesis or difference synthesis. It is customary in crystallography to denote the numerical calculation of the Fourier transformation from the reciprocal to the direct space as a "synthesis". Thus the term "Patterson synthesis" is an expression frequently used). Supposing some part of the structure, denoted ϱ_c, is already known. From this partial model, the Fourier transform can be calculated and we get

$$F_c(\mathbf{h}) = \mathsf{F}[\varrho_c(\mathbf{r})]$$

where the subsript c indicates that these quantities are "calculated structure factors" from a given model, Expressed as a sum after (3.18), $F_c(\mathbf{h})$ is given by

$$F_c(\mathbf{h}) = \sum_{j=1}^{N_c} f_j \, e^{2\pi i \mathbf{h}\mathbf{r}_j} \, e^{-T_j} \tag{5.20}$$

with $N_c < N$ indicating that the portion of N_c among N atoms is known. The principle of difference synthesis is based on the assumption that if N_c does not differ too much from N, the phase $\varphi(F_c)$ [being known from (5.20)] is approximately that of F itself. With the phases of F_c assigned to F_{rel} we have obtained phases for the experimentally derived F's. A scaling of the F_{rel}'s could be obtained from the result of the Wilson plot, but a more precise one is given by the condition

$$\langle |F_c| \rangle = \langle |F| \rangle \tag{5.21}$$

Using (5.21) the quotient

$$\frac{\Sigma_h |F_c(\mathbf{h})|}{\Sigma_h F_{rel}(\mathbf{h})} = c$$

gives an improved scale factor c, which must now be denoted "F_{rel} scale factor" since it acts on F_{rel}. The quantity $|F_o|$, defined by

$$|F_o| = c F_{rel} \tag{5.22}$$

is named "observed structure amplitude", since it is derived directly from experimental observations. Now with F_o and F_c being introduced, we can express the difference electron density calculation as

$$\Delta\varrho(\mathbf{r}) = \mathsf{F}^{-1}[F_o(\mathbf{h}) - F_c(\mathbf{h})] \tag{5.23}$$

with $\varphi[F_o(\mathbf{h})]$ being approximated by $\varphi[F_c(\mathbf{h})]$. From the arithmetic of Frourier transforms, it follows that

$$\Delta\varrho(\mathbf{r}) = \mathsf{F}^{-1}[F_o(\mathbf{h})] - \mathsf{F}^{-1}[F_c(\mathbf{h})]$$

Let us denote by $\varrho_o(\mathbf{r})$ the "observed electron density" being that given by the experimental data together with the true phases. Note that ϱ_o is not equal to the true ϱ, which of course can never be obtained (since neither can all F's be measured nor can experimental errors be completely excluded).

Moreover, let us define

$$\varrho_c(\mathbf{r}) = \mathsf{F}^{-1}[F_c(\mathbf{h})]$$

and approximately

$$\varrho_o(\mathbf{r}) \approx \mathsf{F}^{-1}[F_o(\mathbf{h})]$$

Note that for ϱ_o, only "$\approx$" can be written, since the phases of F_o are only approximately determined. Then we get

$$\Delta\varrho(\mathbf{r}) \approx \varrho_o(\mathbf{r}) - \varrho_c(\mathbf{r}) \tag{5.24}$$

which means that by application of (5.23) we get an electron density distribution for which the known atoms are removed and the remaining residue should show the missing part of the structure.

In practice, however, the success of using this difference Fourier technique will depend on the amount and the accuracy of the known part of the structure, since it is on these parameters that the validity of approximating $\varphi(F_o)$ by $\varphi(F_c)$ depends. Unfortunately, it cannot be determined in advance whether a structural fragment is sufficient to ensure the success of this difference Fourier technique. For this reason, the use of a *heavy* atom position from the Patterson function is generally necessary for the further progress of structure determination. It is evident that in the sum for F,

$$F(\mathbf{h}) = \underbrace{\Sigma_{\text{heavy}}\, f_h\, e^{2\pi i \mathbf{h}\mathbf{r}_h}\, e^{-T_h}}_{F_c} + \Sigma_{\text{light}}\, f_l\, e^{2\pi i \mathbf{h}\mathbf{r}_l}\, e^{-T_l}$$

the weight of F_c will increase and thus its phase will be close to that of F if $f_h \gg f_l$, which is the definition of a heavy-atom structure.

Before trying to identify the heavy-atom portion of the structure by interpretation of the Patterson function, the question should be considered whether the heavy-atom part of the structure is "heavy" enough for a successful structure determination. For that purpose, a rule of thumb is given by the so-called "Sim quotient"; i.e.

$$Q = \frac{\Sigma_h Z_h^2}{\Sigma_l Z_l^2} \tag{5.25}$$

where the Z's are the atomic numbers of the atoms in the unit cell and the summation over h in the numerator is over all atoms already known, while the sum over l in the denominator is taken over all unknown atoms. If

$$Q > 1 \tag{5.26}$$

then the application of the heavy atom method is likely to be successful. However, with very precise diffractometer data, the criterion (5.26) can be relaxed. There is a good chance of success with the heavy atom method if the sum of the squares of the heavy atom atomic numbers is not significantly less than that of the light atoms. A value for Q of 0.5 still gives a reasonable chance of success.

Our test structure KAMTRA, formula $C_4H_5O_6K$, is a good example for testing this rule. With the potassium atom as the heavy atom, we get a Sim quotient:

$$Q = \frac{19^2}{4 \times 6^2 + 6 \times 8^2 + 5} = \frac{361}{144 + 384 + 5} = \frac{361}{533} \approx 0.7$$

As mentioned above, this value might be sufficient for the heavy atom method to be successful. We shall try it, but first we have to learn more about some special properties of the Patterson function which derive from the space group symmetry.

5.2.3 *Harker Sections*

It was first pointed out by Harker (1936) that space group symmetry results in special sections of a Patterson synthesis which can be interpreted as projections of the structure onto a plane or a line. These sections, named Harker planes or lines, or generally Harker sections, are very useful for the interpretation of a Patterson function.

Let us discuss, for example, the space group $P2_1/c$. The symmetry operation representing the screw axis is

$$-x,\ 1/2 + y,\ 1/2 - z$$

Every atom having the vector $\mathbf{r} = (x, y, z)$ has its equivalent in $\mathbf{r}' = (-x, 1/2 + y, 1/2 - z)$. Since the Patterson function contains the difference vectors between each pair of atoms, the vector

$$\mathbf{u}(\mathbf{r}) = \mathbf{r} - \mathbf{r}' = (2x, -1/2, 2z - 1/2) = (u, -1/2, w)$$

will be present and this is true for each atom of the structure. From the periodicity of the crystal lattice, +1 can be added to the y- and z-component and we get

$$\mathbf{u}(\mathbf{r}) = (2x, 1/2, 2z + 1/2) = (u, 1/2, w)$$

This result can be interpreted as a projection of the structure on the plane $y = 1/2$ in twice the original scale and shifted in z by 1/2. On this Harker plane, a maximum at $(u, 1/2, w)$ gives the atomic coordinates

$$x = u/2$$
$$z = (w - 1/2)/2$$

Since this is true for each atom, in principle a complete two-dimensional projection of the structure can be obtained.

There is another Harker section for the space group $P2_1/c$ and that is a Harker line caused by the glide plane symmetry. The difference vector of

$$\mathbf{r} = (x, y, z)$$

and

$$\mathbf{r}' = (x, 1/2 - y, 1/2 + z)$$

is

$$\mathbf{u}(\mathbf{r}) = (0, 2y - 1/2, -1/2) = (0, 2y + 1/2, 1/2)$$

This can be interpreted as a projection of the structure on the line (0, v, 1/2). Generally, the projection of a complete structure on a single line is not very helpful but in combination with a Harker plane, such a line projection may give useful information. If, for instance, the x and z coordinates of a heavy-atom position have been determined from a Harker plane, it may be possible to obtain the third coordinate from the Harker line.

Although a Harker plane may contain a projection of the complete structure and information about the third dimension can be derived from Harker lines, in practice, there are difficulties, because Harker maxima may be superimposed by other vectors which correspond to vectors between atoms which are not symmetry related. Nevertheless, Harker sections should always be inspected very carefully. In spite of some ambiguities, their interpretation is often possible if the geometry of the structure is rather simple (expecially for planar molecules) or if a single heavy-atom position has to be determined.

It can be generalized, from the example of space group $P2_1/c$, that symmetry elements derived from rotation axes will result in Harker planes while symmetry elements derived from mirror planes cause Harker lines. If, for example, more than one rotational symmetry element is present, we have more than one Harker plane and the information derived from one Harker plane may be completed and confirmed by that from the others. Therefore, in space groups of higher symmetry, the information of all the Harker planes and lines may permit the interpretation of a significant part of the structure. These days Harker sections are particularly useful in protein crystallography where it is necessary for phase determination to work on several heavy-atom derivatives of one protein. Harker sections are then an important aid for verifying the locations of heavy atoms.

A further symmetry-related property should be noted here, although its application is more limited. If the space group is centrosymmetric, every vector $\mathbf{r} = (x, y, z)$ has its equivalent in $\mathbf{r}' = (-x, -y, -z)$ and we obtain a Patterson vector $\mathbf{u}(\mathbf{r})$ for every atomic vector $\mathbf{r}$

$$\mathbf{u}(\mathbf{r}) = (2x, 2y, 2z)$$

It follows that the vector $2\mathbf{r}$ of every atomic vector $\mathbf{r}$ is present in the Patterson synthesis. But usually these special vectors can only be resolved from one another for a heavy atom vector. Such a vector does, however, provide a check whether the interpretation of the Harker sections was correct.

5.2.4 *Numerical Calculation of Fourier Syntheses*

The computation of a Patterson function is usually done by a program which is able to do all types of Fourier inverse transforms necessary in single crystal analysis. Since crystallographers call this process the "calculation of a Fourier synthesis", a program of that kind is called a "Fourier synthesis program". The numerical formalism, the general aspects of preparing the input data, and how to obtain a suitable output are the same, so we shall discuss here the method of numerical calculation for a Fourier synthesis.

The calculation is based on the integral

$$R(\mathbf{r}) = \int_{V^*} G(\mathbf{b})\, e^{-2\pi i \mathbf{br}}\, dV^*$$

where G (**b**) is a general function of reciprocal space, which we shall replace later by the different types of actual functions. In the case of single crystals, it reduces, after (3.19), to the sum

$$R(\mathbf{r}) = (1/V)\, \Sigma_h\, G(\mathbf{h})\, e^{-2\pi i \mathbf{hr}}$$

or with

$$G(\mathbf{h}) = |G(\mathbf{h})|\, e^{2\pi i \varphi(\mathbf{h})}, \quad \varphi(\mathbf{h}) = \text{phase of } G(\mathbf{h}),$$

$$R(\mathbf{r}) = (1/V)\, \Sigma_h |G(\mathbf{h})|\, e^{2\pi i [\varphi(\mathbf{h}) - \mathbf{hr}]}$$

To avoid complex arithmetic, Friedel pairs of reflections are coupled. Provided that

$$|G(-\mathbf{h})| = |G(\mathbf{h})| \quad \text{and} \quad \varphi(-\mathbf{h}) = -\varphi(\mathbf{h})$$

we get

$$|G(\mathbf{h})|\, e^{2\pi i [\varphi(\mathbf{h}) - \mathbf{h}]} + |G(-\mathbf{h})|\, e^{2\pi i [\varphi(-\mathbf{h}) + \mathbf{h}]} = 2|G(\mathbf{h})| \cos 2\pi [\mathbf{hr} - \varphi(\mathbf{h})]$$

Hence,

$$R(\mathbf{r}) = (2/V) \sum_{h=0}^{\infty} \sum_{k=-\infty}^{+\infty} \sum_{l=-\infty}^{+\infty} |G(\mathbf{h})| \cos 2\pi [hx + ky + lz - \varphi(hkl)] \qquad (5.27)$$

The only reflection having no Friedel pair is the 000 reflection. The intensity of the 000 reflection and therefore | F (000) | can never be obtained experimentally, since it had to be taken at $\theta = 0°$. However, F (000) can be calculated theoretically. From formula (3.18) for F (**h**), we get for **h** = **0**

$$F(\mathbf{0}) = \sum_{j=1}^{N} f_j(0)\, e^0$$

Since for $s = \sin\theta/\lambda = 0$ the f_j's are equal to the atomic number Z_j of contributing atoms, we get

$$F(000) = \sum_{j=1}^{N} Z_j \qquad (5.28)$$

In other words, the magnitude of F (000) is given by the number of electrons present in the unit cell [Note that (5.28) is a general property of Fourier transforms, it was already derived in 1.2.1, formula (1.36)]. It follows from (5.28) that the phase of F (000) is also known and is equal to zero. We can say that F (000) is the only structure factor the magnitude *and* phase of which are originally known. It is the only reflection which can never be measured experimentally. In order to avoid a separate handling of the 000 reflection in the Fourier sum (5.27), it is usual to enter $|G(\mathbf{h})|/2$ into the sum if $\mathbf{h}=\mathbf{0}$, but note that the Fourier cofficient for $\mathbf{h}=\mathbf{0}$ is of minor relevance in the sum (5.27). Since the trigonometric term reduces to $+1$, $|G(\mathbf{0})|$ acts only as a constant which can be omitted in some cases.

We will now discuss the different functions that are used for G and the physical interpretation of the resulting distribution $R(\mathbf{r})$ of direct space.

(1) $|G(\mathbf{h})| = |F_o(\mathbf{h})|, \quad \varphi(\mathbf{h}) = \varphi[F_c(\mathbf{h})]$.

This is called an "F_o-Fourier" synthesis and the result is the observed electron density distribution $\varrho_o(\mathbf{r})$, with reservations arising from possible uncertainties of the F_c-phases. The F (000) reflection may or may not be included in the F_o-Fourier calculation. If it is icluded, a rather accurate scaling of the F_o's must be provided.

(2) $G(\mathbf{h}) = F_c(\mathbf{h})$.

The result is a so-called "F_c-Fourier" synthesis representing the calculated electron density $\varrho_c(\mathbf{r})$. Since the F_c's are scaled properly, F (000) should always be included.

(3) $|G(\mathbf{h})| = |F_o| - |F_c|, \quad \varphi(\mathbf{h}) = \varphi(F_c)$.

From this calculation a difference synthesis, as discussed in 5.2.2, is obtained. F (000) is not used, since it cancels by the subtraction.

If the F_c part of the structure factor is obtained from only a small structural fragment, it seems likely that the phase of F_c will be of low precision if the magnitude of $|F_c|$ is significantly smaller than that of $|F_o|$. For example, in a centric structure with a reflection having $|F_o| = 100$ and $F_c = -80$, it is unlikely that the atoms not yet included in the F_c-calculation will all contribute with large enough positive terms to give a final $F_c = +100$. It can therefore be assumed that "$-$" is the correct sign for this reflection. If, on the other hand, F_c has only a small value, say -10, nothing can be assumed about the final sign of this reflection. It is therefore recommended that this reflection is excluded from the Fourier summation. For that purpose, a so-called "rejection ratio", j, is used, so that reflections with

$$|F_c| - j|F_o| < 0 \tag{5.29}$$

are excluded. In the preliminary stages of a structure determination, $j = 0.5$ is a good choice which can be reduced to zero as the structure approaches completion. A rejection ratio may also be used in an F_o-Fourier synthesis calculation.

(4) $G(\mathbf{h}) = F_{rel}^2$.

Since F_{rel}^2 is proportional to FF* [see (5.8)], this calculation produces a Patterson synthesis. Since the F_{rel}'s are usually unscaled, it makes no physical sense to include F(000) in the calculation. The lack of a proper scaling of the F_{rel}'s acts only on the scaling of the Patterson function. However, since the user is only interested in the positions of maxima and their relative weight, this is of no significance.

(5) $G(\mathbf{h}) = E^2(\mathbf{h}) - 1$.

From this type of Fourier coefficient, we get a so-called "vector map" $V(\mathbf{r})$. From the properties of this synthesis, it is a "sharpened Patterson synthesis" with the origin removed. The interpretation of the Patterson function using F_{rel}^2 is complicated by overlap of two or more maxima, caused by the fact that the atomic maxima and thus difference vector maxima have a finite extension. Since the E-values relate to atoms without thermal motion, it follows that the maxima derived from a calculation using E^2 values should have less peak overlap. These peaks are said to be "sharpened" relative to those of a normal Patterson synthesis. The large maximum at the origin also obscures the analysis of difference vectors of short length. By using $E^2 - 1$ instead of E^2 as Fourier coefficients, the origin peak can be removed,

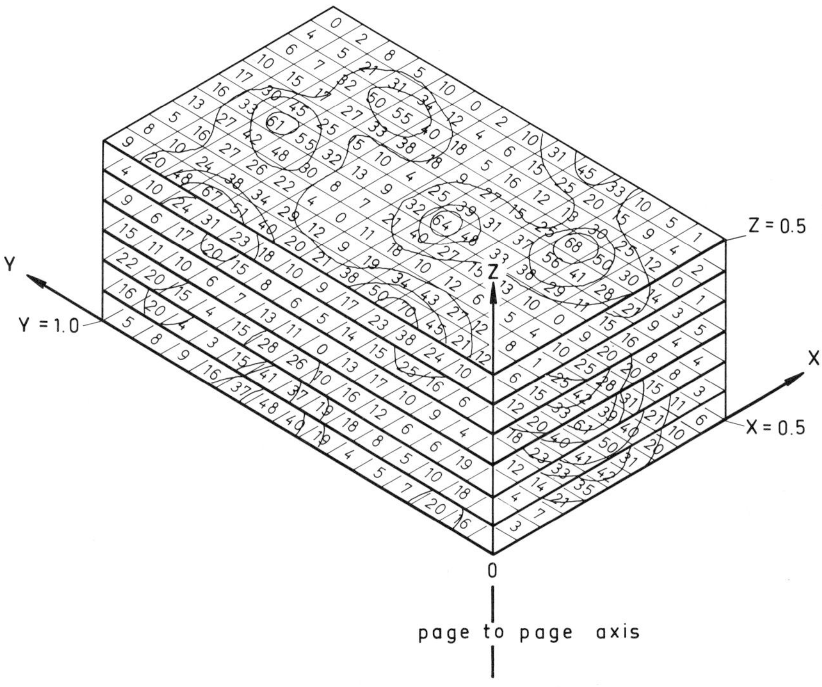

Fig. 5.6. Sub-division of a unit cell by a three-dimensional grid.

$$V(\mathbf{r}=0) = (2/V)\,\Sigma_h\big(E^2(\mathbf{h}) - 1\big)\cos(0)$$
$$= (2/V)\,[\{\Sigma_h E^2(\mathbf{h})\} - M]$$

where M is the total number of reflections. Since it follows from (5.18) that

$$\Sigma_h E^2(\mathbf{h}) = M,$$

we get

$$V(\mathbf{r}=0) = 0$$

which means that the large origin maximum usually present in a Patterson synthesis is removed.

(6) $|G(\mathbf{h})| = |E(\mathbf{h})|$, $\varphi(\mathbf{h})$ derived from "Direct Methods".

This type of Fourier synthesis will be used frequently in the application of "Direct Methods" (to be discussed in section 5.3). The results is an electron density distribution having sharper maxima than when $|F_o|$'s are used.

For the calculation of a Fourier synthesis, a three-dimensional grid will be used for subdividing the unit cell into "grid points" (see Fig. 5.6). The calculation of $R(\mathbf{r})$ is then performed for every grid point of a specified section of the unit cell. The grid can either be chosen by the user or will be provided automatically by the Fourier program. The number of grid points in each direction should be chosen in relation to the length of the axes and the resolution of the structure determination. As already pointed out, the resolution cannot be better than half the wavelength of the radiation used – that means ≈ 0.8 Å for CuKα – it is not sensible therefore to choose more grid points than necessary for this resolution. A choice of three grid points/Å is reasonable.

If the three cell directions are subdivided by N, M and P points, a position vector $\mathbf{r}$ having the fractional coordinates (x, y, z) has the grid coordinates

$$n = [xN], \quad m = [yM], \quad p = [zP] \tag{5.30}$$

where [ ] means take the nearest integer. On the other hand, x, y, z are obtained from the fractions n/N, m/M, p/P.

It is not always necessary to calculate the Fourier synthesis for the whole unit cell, and the calculations can be restricted to a part of the unit cell that is symmetry independent. Since a Patterson synthesis is always centrosymmetric, more than half a unit cell is never necessary.

The output map of a Fourier synthesis usually consists of two-dimensional sections perpendicular to one of the unit cell directions. Each section, or page, consists of the density results for a fixed grid value in that direction (Fig. 5.6). By contouring equidensity points, the density distribution may be illustrated graphically.

Several output options, such as the correction for oblique axes, are provided in every well-organized computer program. The actual details of the output are given in the program descriptions. Moreover, except for protein structure analysis, it is now customary to calculate a Fourier synthesis and keep the output only on the mass storage of the computer and dispense with any printed output. In this case, the results are analyzed by a peak-searching program which determines all peaks above a given limit, eliminates the symmetry-dependent peaks, sorts the unique peaks with respect to their height, and prints out the atomic coordinates of a specified number of the highest maxima. This technique saves paper and printer time and should be applied when no more information than peak positions ans relative height are desired.

5.2.5 *Application of Heavy Atom Method to KAMTRA*

Now let us apply the "heavy atom method" to an example. The $P2_1 2_1 2_1$ space group of KAMTRA is very favorable for the application of Patterson methods, since from this symmetry three Harker planes can be derived. The following four symmetry operations

(1) x, y, z (2) $1/2 - x, -y, 1/2 + z$

(3) $1/2 + x, 1/2 - y, -z$ (4) $-x, 1/2 + y, 1/2 - z$

generate three Harker planes, given by

(1)–(2): $H_1(u, v, 1/2)$: $(1/2 + 2x, 2y, 1/2)$

(1)–(3): $H_2(1/2, v, w)$: $(1/2, 1/2 + 2y, 2z)$

(1)–(4): $H_3(u, 1/2, w)$: $(2x, 1/2, 1/2 + 2z)$

(A constant $-1/2$ is always replaced by $+1/2$, since $+1/2$ and $-1/2$ differ only by a translation of one unit cell.)

For an inspection of these planes we have to calculate the Patterson function for the three sections given by $u = 1/2$, $v = 1/2$, $w = 1/2$. With the lattice constants of 7.8, 10.6 and 7.6 Å, a grid of 28th in the x-direction, 32nd in the y-direction, and 23rd in the z-direction gives approximately three grid points per Angstrom in all three directions. The results are shown in Fig. 5.7. In all three sections, four high maxima are found which are all symmetry-related so that only one independent high maximum is present. We assume that it is the potassium-potassium vector.

Taking H_3, the position (in grid coordinates) is at

$$u = 18.6, \quad w = 19.4 \quad \text{height: } 939,$$

it follows that the x- and z-coordinates of the potassium position are

$$x = u/2 = 9.3$$

$$z = (w - 1/2)/2 = 3.95$$

Now on H_2, one of the four maxima should be at $w = 2z = 7.9$. Indeed one maximum, height 932, is at

$$v = 18.2, \quad w = 7.7$$

With the uncertainties of interpolation, we can regard $w = 7.7$ as equal to $w = 7.9$. So we get from this Harker plane

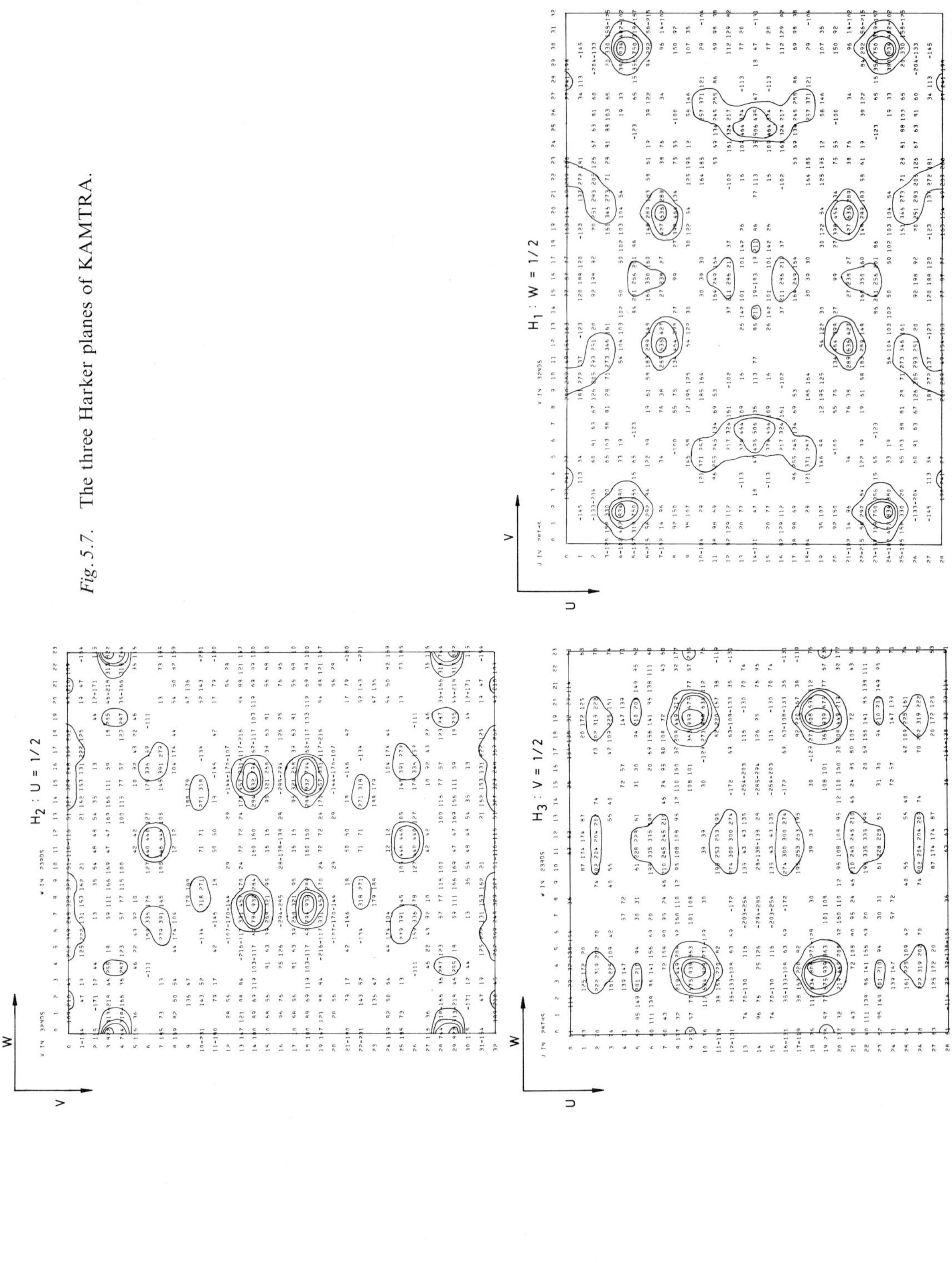

Fig. 5.7. The three Harker planes of KAMTRA.

$$y = (v - 1/2)/2 = 1.1, \quad z = w/2 = 3.85$$

so we now have all three grid coordinates of the potassium position:

$$x = 9.3; \quad y = 1.1; \quad z = 3.9 \quad \text{(by averaging)}$$

We should then observe a high maximum on H_1 at

$$u = 14 + 18.6 = 32.6 = 4.6 \quad \text{(from the periodicity)}$$

and $v = 2.2$.

Indeed we find this maximum, height 834, at $u = 4.4$ and $v = 2.2$. This consistency in the three section indicates that the potassium position vector is correctly found. Its fractional coordinates are

$$x = 9.3/28 = 0.332$$

$$y = 1.1/32 = 0.034$$

$$z = 3.9/23 = 0.170$$

No more information is likely to be obtained from the Patterson synthesis since all further high maxima correspond to K-O or K-C vectors, of which there are such a large number that an interpretation is impossible. We must therefore complete the structure determination by a difference synthesis. We shall describe this process in the section concerned with refinement (see 6.4.1).

At this stage, we have established that the heavy-atom method has been successful for locating the potassium position with certainty.

5.3 Direct Methods

Up to the early sixties, the principal method of structure determination was by interpretation of the Patterson function. As a consequence, a great number of chemical compounds, especially the so-called "light atom structures" of organic chemistry, were very difficult or impossible to investigate by the method of single crystal analysis. The development of the so-called "Direct Methods" of phase determination has removed this limitation.

The name "Direct Methods" is derived from the fact that the phases of the structure factors are derived from the magnitudes of the F's rather than indirectly by an interpretation of the Patterson function. However, extensive numerical calculations are necessary to apply this technique and the development of direct methods was only possible because of the rapid progress in computer technology made in the 1960's.

The detailed results of the theory of direct methods have been published in a large number of papers which use a complicated mathematical formalism. For our purposes, it is sufficient to present the fundamental ideas of that method and the formulae mostly used in the practical application. A short and not too detailed representation of the principles of direct methods must be accompanied by some simplifications. For the practical structure analyst, it is only necessary to know about the main principles and to be familiar with their application.

5.3.1 *Fundamental Formulae*

The principles of direct methods are based on the fact that the experimental data, that is the $|F_{rel}|$'s or the magnitudes of F_o's, already contain information about the phases of the structure factors. This was first pointed out by Harker and Kasper in their so-called "Harker-Kasper inequalities" [Harker & Kasper, Acta Cryst. *1*, 70 (1948)].

By application of the Cauchy-Schwarz inequality

$$|\int f(x)\, g(x)\, dx|^2 \leqq \left(\int |f(x)|^2 dx\right) \left(\int |g(x)|^2 dx\right)$$

to the expression for the structure factor, some remarkable results can be obtained. With

$$F(\mathbf{h}) = \int_V \varrho(\mathbf{r})\, e^{2\pi i \mathbf{hr}}\, dV,$$

supposing $\varrho(\mathbf{r}) \geq 0$ for all $\mathbf{r}$, and

$$f(\mathbf{h}) = \varrho(\mathbf{r})^{1/2}$$
$$g(\mathbf{h}) = \varrho(\mathbf{r})^{1/2}\, e^{2\pi i \mathbf{hr}}$$

we get

$$|F(\mathbf{h})|^2 \leq \left(\int_V \varrho(\mathbf{r})\, dV\right) \left(\int_V \varrho(\mathbf{r})\, [|e^{2\pi i \mathbf{hr}}|]^2\, dV\right)$$

Since

$$|e^{2\pi i \mathbf{hr}}|^2 = 1$$

and

$$\int \varrho(\mathbf{r})\, dV = Z,$$

with $Z = F(000)$ (see 5.28) being equal to the number of electrons in the unit cell, we get

$$|F(\mathbf{h})| \leqq F(000) \tag{5.31}$$

which is, of course, a trivial result. However, with the additional provision of special symmetry elements present in the unit cell, some important non-trivial results can be derived. Let us assume, for example, that an inversion center is present. Then $F(\mathbf{h})$ reduces to

$$F(\mathbf{h}) = \int_{V/2} \varrho(\mathbf{r})\, \left(e^{2\pi i \mathbf{hr}} + e^{-2\pi i \mathbf{hr}}\right) dV$$

$$= 2 \int_{V/2} \varrho(\mathbf{r}) \cos 2\pi \mathbf{hr}\, dV$$

Choosing f and g as above, we get

$$|F(\mathbf{h})|^2 \leq 2Z \int_{V/2} \varrho(\mathbf{r})\, (\cos 2\pi \mathbf{hr})^2\, dV,$$

with $\cos^2\alpha = 1/2 + 1/2\cos 2\alpha$, we get

$$|F(\mathbf{h})|^2 \leq 2Z\Big(\int_{V/2} (\varrho/2)\,dV + \int_{V/2} (\varrho/2)\cos 2\pi(2\mathbf{h})\mathbf{r}\,dV\Big) = Z/2\,\big(Z + F(2\mathbf{h})\big)$$

or with $Z = F(000)$,

$$|F(\mathbf{h})|^2 \leq F(000)\,[F(000)/2 + F(2\mathbf{h})/2]$$

Defining

$$u(\mathbf{h}) = \frac{F(\mathbf{h})}{F(000)}$$

we get finally

$$|u(\mathbf{h})|^2 \leq 1/2 + u(2\mathbf{h})/2 \qquad (5.32)$$

and with using u instead of F (5.31) reduces to

$$|u(\mathbf{h}) \leq 1$$

From (5.32) information about a phase, or, since we have a centrosymmetric problem, about a sign, can be derived from the magnitude of the u's and the $|u|$'s are derived directly from the experimental data. Let us suppose that we have two reflections **h** and 2**h** with their u-values having large magnitudes, say both 0.6. Then (5.32) is only satisfied for $u(2\mathbf{h}) = +0.6$ (since $|u(\mathbf{h})|^2 = 0.36$ and $1/2 + (1/2)0.6 = 0.8$, while $1/2 - (1/2)0.6 = 0.2$), so the sign of $F(2\mathbf{h})$ must be $+$. However, a definite conclusion from (5.32) can only be deduced for a limited number of reflections. If, for instance, both reflections have small magnitudes of u, say 0.2 or so, no information can be derived, since in this case $+u(2\mathbf{h})$ as well as $-u(2\mathbf{h})$ satisfy this equation.

Although further Harker-Kasper inequalities can be derived for other symmetry elements, their practical use is limited, since reflections with large u-magnitudes are usually rare and hence an insufficient number of phases can be determined. Therefore, Harker-Kasper inequalities are no longer applied in practical structure analysis. Nevertheless, this first recognition of a relation between magnitudes and phases was an important result, which has been developed into more powerful methods of direct phase determination. A large number of investigations on that subject have been initiated, of which one of the earlier important results was the Sayre equation developed in 1952 [Sayre, D., Acta Cryst. *5*, 60 (1952)]. It is one of the basic formulae of "Direct Methods".

A simple derivation of Sayre's equation can be obtained using a simplified structural model. Let us use again the model introduced in 5.1.2 with point-shaped atoms of unique density having no thermal motion, so that there is no overlap between pairs of atoms. For this model

$$\varrho(\mathbf{r}) = [\varrho(\mathbf{r})]^2$$

The Fourier transform of $\varrho(\mathbf{r})$ is then denoted by $G(\mathbf{h})$ (see 5.1.2). Since the Fourier transform of a product can be expressed by a convolution operation (Convolution Theorem, 1.2.1), we get

$$\mathbf{F}[\varrho(\mathbf{r})] = \{\mathbf{F}[\varrho(\mathbf{r})]\}^{\frown 2}$$

or, by using $G(\mathbf{h})$ for the Fourier transform,

$$G(\mathbf{h}) = \int_{V^*} G(\mathbf{h}')\, G(\mathbf{h}-\mathbf{h}')\, dV^*$$

Since it was shown in 5.1.2 that the E-values are closely related to that model, and since for single crystals the integral over the reciprocal space can be replaced by a sum, we get

$$E(\mathbf{h}) = T\, \Sigma_{\mathbf{h}'}\, E(\mathbf{h}')\, E(\mathbf{h}-\mathbf{h}') \tag{5.33}$$

(5.33) is the *Sayre Equation*, which is a key formula in the theory of "Direct Methods". The non-negative factor T does not affect its application, since it can be calculated. Although we have derived this equation from a very special model, it is valid generally. For a real structural model, only another factor T has to be introduced. From Sayre's equation, two formulae to be applied in actual phase determination will be derived, depending on whether the structure is centrosymmetric or acentric.

It must be noted at this point that the procedure in practical direct phase determination is different in the centric and acentric case. This is because the centric structure factors have phases restricted to 0 or π. Since it is, in general, less difficult to decide between two possible values than to determine a phase out of the whole range from 0 to 2π, it has become customary, although not necessary, to handle these two cases separately. If it could not be decided unambiguously by space group determination whether a center of symmetry is present or not, the distribution and the averages of E-values discussed in 5.1.2 (Table 5.1a) can be used as further criteria.

In the centrosymmetric case, the E-values have a sign of + or −. Sayre's equation can then be interpreted as follows. For reflections $\mathbf{h}$ with $|E(\mathbf{h})|$ being sufficiently large, it is likely that the sum on the right side of (5.33) will contain more terms $E(\mathbf{h}')$ $E(\mathbf{h}-\mathbf{h}')$ having the same sign as $E(\mathbf{h})$ itself, than terms of opposite sign. Otherwise equation (5.33) could not hold. This is especially true for those terms for which $|E(\mathbf{h}')|$ *and* $|E(\mathbf{h}-\mathbf{h}')|$ are large, since they are the major contributors to the sum. So there exists a more than 50% probability that for large E-values,

$$s(\mathbf{h}) = s(\mathbf{h}')\, s(\mathbf{h}-\mathbf{h}')$$

where $s(\mathbf{h})$ denotes the sign of $E(\mathbf{h})$. This equation remains valid if on the left side $\mathbf{h}$ is replaced by $-\mathbf{h}$ (since $s(\mathbf{h}) = s(-\mathbf{h})$). Setting $-\mathbf{h} = \mathbf{h}_1$, $\mathbf{h}' = \mathbf{h}_2$ and $\mathbf{h}-\mathbf{h}' = \mathbf{h}_3$, we get finally

$$s(\mathbf{h}_1) = s(\mathbf{h}_2)\, s(\mathbf{h}_3)$$

or

$$s(\mathbf{h}_1)\, s(\mathbf{h}_2)\, s(\mathbf{h}_3) = 1 \tag{5.34}$$

if the three reflections $\mathbf{h}_1$, $\mathbf{h}_2$, $\mathbf{h}_3$ satisfy the equation

$$\mathbf{h}_1 + \mathbf{h}_2 + \mathbf{h}_3 = 0 \tag{5.35}$$

Reflection triplets for which (5.35) holds are said to be related by a Σ_2-relation. As we shall see, these Σ_2-relations play an important role with all applications of direct methods.

From the derivation of (5.34), it is clear that this equation cannot hold exactly, so that instead of writing "=" it would be better to write "≈". Fortunately, the probability that (5.34) is valid for a given structure can be calculated, as was first done by Cochran and Woolfson [Cochran, W. and Woolfson, M. M., Acta Cryst. *8*, 1 (1955)]. Setting

$$n_j = \frac{f_j}{\sum_{j=1}^{N} f_j} \tag{5.36}$$

with the f_j being the atom form factors of the N atoms of the unit cell, and

$$\sigma_k = \sum_{j=1}^{N} n_j^k \tag{5.37}$$

the probability, p, that (5.34) holds can be expressed by

$$p = 1/2 + (1/2)\tanh\left[\frac{\sigma_3}{\sigma_2^{3/2}}\,|E(\mathbf{h}_1)\,E(\mathbf{h}_2)\,E(\mathbf{h}_3)|\right] \tag{5.38}$$

For an easier interpretation, let us transform (5.38) to the case of N atoms of equal type. Then

$$n_j = n = \frac{f}{Nf} = \frac{1}{N}$$

$$\sigma_k = N\,\frac{1}{N^k} = \frac{1}{N^{k-1}}$$

$$\frac{\sigma_3}{\sigma_2^{3/2}} = \frac{1/N^2}{1/N\sqrt{N}} = \frac{1}{\sqrt{N}}$$

and we get

$$p = 1/2 + (1/2)\tanh\left[\frac{1}{\sqrt{N}}\,|E(\mathbf{h}_1)\,E(\mathbf{h}_2)\,E(\mathbf{h}_3)|\right] \tag{5.39}$$

Now we can interpret the sign relationship (5.34) together with (5.38) or (5.39) as follows. If we have three reflections connected by a Σ_2-relation (5.35), and if the signs of two of them are known, then the sign of the third reflection can be deduced from (5.34) with a probability given by (5.38). This probability increases with the magnitudes of contributing E's, but decreases with the number of atoms in the unit cell, thus with the size of the structure, as shown by (5.39). Several questions arise with this interpretation of (5.34). The first is, where to obtain the known signs so that the sign relationship can be applied? This question will be discussed in 5.3.2. As we shall see, it will be possible to obtain the required starting set of known signs. Another question is, are the probabilities calculated from (5.38) large enough so that sign determinations from (5.34) are valid? To get an impression of the magnitude of p, we have calculated p from (5.39) [which does not differ too much from (5.38)] for various values of N and for different magnitudes of E, assuming $|E(\mathbf{h}_1)| = |E(\mathbf{h}_2)| = |E(\mathbf{h}_3)| = E$. The results, given in Table 5.2, show that for a medium-sized structure with N = 120 atoms in the unit cell, p is greater than 95% only if all three contributing reflections have E-values greater than 2.5.

Table 5.2. Probabilities for the Sign Relationship Calculated for Various Parameters E and N.

E	N = 40	80	120	160	200
3.0	1.00	1.00	0.99	0.99	0.98
2.5	0.99	0.97	0.95	0.92	0.90
2.0	0.93	0.86	0.81	0.78	0.76
1.5	0.74	0.68	0.65	0.63	0.62
1.0	0.58	0.55	0.54	0.54	0.53

As follows from Table 5.1, only about 1% of all reflections have E-values of that magnitude in centrosymmetric structures. If we have, for instance, a total of 3000 reflections, only 30 will have E-values above 2.5, and out of these, only a small number of Σ_2-relationships will result. Although relations of that high probability are infrequently obtained, it is dangerous to accept lower probabilities. Note that already a probability of 90% means that from 10 Σ_2-relations, one sign determination is wrong. For a structure with hundreds of reflections, hundreds of Σ_2-relations must be applied, therefore it is clear that a significant number of sign determinations will be wrong. Furthermore, every sign determined from a Σ_2-relationship may be used as input in a further relation, so that a wrong sign determined in the first stages of phase determination will be propagated throughout the whole phase determination process.

If, in the latter stages of sign determination, a large number of signs is known, it can happen that an unknown reflection is contained in more than one Σ_2-relation. Then, of course,

$$s(\mathbf{h}) = \Sigma_{h'}\, s(\mathbf{h}')\, s(\mathbf{h} - \mathbf{h}') \tag{5.40}$$

is a better approximation of Sayre's equation, and in this case the probability for the sign of **h** to be "+" is given by

$$p_+(\mathbf{h}) = 1/2 + (1/2)\tanh\left[\frac{\sigma_3}{\sigma_2^{3/2}}\,|E(\mathbf{h})|\,\Sigma_{h'}\,E(\mathbf{h}')\,E(\mathbf{h} - \mathbf{h}')\right] \tag{5.41}$$

while the probability $p_-(\mathbf{h})$ of a sign to be "−" is given by

$$p_-(\mathbf{h}) = 1 - p_+(\mathbf{h}) \tag{5.42}$$

For phase determination in the acentric case, a further formula can be derived from Sayre's equation. Separating (5.33) into its real and imaginary part, we get

$$|E(\mathbf{h})| \sin\varphi(\mathbf{h}) = T\,\Sigma_{h'} |E(\mathbf{h}')\,E(\mathbf{h}-\mathbf{h}')| \sin[\varphi(\mathbf{h}') + \varphi(\mathbf{h}-\mathbf{h}')]$$

and

$$|E(\mathbf{h})| \cos\varphi(\mathbf{h}) = T\,\Sigma_{h'} |E(\mathbf{h}')\,E(\mathbf{h}-\mathbf{h}')| \cos[\varphi(\mathbf{h}') + \varphi(\mathbf{h}-\mathbf{h}')]$$

By division, we get

$$\tan\varphi(\mathbf{h}) = \frac{\Sigma_{h'} |E(\mathbf{h}')\,E(\mathbf{h}-\mathbf{h}')|\ \sin[\varphi(\mathbf{h}') + \varphi(\mathbf{h}-\mathbf{h}')]}{\Sigma_{h'} |E(\mathbf{h}')\,E(\mathbf{h}-\mathbf{h}')| \cos[\varphi(\mathbf{h}') + \varphi(\mathbf{h}-\mathbf{h}')]} \tag{5.43}$$

This is the well-known "tangent formula" derived by Karle and Hauptman in 1956 [Hauptman, H. and Karle, J., Acta Cryst. *9*, 635 (1956)]. Just as (5.34) is the key formula for phase determination in the centric case, the tangent formula is the key formula for phase determination in the acentric case.

The process for phase determination by application of (5.43) is in principle the same as in the centric case. Σ_2-relations have to be found and if the phases of two reflections are known, this relation can be used as input to (5.43) to determine the third. Note that (5.43) holds exactly only if the summation is taken over *all* $\mathbf{h}'$. In practical application, however, only a few terms can be used. In the first stages of phasing it may happen that only one term is available. Then the result of (5.43) is only a first approximation of $\varphi(\mathbf{h})$, but after having expanded the phase determination to a larger set, a re-phasing process may be started using more terms in the tangent sums. By an iterative expansion and re-phasing process, convergence to a set of sufficient correct phases can be obtained in many cases. Again it is true that the major contributors to the sums in (5.43) are those terms having large $|E|$-values, so that a start with relations which include reflections of large $|E|$-values is made, as for the centric case.

Summarizing all the theoretical results of direct methods, we can state that a practical application is possible if; first, the structure is not too large, second, a set of known phases can be obtained, and third, this set can be used in a sufficient number of Σ_2-relations between reflections of large E-values for the determination of additional phases. From all the aspects discussed above, it can be noted that phase determination by direct methods is not a guaranteed success. However, very powerful programs, such as MULTAN and MAGLAN, using subtle procedures, have been developed and are distributed worldwide. Experience in the last few years has shown that, in spite of the fact that all the formulae are approximations, direct methods are by far the most powerful general method of phase determination presently available. Provided that a set of reasonably accurate intensity data is available, centric structures will be solved nearly without exception, if the number of atoms in the asymmetric unit is not larger than 100 (exluding hydrogens). For acentric structures of up to 100 atoms, more failures may occur, but in principle, the same is true as for the centric case.

5.3.2 *Origin Definition, Choice of Starting Set*

It was shown in the last section that the application of direct methods requires a set of reflections with known phases. Fortunately, a set of up to three known phases can be obtained by the definition of the unit-cell origin. As was pointed out in the discussion of crystal symmetry, the origin of a unit cell is chosen to be in a special position relative to the symmetry elements. For instance ,in a centrosymmetric cell, the origin must always coincide with the inversion center. In every centric cell, inversion centers are not only present at (0, 0, 0) but also in seven other positions, so that a total of eight centers of symmetry can be found. They are located at (see Fig. 5.8)

(0, 0, 0), (1/2, 0, 0), (0, 1/2, 0), (0, 0, 1/2)
(1/2, 1/2, 0), (1/2, 0, 1/2), (0, 1/2, 1/2), (1/2, 1/2, 1/2)

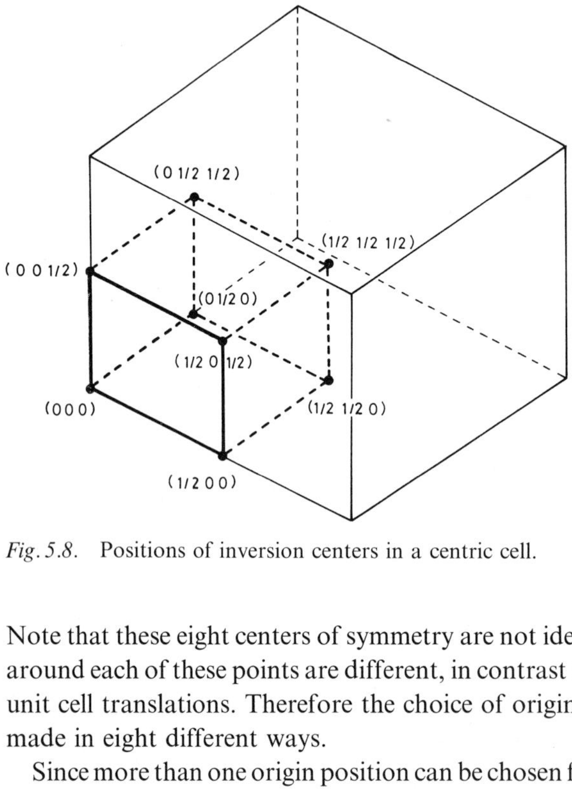

Fig. 5.8. Positions of inversion centers in a centric cell.

Note that these eight centers of symmetry are not identical, since the structural motifs around each of these points are different, in contrast to centers of symmetry related by unit cell translations. Therefore the choice of origin at a center of symmetry can be made in eight different ways.

Since more than one origin position can be chosen for non-centric structures also, we must discuss the general influence of origin definition on the structure factor expression. If 0 and 0′ are two origins with $\Delta\mathbf{r}$ being the origin shift vector (see Fig. 5.9), every point P can be expressed by a vector $\mathbf{r}$ relative to 0 and a vector $\mathbf{r}'$ relative to 0′ with

$$\mathbf{r} = \Delta\mathbf{r} + \mathbf{r}'$$

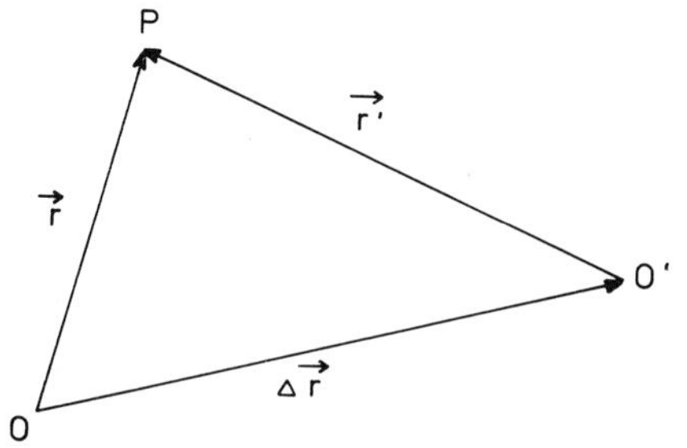

Fig. 5.9. Relationship between the position vectors **r** and **r**′ and the origin shift vector $\Delta\mathbf{r}$.

The structure factor expressions are

$$F(\mathbf{h}) = |F|\, e^{i}\varphi = \sum_{j=1}^{N} f_j\, e^{2\pi i \mathbf{h}\mathbf{r}_j}$$

and

$$F'(\mathbf{h}) = |F'|\, e^{i\varphi'} = \sum_{j=1}^{N} f_j\, e^{2\pi i \mathbf{h}\mathbf{r}'_j}$$

$$= \sum_{j=1}^{N} f_j\, e^{2\pi i \mathbf{h}(\mathbf{r}_j - \Delta\mathbf{r})} = e^{-i(2\pi \mathbf{h}\Delta\mathbf{r})}\, F(\mathbf{h})$$

$$= |F|\, e^{i(\varphi - \Delta\varphi)}$$

with

$$\Delta\varphi = 2\pi \mathbf{h}\Delta\mathbf{r}. \tag{5.44}$$

It follows that F and F′ have equal amplitudes, but the choice of different origin causes a phase shift $\Delta\varphi$, given by (5.44). This result is important and requires interpretation. An immediate consequence is that no phase determination is possible unless the origin is specified; otherwise the phases are ambiguous. In this connection, we have to ask why tangent formula (5.43) or the sign-relationship (5.34) can be valid since they propose phase values with no provisions for defining an origin. In fact, this is only possible because the reflections involved in these two formulas satisfy Σ_2-relations and, moreover, it can be shown that the phase sum $\varphi_1 + \varphi_2 + \varphi_3$ of three reflections involved in a Σ_2-relation is a so-called "structure invariant".

Phases, or combinations of phases, are said to be "structure invariants" if they do not change with origin transformations. A second notation used in this connection is that of a "structure semi-invariant". If a phase or a combination of phases is invariant with respect to origin shifts for a given structure, then we have a "structure semi-invariant". Every structure invariant is a semi-invariant, but the inverse generally does not hold. These related notations can be explained by two examples. In the space group P4/m, the reflection (111) is invariant if, as is conventional, the origin is placed only on those inversion centers which are on the 4-fold axis. In other space groups, e.g. in $P\bar{1}$, this reflection depends on the choice of origin, hence it is a semi-invariant for a structure in P4/m, if the origin is chosen only on the 4-fold axis, but it is not a structure invariant.

On the other hand, the phase sum $\varphi(\mathbf{h}_1) + \varphi(\mathbf{h}_2) + \varphi(\mathbf{h}_3)$ is invariant if the three reflections satisfy $\mathbf{h}_1 + \mathbf{h}_2 + \mathbf{h}_3 = \mathbf{0}$, and this is true for all space groups. This can easily be seen from the phase shift calculated above. If F(**h**) and F′(**h**) are related to 0 and 0′, we get $(\varphi_i = \varphi(\mathbf{h}_i))$,

$$F(\mathbf{h}_1)\,F(\mathbf{h}_2)\,F(\mathbf{h}_3) = |F(\mathbf{h}_1)|\,|F(\mathbf{h}_2)|\,|F(\mathbf{h}_3)|\,e^{i(\varphi_1+\varphi_2+\varphi_3)}$$

and

$$F'(\mathbf{h}_1)\,F'(\mathbf{h}_2)\,F'(\mathbf{h}_3) = |F'(\mathbf{h}_1)|\,|F'(\mathbf{h}_2)|\,|F'(\mathbf{h}_3)|\,e^{i(\varphi'_1+\varphi'_2+\varphi'_3)}$$

$$= |F(\mathbf{h}_1)|\,|F(\mathbf{h}_2)|\,|F(\mathbf{h}_3)|\,e^{i[\varphi_1+\varphi_2+\varphi_3-2\pi\Delta\mathbf{r}(\mathbf{h}_1+\mathbf{h}_2+\mathbf{h}_3)]}$$

Since $\mathbf{h}_1 + \mathbf{h}_2 + \mathbf{h}_3 = \mathbf{0}$, we obtain

$$\varphi_{1'} + \varphi_{2'} + \varphi_{3'} = \varphi_1 + \varphi_2 + \varphi_3$$

hence $\varphi_1 + \varphi_2 + \varphi_3$ is a structure invariant. This property is the reason why phase determination is possible from the sign relationship (5.34) and the tangent formula, since these two formulae define the phases of the structure invariants which is independent of the choice of origin.

To understand how origin definition can be used to get known phases, let us study the problem for the space group $P\bar{1}$. This space group has the eight inversion centers as shown in Fig. 5.8, each of which may be chosen as the origin. Since only π or 2π are possible phase shifts (we have a centric cell) it is of interest only whether the scalar product $\mathbf{h}\Delta\mathbf{r}$ is an odd or an even multiple of 1/2. So for phase-shift considerations, the actual values of reflection indices are of no consequence. It is only necessary to know whether they are odd or even. Therefore, all reflections can be put into eight categories, denoted by eee, oee, eoe, eeo, ooe, oeo, eoo, ooo (e = even, o = odd). These categories are named *parity groups*. Every pair of reflections belonging to the same category have the same properties with respect to a phase shift. Now we can calculate the phase shift for each category if the origin is transferred from (0, 0, 0) to one of the other possible positions. The results are given in Table 5.3.

A phase shift 0 means that the sign of a reflection is not changed, a shift π indicates a change to opposite sign. The reflection 111, for instance, belonging to the parity group ooo will change its sign if the origin is shifted by 1/2 in each axial direction or if shifted by (1/2, 1/2, 1/2). For the other origin shifts, its sign will remain unchanged. On the other hand, a reflection of parity group eee is completely determined by the structure and does *not* depend on any origin shift, as can be seen from the first column which shows that all shifts are zero. All eee reflections are therefore structure semi-invariants for this space group, but, as shown in Table 5.3, this is the only parity group having this property.

The results listed in Table 5.3 can be used for origin definition as follows. For a given structure, the reflections have a fixed sign distribution with respect to the structure itself and a special origin. Although no sign is known, we can fix the sign of a non-semi-

Table 5.3. Phase Shifts in P$\bar{1}$.

		Phase Shift for Reflection Category							
No.	Origin shift	eee	oee	eoe	eeo	ooe	oeo	eoo	ooo
1	0	0	0	0	0	0	0	0	0
2	1/2, 0, 0	0	π	0	0	π	π	0	π
3	0, 1/2, 0	0	0	π	0	π	0	π	π
4	0, 0, 1/2	0	0	0	π	0	π	π	π
5	1/2, 1/2, 0	0	π	π	0	0	π	π	0
6	1/2, 0, 1/2	0	π	0	π	π	0	π	0
7	0, 1/2, 1/2	0	0	π	π	π	π	0	0
8	1/2, 1/2, 1/2	0	π	π	π	0	0	0	π

invariant reflection (e.g. of a reflection not belonging to the eee parity group) to a value of our choice (see Fig. 5.10). This can be, for example, a reflection of the parity group ooo for which we define the sign value, a. The sign will remain a if we transform the structure by one of the origin shifts, 1, 5, 6, or 7 (Table 5.3). Let b be the sign of a second reflection of, say, eoo parity group relative to the origin 1. Since we do not know b, the sign of this second reflection can be chosen either as +b or −b. If we choose +b we can allow an origin shift 1 or 7 (see Fig. 5.10a and b). If −b was chosen instead of b, an origin shift 5 or 6 has to be applied. Now let us proceed with a third reflection and

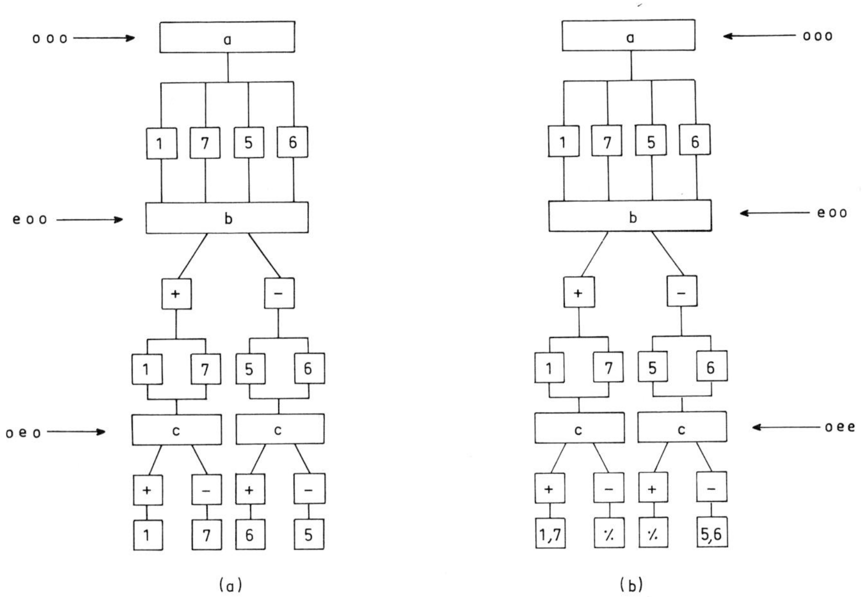

Fig. 5.10. Reflection categories and origin shifts:
(a) allowed combination; (b) forbidden combination.

assume it to be taken from the parity group oeo (see Fig. 5.10a). For the initial phase c, we can choose +c or −c. If we are in the path of origin shifts 1 or 7 (+b was chosen), +c would make no change of origin necessary, which means we stay at origin 1. If we choose −c, we must change to origin 7. In the path of origin shifts 5 or 6 (−b was chosen), a choice of +c requires the origin 6, whereas if −c was chosen, origin 5 must be used.

Therefore for every sign combination of the second and third reflection relative to the first, there is only one origin which is consistent with a given sign combination. However, the sign of a fourth reflection can no longer be chosen arbitrarily. Since it can have two possible sign values and the origin shift is already uniquely fixed by three reflections, it cannot be ensured that the origin shift necessary for the fourth reflection is identical to that previously fixed. To demonstrate this with the example discussed above, let us suppose that no. 6 was the origin shift obtained for three reflections. Suppose a fourth reflection of parity group ooe had the initial sign d. Since we are obliged to apply origin shift 6, d becomes −d and is therefore fixed. However, since we do not know this sign, we cannot select the correct one and thus a choice is impossible. The same holds for a reflection of any other parity group, and for those having a phase shift 0 for origin shift 6.

Before discussing the final interpretations of the example of Fig. 5.10a, let us deal with the example of Fig. 5.10b. It differs from that of Fig. 5.10a only by the choice of the third reflection, which is taken out of the parity group oee instead of oeo. If the third reflection has the original sign c and if we are in the path of origin shift 1 or 7, (+b was selected), +c would permit these origin shifts but −c would not (see Table 5.3), since neither 1 nor 7 could change a sign. In path 5 or 6, we have a similar situation. In this case, neither 5 nor 6 are consistent with +c, so that for both choices of c it is impossible to have all sign combinations by means of origin shifts.

It follows that each choice of sign of three reflections of the parity groups ooo, eoo and oeo can be realized by a uniquely determined origin shift, so that this definition of signs can be interpreted as an implicit definition of one of the possible origins. In the case of the parity groups ooo, eoo and oee, the origin is either determined ambiguously (+c in path 1, 7 or −c in path 5, 6) or its definition is impossible (−c in path 1, 7 or +c in path 5, 6). Therefore, this combination of parity groups cannot be used for fixing the origin.

The question whether a combination of parity groups can be used for selecting an origin can, in principle, be decided as shown in the examples. However, this procedure can be expressed more simply by a method which has the advantage that it is valid also for other space groups. For each parity group, a vector **p** is introduced with components which have values 0 or 1, with 0 representing an even parity, 1 an odd parity. The eight vectors are then

$$\mathbf{p}_1 = \begin{pmatrix} 0 \\ 0 \\ 0 \end{pmatrix} \text{ for eee;} \qquad \mathbf{p}_2 = \begin{pmatrix} 1 \\ 0 \\ 0 \end{pmatrix} \text{ for oee;} \qquad \mathbf{p}_3 = \begin{pmatrix} 0 \\ 1 \\ 0 \end{pmatrix} \text{ for eoe;}$$

$$\mathbf{p}_4 = \begin{pmatrix} 0 \\ 0 \\ 1 \end{pmatrix} \text{ for eeo;} \qquad \mathbf{p}_5 = \begin{pmatrix} 1 \\ 1 \\ 0 \end{pmatrix} \text{ for ooe;} \qquad \mathbf{p}_6 = \begin{pmatrix} 1 \\ 0 \\ 1 \end{pmatrix} \text{ for oeo;}$$

$$\mathbf{p}_7 = \begin{pmatrix} 0 \\ 1 \\ 1 \end{pmatrix} \text{ for eoo;} \qquad \mathbf{p}_8 = \begin{pmatrix} 1 \\ 1 \\ 1 \end{pmatrix} \text{ for ooo.}$$

Although the same vectors are obtained, another explanation of the $\mathbf{p}_i$ can be given. Calculate for every vector $\mathbf{h} = (hkl)$, a vector $\mathbf{h}(m)$ which is obtained by reduction of the hkl's modulus 2

$$\mathbf{h}(m) = \mathbf{h} \bmod \mathbf{m} \tag{5.45}$$

with

$$\mathbf{m} = \begin{pmatrix} 2 \\ 2 \\ 2 \end{pmatrix}$$

(The reduction of a modulus b is defined in the following way: Divide a by b in the sense of an integer division. Then the residue is the desired quantity, e.g. $12 \bmod 5 = 2$, $4 \bmod 2 = 0$, $5\pi \bmod \pi = \pi$. It can be expressed analytically by $a \bmod b = a - [a/b]\, b$, where $[./.]$ indicates integer division. The technique of reduction modulus 2 is frequently used to describe the properties of odd or even integers. If n is odd, $n \bmod 2 = 1$; otherwise $n \bmod 2 = 0$.)

One of the eight vectors $\mathbf{p}_i$ can be uniquely assigned to each reflection $\mathbf{h} = (hkl)$. Either the parity group of $\mathbf{h}$ has to be determined or $\mathbf{h}(m)$ has to be calculated from (5.45). Doing this calculation for a few examples:

(a) $\mathbf{h} = \begin{pmatrix} 4 \\ 2 \\ 0 \end{pmatrix}$, parity group eee,

hence $\mathbf{p} = \mathbf{p}_1 = \begin{pmatrix} 0 \\ 0 \\ 0 \end{pmatrix}$

Calculation of $\mathbf{h}(m)$ leads to the same result:

$$h(m) = \begin{pmatrix} 4 \bmod 2 \\ 2 \bmod 2 \\ 0 \bmod 2 \end{pmatrix} = \begin{pmatrix} 0 \\ 0 \\ 0 \end{pmatrix}$$

(b) $\mathbf{h} = \begin{pmatrix} 5 \\ -3 \\ 6 \end{pmatrix}$, parity group ooe, $\quad \mathbf{p} = \mathbf{p}_5 = \begin{pmatrix} 1 \\ 1 \\ 0 \end{pmatrix}$,

$$\mathbf{h}(m) = \begin{pmatrix} 5 \bmod 2 \\ -3 \bmod 2 \\ 6 \bmod 2 \end{pmatrix} = \begin{pmatrix} 1 \\ 1 \\ 0 \end{pmatrix}$$

(c) $\mathbf{h} = \begin{pmatrix} -1 \\ -1 \\ 3 \end{pmatrix}$, parity group ooo, $\quad \mathbf{p} = \mathbf{p}_8 = \begin{pmatrix} 1 \\ 1 \\ 1 \end{pmatrix}$,

$$\mathbf{h}(\mathrm{m}) = \begin{pmatrix} -1 \bmod 2 \\ -1 \bmod 2 \\ 3 \bmod 2 \end{pmatrix} = \begin{pmatrix} 1 \\ 1 \\ 1 \end{pmatrix}$$

Hence it can be decided for the three reflections, whether or not they are suitable for the choice of origin by the following rule:

The reflections $\mathbf{h}_1$, $\mathbf{h}_2$, $\mathbf{h}_3$ can be used for origin definition if and only if their parity vectors are linearly independent; or if the determinant consisting of the vectors $\mathbf{h}_1(\mathrm{m})$, $\mathbf{h}_2(\mathrm{m})$, $\mathbf{h}_3(\mathrm{m})$ is equal to ± 1:

$$D[\mathbf{h}(\mathrm{m})] = \begin{vmatrix} h_1(\mathrm{m}) & k_1(\mathrm{m}) & l_1(\mathrm{m}) \\ h_2(\mathrm{m}) & k_2(\mathrm{m}) & l_2(\mathrm{m}) \\ h_3(\mathrm{m}) & k_3(\mathrm{m}) & l_3(\mathrm{m}) \end{vmatrix} = \pm 1 \tag{5.46}$$

The examination whether a reflection triplet satisfies this condition can be done geometrically or analytically. If the geometrical way is preferred, the eight parity vectors should be arranged at the edges of a cube as shown in Fig. 5.11a. Then it must be checked whether the three parity vectors belonging to a reflection triplet are coplanar or not. Only if they are not all situated in one plane are they suitable for origin fixing. Fig. 5.11b shows an example of an impossible combination of parity groups, while in Fig. 5.11c, the parity vectors are non-coplanar, hence suitable for choice of origin.

For the analytical verification of (5.46) the reflection indices must be taken modulus 2. Let us suppose for example that we have three reflections of the parity groups used in Fig. 5.11b, say

$$\mathbf{h}_1 = (2\ \ -3\ \ 4), \qquad \mathbf{h}_2 = (1\ \ 0\ \ 3), \qquad \mathbf{h}_3 = (-7\ \ 5\ \ 11)$$

then we get

$$D[\mathbf{h}(\mathrm{m})] = \begin{pmatrix} 0 & 1 & 0 \\ 1 & 0 & 1 \\ 1 & 1 & 1 \end{pmatrix} = 0 \neq \pm 1$$

Having, on the other hand, three reflections of the parity groups as shown in Fig. 5.11c, say

$$\mathbf{h}_1 = (2\ \ 2\ \ 5), \qquad \mathbf{h}_2 = (2\ \ 7\ \ -4), \qquad \mathbf{h}_3 = (1\ \ 1\ \ -6)$$

we obtain

$$D[\mathbf{h}(\mathrm{m})] = \begin{pmatrix} 0 & 0 & 1 \\ 0 & 1 & 0 \\ 1 & 1 & 0 \end{pmatrix} = -1$$

As with the geometrical construction, the second reflection triplet can be taken for the choice of origin, but not the first.

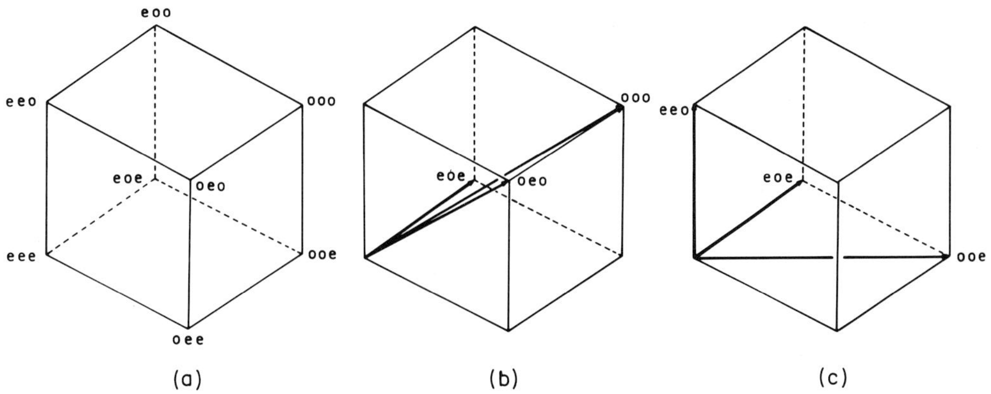

Fig. 5.11. (a) Geometrical representation of the parity vectors in a cube;
(b) an invalid combination of parity groups;
(c) an acceptable combination for origin-fixing.

Since the only property of $P\bar{1}$ used in this discussion was the choice of the origin at the eight points shown in Fig. 5.8, it follows that this rule holds for every space group which permits these eight points to be defined as origins. This condition holds for a large number of space groups, that is, all primitive centric space groups in the triclinic, monoclinic and orthorhombic system, and all primitive acentric space groups of the crystal class 222. Since many inorganic and most of the organic structures crystallize in one of these space groups, the rules derived above can be used for many structure determinations.

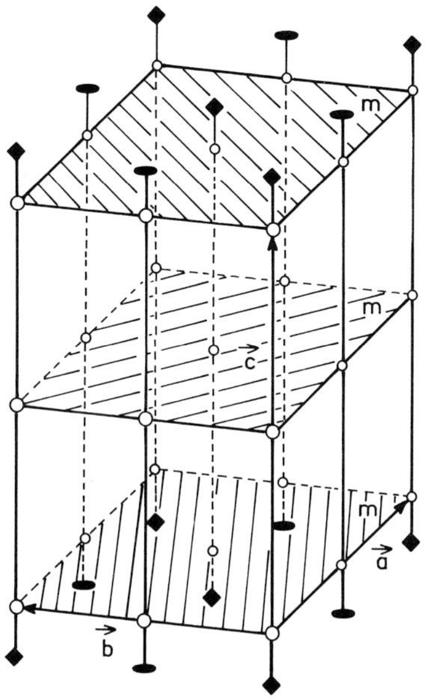

Fig. 5.12. Distribution of the symmetry elements in space group P4/m.

For the space groups not included in the set described above, a generalization of the origin fixing rule has to be given. Since all the fundamental aspects have already been discussed, it is sufficient to proceed with other space groups by indicating the major properties to be observed when generalizing the origin-fixing procedure. Let us, for instance, consider the symmetry elements in the space group P4/m, which are shown in Fig. 5.12. Although the inversion centers are placed at the same points as in $P\bar{1}$ the situation is, nevertheless, different. Some of the inversion centers are associated with 2-fold axes, others with 4-fold axes. Since the origin must be situated on the symmetry element with the highest symmetry, we can choose the origin on the 4-fold axes only. These positions are

$$(0,0,0), \quad (1/2,1/2,0), \quad (0,0,1/2), \quad (1/2,1/2,1/2)$$

If we calculate the phase shifts for all reflection types relative to these origins (by use of 5.44), we find that the phase shift depends only on whether for a given $\mathbf{h} = (hkl)$, $h + k$ is even or odd, or on l being even or odd. So for this space group, we can subdivide all reflections into four categories:

(1) ee for $h + k$ even, l even
(2) eo for $h + k$ even, l odd
(3) oe for $h + k$ odd, l even
(4) oo for $h + k$ odd, l odd

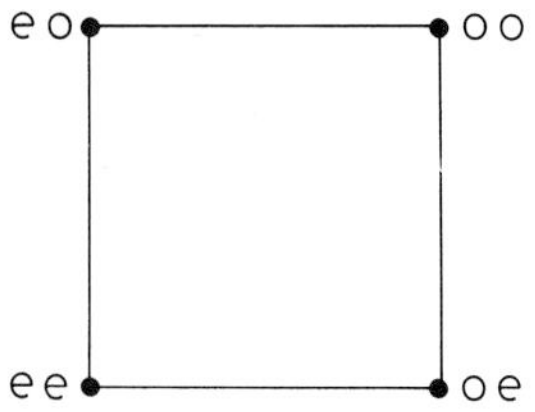

Fig. 5.13. Two-dimensional parity vectors represented in a square.

Examples: $\mathbf{h} = (1\ 1\ 4)$ belongs to the ee category, $\mathbf{h} = (0\ 3\ 2)$ to the oe category, and $\mathbf{h} = (1\ 4\ 3)$ to the oo category. Using the same arguments as for $P\bar{1}$, we find the analogous rule for the choice of the origin-defining reflections. If expressed graphically by a square (Fig. 5.13), its four edges represent four two-dimensional vectors (0, 0), (1, 0), (0, 1), and (1, 1), each being associated to one category. As in $P\bar{1}$, linearly independent vectors indicate a valid choice of origin-defining reflections. Since in two dimensions only two vectors can be linearly independent, the rule for origin-fixing in a space group of that type is: Two reflections can be chosen for origin-fixing, if their category vectors (Fig. 5.13) are linearly independent. The same can be expressed analytically: Calculate for every reflection (hkl) a *two-dimensional* vector $\mathbf{h(m)}$ by

$$\mathbf{h}(m) = \begin{pmatrix} h+k \\ l \end{pmatrix} \bmod \mathbf{m} \tag{5.47}$$

with $\mathbf{m} = \begin{pmatrix} 2 \\ 2 \end{pmatrix}$

Then two reflections $\mathbf{h}_1$, $\mathbf{h}_2$ can be used for origin definition, if and only if the determinant consisting of the vectors $\mathbf{h}_1(m)$ and $\mathbf{h}_2(m)$ is equal to ± 1:

$$D[\mathbf{h}(m)] = \begin{vmatrix} (h_1+k_1)(m) & l_1(m) \\ (h_2+k_2)(m) & l_2(m) \end{vmatrix} = \pm 1 \tag{5.48}$$

Example: The pair $\mathbf{h}_1 = (0\ 3\ 2)$ and $\mathbf{h}_2 = (1\ 4\ 3)$ is a valid choice, since $\mathbf{h}_1(m) = (1\ 0)$, $\mathbf{h}_2(m) = (1\ 1)$ and

$$D[\mathbf{h}(m)] = \begin{vmatrix} 1 & 0 \\ 1 & 1 \end{vmatrix} = 1$$

An invalid choice would be given by $\mathbf{h}_1 = (1\ 1\ 4)$ and $\mathbf{h}_2 = (0\ 3\ 2)$, since we get $\mathbf{h}_1(m) = (0\ 0)$, $\mathbf{h}_2(m) = (1\ 0)$ and

$$D[\mathbf{h}(m)] = \begin{vmatrix} 0 & 0 \\ 1 & 0 \end{vmatrix} = 0$$

The main difference from the space group $P\bar{1}$ is given by the conclusion that only two instead of three reflections can be used for origin definition. This is a direct consequence of reducing the possible origin positions and it therefore seems reasonable that for higher symmetric space groups, the number of selected reflections may be reduced to one or even zero.

Now we will find a general procedure for obtaining the $D[\mathbf{h}(m)]$ determinant for the space groups discussed above ($P\bar{1}$ and $P4/m$). For that purpose, it is only necessary to introduce a matrix M and a vector $\mathbf{m}$ which allows the calculation of $\mathbf{h}(m)$ by

$$\mathbf{h}(m) = M\mathbf{h} \bmod \mathbf{m} \tag{5.49}$$

For $P\bar{1}$, we get equation (5.45), with

$$M = \begin{pmatrix} 1 & 0 & 0 \\ 0 & 1 & 0 \\ 0 & 0 & 1 \end{pmatrix} \quad \text{and} \quad \mathbf{m} = \begin{pmatrix} 2 \\ 2 \\ 2 \end{pmatrix}$$

For $P4/m$, we get equation (5.47), with

$$M = \begin{pmatrix} 1 & 1 & 0 \\ 0 & 0 & 1 \\ 0 & 0 & 0 \end{pmatrix} \quad \text{and} \quad \mathbf{m} = \begin{pmatrix} 2 \\ 2 \\ 0 \end{pmatrix}$$

(Agreement should be made to neglect zero components of $\mathbf{h}(m)$.)

The matrix M is the matrix of semi-invariants; the vector **m** is called the semi-invariant modulus. These two quantities can be derived for every space group. By generalization of what we have found for $P\bar{1}$ and for P4/m, we can express the following rule for origin-definition, which is valid for *all* space groups.

The number n of reflections used for origin-fixing is given by the number of non-zero components of the vector $\mathbf{h}(m)$, which can be obtained from the matrix M of semi-invariants and the semi-invariant modulus **m** after (5.49). A set of n reflections $\mathbf{h}_1, \ldots \mathbf{h}_n$ is a valid choice of origin-defining reflections, if the determinant of vectors $\mathbf{h}_1(m), \ldots \mathbf{h}_n(m)$ is equal to ± 1,

$$D[\mathbf{h}(m)] = \pm 1 \tag{5.50}$$

The quantities needed in (5.49) were calculated by Karle & Hauptman and have been published in four papers [Hauptman, H. and Karle, J., (1953), ACA Monograph No. 3, Pittsburgh: Polycrystal Book Service, PA, USA; Acta Cryst. *9*, 45 (1956); Acta Cryst. *12*, 93 (1959); Acta Cryst. *14*, 217 (1961)]. They are now summarized in Vol. IV of the International Tables (1974).

Two further remarks should be added. While the semi-invariant moduli contain only the number 2 as components for all centric space groups, this does not hold for other space groups. Since the number 2 was caused by origin positions having as their only possible values, 0 or 1/2, we can see that other integers will result for other origin shifts (c.g. a 3 for a shift of 1/3). Sometimes, however, a zero is a component of the semi-invariant modulus, as, for example, in $P2_1$, where

$$\mathbf{m} = \begin{pmatrix} 2 \\ 0 \\ 2 \end{pmatrix}$$

A zero is used if for a given direction in the unit cell the origin can be fixed arbitrarily. This is generally the case in any of the 68 space groups belonging to the 10 polar classes (see Table 3.7 in 3.3.1.) In $P2_1$, this is the y-direction, which is that of the 2_1-axis. Since the origin in that direction is not fixed, the phase shift depends on the actual value of the corresponding reflection index, so that this index cannot be reduced relative to a modulus. Since the reduction of a quantity Z modulus zero is defined by taking Z itself, a zero is the component of the semi-invariant moduli in all cases when the origin shift is not restricted to a fixed value.

Major difficulties usually arise if origin-fixing has to be applied to centered space groups. Care has to be taken that for these cases the information needed for (5.49) is listed in the International Tables referred to a corresponding primitive cell. Using the transformation matrices,

$$C \rightarrow P\colon\ S = \begin{pmatrix} 1 & 1 & 0 \\ 1 & -1 & 0 \\ 0 & 0 & -1 \end{pmatrix}$$

$$A \rightarrow P:\ S = \begin{pmatrix} -1 & 0 & 0 \\ 0 & -1 & 1 \\ 0 & 1 & 1 \end{pmatrix}$$

$$I \rightarrow P:\ S = \begin{pmatrix} 0 & 1 & 1 \\ 1 & 0 & 1 \\ 1 & 1 & 0 \end{pmatrix} \qquad (5.51)$$

$$F \rightarrow P:\ S = \begin{pmatrix} -1 & 1 & 1 \\ 1 & -1 & 1 \\ 1 & 1 & -1 \end{pmatrix}$$

however, it is possible to carry out the origin-fixing, and thus the complete phase determination in the centered space group. Note that the matrices given above are defined as transformations of coordinates $\mathbf{r}_c$ of the centered cell to coordinates $\mathbf{r}_p$ of the primitive cell, $\mathbf{r}_p = S\mathbf{r}_c$. As shown in 1.1.5, the indices $\mathbf{h}_c$ of the centered cell are then obtained from the indices $\mathbf{h}_p$ of the primitive cell by the transposition matrix S'. Since $S = S'$ for the matrices listed above, we get $\mathbf{h}_c = S\mathbf{h}_p$.

For example, in space group C2, M and $\mathbf{m}$ are

$$M = \begin{pmatrix} 1 & -1 & 0 \\ 0 & 0 & 1 \\ 0 & 0 & 0 \end{pmatrix} \qquad \mathbf{m} = \begin{pmatrix} 0 \\ 2 \end{pmatrix}$$

so that application of (5.49) leads to

$$\mathbf{h}(m) = \begin{pmatrix} h - k \\ 1 \end{pmatrix} \quad \mathrm{mod} \begin{pmatrix} 0 \\ 2 \end{pmatrix}$$

where hkl refers to the primitive cell. To satisfy $D[\mathbf{h}(m)] = \pm 1$, we could, for instance, choose two reflections $\mathbf{h}_{1p} = (h_1 k_1 o)$, $\mathbf{h}_{2p} = (h_2, h_2 - 1, e)$, with o and e being arbitrary odd and even numbers. Then we get

$$D[\mathbf{h}(m)] = \begin{vmatrix} h_1 - k_1 & 1 \\ 1 & 0 \end{vmatrix} = -1$$

The indices in the centered cell are then obtained by the transformation

$$\mathbf{h}_{1c} = \begin{pmatrix} 1 & 1 & 0 \\ 1 & -1 & 0 \\ 0 & 0 & -1 \end{pmatrix} \begin{pmatrix} h_1 \\ k_1 \\ o \end{pmatrix} = \begin{pmatrix} h_1 + k_1 \\ h_1 - k_1 \\ -o \end{pmatrix} = \begin{pmatrix} h_{1c} \\ k_{1c} \\ l_{1c} \end{pmatrix}$$

$$h_{2c} = \begin{pmatrix} 1 & 1 & 0 \\ 1 & -1 & 0 \\ 0 & 0 & -1 \end{pmatrix} \begin{pmatrix} h_2 \\ h_2 - 1 \\ e \end{pmatrix} = \begin{pmatrix} 2h_2 - 1 \\ 1 \\ -e \end{pmatrix} = \begin{pmatrix} h_{2c} \\ k_{2c} \\ l_{2c} \end{pmatrix}$$

with the following relations

$$h_1 + k_1 = h_{1c}$$
$$h_1 - k_1 = k_{1c}$$
$$2h_2 - 1 = h_{2c}$$

or

$$h_1 = (h_{1c} + k_{1c})/2$$
$$k_1 = (h_{1c} - k_{1c})/2$$
$$h_2 = (h_{2c} + 1)/2$$

Because of the C-centering, all reflections $\mathbf{h}_c$ satisfy $h_c + k_c = \text{even}$, so that for every reflection pair $\mathbf{h}_{1c}$ and $\mathbf{h}_{2c}$, an integer solution for h_1, k_1, h_2 exists, provided that h_{2c} is odd. That means, every reflection pair

$$\mathbf{h}_{1c} = \begin{pmatrix} h_{1c} \\ k_{1c} \\ o \end{pmatrix} \quad \text{and} \quad \mathbf{h}_{2c} = \begin{pmatrix} h_{2c} \\ 1 \\ e \end{pmatrix}, \; h_{2c} \text{ odd}$$

may be chosen for fixing the origin in the centered space group C2, since the transforms to the primitive cell satisfy (5.50). Then the phase determination procedure can be carried out in the centered cell.

In addition to the problem of origin-fixing, a further aspect has to be discussed in the case of acentric space groups. In this case, the chirality of a structure is unknown, since it follows from Friedel's law that two structures which differ only by a center of symmetry are not distinguishable from the reflection intensities. If we denote the two

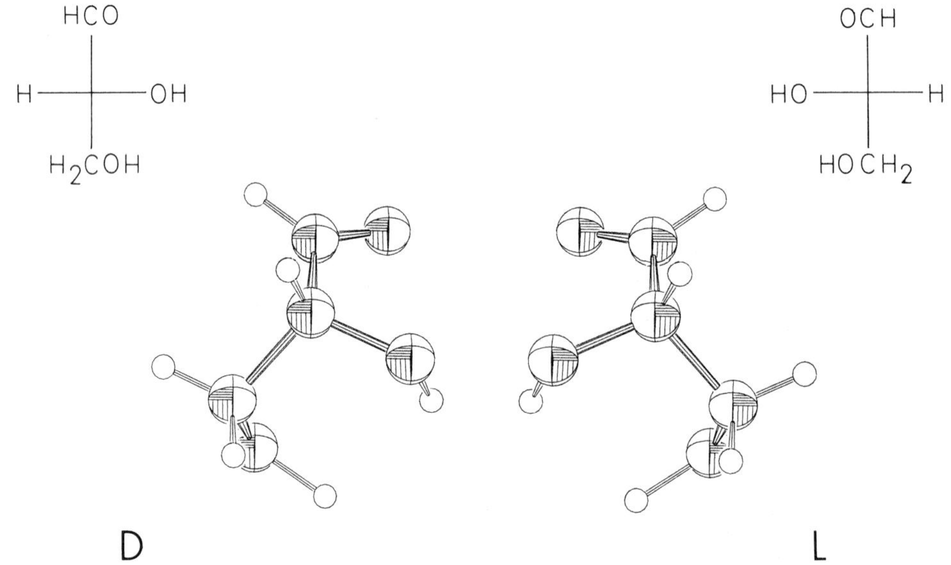

Fig. 5.14. D- and L-glycerinaldehyde; example of a pair of enantiomorphic structures.

structure types as left-handed (L) and right-handed (R) structures (for an example see Fig. 5.14), their atomic positions are related by

$$\mathbf{r}_j(R) = -\mathbf{r}_j(L) \qquad j = 1, \ldots N \tag{5.52}$$

if N is the number of contributing atoms and $\mathbf{r}_j$ the vector of the j th atom. From the definition of the structure factor if follows that

$$F_R(\mathbf{h}) = \sum_{j=1}^{N} f_j\, e^{2\pi i \mathbf{h}\mathbf{r}_j(R)} = \sum_{j=1}^{N} f_j\, e^{-2\pi i \mathbf{h}\mathbf{r}_j(L)} = F_L^*(\mathbf{h}) = F_L(-\mathbf{h})$$

Hence

$$|F_R(\mathbf{h})| = |F_L(\mathbf{h})|$$

and (5.53)

$$\varphi_R(\mathbf{h}) = -\varphi_L(\mathbf{h})$$

It follows that, if optically active compounds such as amino acids, carbohydrates, or any natural products are investigated, the X-ray analysis gives no information whether the L or the D form is present, provided that Friedel's law holds (note that under certain conditions when Friedel's law is no longer valid, (5.53) does not hold, and a decision between the D and L form can be made. We shall discuss this subject in the section on "anomalous dispersion", see 6.2.3).

It follows from (5.53) that, in spite of fixing the origin, two phase sets can exist for an acentric structure [for centric structures with $\varphi(\mathbf{h})$ being restricted to 0 or π, $\varphi(\mathbf{h}) = -\varphi(\mathbf{h})$ always holds], one corresponding to the left-handed structure and the other to the right-handed structure. At the stage of phase determination, a decision in favor of one enantiomorphic form has to be made. This can, in principle, be done by restricting the phase of a reflection which is significantly different from 0 or π to one-half of the permissible range, say 0 to π. The problem, however, for originally unknown phases, is that it is difficult to know whether a phase is clearly different from 0 or π. In certain space groups, as for instance in $P2_12_12_1$, special reflection groups exist in which phases are restricted to $\pi/2$ or $(3/2)\pi$, due to the symmetry. In this case, the enantiomorph-fixing can be done by assigning one of these restricted phase values to one reflection of that group. However, care must be taken that this assignment is not compensated for by an origin shift. For example: in the course of the structure determination of 2, 6-anhydro-β-D-fructofuranose [Dreissig, W. and Luger, P., Acta Cryst. *B29*, 1409 (1973)], space group $P2_12_12_1$, we chose the following reflections for fixing the origin: indices 9 2 0, 13 7 0, 5 0 7 (these reflections were 4th, 15th and 55th, respectively, in the sequence of descending E-values). In this space group, $\mathbf{h}(m)$ is obtained by

$$\mathbf{h}(m) = \begin{pmatrix} 1 & 0 & 0 \\ 0 & 1 & 0 \\ 0 & 0 & 1 \end{pmatrix} \mathbf{h} \bmod \begin{pmatrix} 2 \\ 2 \\ 2 \end{pmatrix}$$

Then we get for this reflection triplet,

$$D(m) = \begin{vmatrix} 1 & 0 & 0 \\ 1 & 1 & 0 \\ 1 & 0 & 1 \end{vmatrix} = 1$$

therefore the origin has been properly chosen. All three origin reflections are restricted to the special phases $\pi/2$ or $(3\pi)/2$. To fix the enantiomorphous form, we selected a further reflection which is restricted to $\pi/2$ or $(3\pi)/2$. Two reflections with relatively high E-values were available: no. 6, 0 5 5 and no. 8, 5 10 0. It has to be noted that no. 8 would be an invalid choice, while no. 6 is valid. This can clearly be seen if the phases of all reflections in question are considered for both enantiomorphous forms (Table 5.4). All phases differ by π for the two enantiomorphous forms. However, the origin-defining reflections do not depend on whether the L or the R-form has been chosen, since there exists one (and only one) origin shift $\Delta\mathbf{r}$ [in our example, $\Delta\mathbf{r} = (1/2, 0, 0)$] which allows the transformation (see 5.44) of the φ(R)-phases to their initial values. This holds also for reflection no. 8, but *not* for reflection no. 6, so that an assignment of $+$ or $-\pi/2$ to the phase of reflection no. 6 is indeed a fixing of the enantiomorphic form. Reflection no. 9 could not be chosen for that purpose, since its phase is restricted to 0 or π and is therefore not suitable for enantiomorph-fixing. For space groups having no favorable reflection groups, a non-zero phase angle may be assigned to one of the origin-fixing reflections, but it must be ensured that this reflection can have a general phase value.

Table 5.4. Reflections for Enantiomorph-fixing for 2,6-Anhydro-β-D-fructofuranose.

No.	Indices	φ(L)	φ(R)	φ after origin shift (1/2, 0, 0)	Suitable for enantiomorph-fixing
4	9 2 0	$\pi/2$	$-\pi/2$	$\pi/2$	(NP)
15	13 7 0	$\pi/2$	$-\pi/2$	$\pi/2$	(NP)
55	5 0 7	$\pi/2$	$-\pi/2$	$\pi/2$	(NP)
6	0 5 5	$\pi/2$	$-\pi/2$	$-\pi/2$	Yes
8	5 10 0	$\pi/2$	$-\pi/2$	$\pi/2$	No
9	0 2 7	π	$-\pi = \pi$	$-\pi = \pi$	Not at all

For the moment, origin and (for some space groups) enantiomorph-fixing reflections are the only ones for which phases can be assumed to be known and which can be used as input to either (5.34) or the tangent formula (5.43). Although we have learned that these starting reflections have to satisfy certain conditions (see 5.50), a large number of these starting sets can be chosen. The choice of the most favorable origin reflections is one of the most important decisions in the course of phase determination, and has great influence on whether the structure determination will be successful or not. Two criteria are important for the choice of a good starting set:

(1) The starting set should include reflections having E-values as high as possible to ensure a high probability for the signs derived from (5.34) (see 5.38 in the centric case).

In the acentric case, sums with high E-values are the major contributors to the tangent sums in (5.43) so that they will mainly determine $\varphi(\mathbf{h})$.

(2) The starting reflections should be involved in as many Σ_2-relations as possible, so that a large number of additional phases can be determined.

The problem of finding a good starting set is given by the fact that *both* (1) and (2) must be satisfied.

For centric structures, a successful sign determination may be obtained using only the origin-fixing reflections as a starting set. However, the number of reflections determined from these starting reflections is generally too small to be sufficient for structure determination, especially for acentric structures. Therefore, it is necessary to include additional reflections in the starting set. These additional reflections are known as "variables". In contrast to the origin- and enantiomorph-fixing reflections, no fixed phase values can be assigned. For centric structures, once the + or − sign has been assigned to every variable reflection, a complete sign determination will be calculated for each choice of sign. Since one of the two signs must be correct, one of the two sign determinations should lead to a correct solution. If n variables are used, 2^n trials have to be calculated, so that the number of trials increases exponentially with the number of variable reflections.

The problem becomes more difficult in acentric phase determinations. Since the phases are unrestricted and can vary over 0 to 2π, it would, in principle, be necessary to use a large number of starting phase values taken from the whole range of data. Fortunately, it has been shown from experience that it is generally sufficient to use only four starting values for a general reflection: these are $\pi/4$; $(3\pi)/4$; $(5\pi)/4$; $(7\pi)/4$. However, for every general reflection used as a variable in the starting set, four trials have to be calculated. If m general reflections are necessary, the number of trials will be 4^m, which means that for acentric structures, great care has to be taken to keep the number of variable reflections as small as possible.

As already mentioned, in certain acentric space groups, the phases of special reflections are restricted to fixed values due to the space group symmetry. Reflections of that type are denoted as "special reflections" with respect to the phase determination. For example, in the space group P222, the structure factor expression can be reduced to

$$F(hkl) = A + iB$$

with

$$A = 4\cos 2\pi hx \cos 2\pi ky \cos 2\pi lz$$
$$B = -4 \sin 2\pi hx \sin 2\pi ky \sin 2\pi lz$$

It follows that

$$B = 0 \text{ if } h = 0 \text{ or } k = 0 \text{ or } l = 0$$

That means that for all reflections of type hk0, h0l, 0kl, F(hkl) is no longer a complex number, hence its phase is restricted to 0 or π, as for centric structures. It is therefore desirable to include special reflections in the starting set since they reduce the number of trials significantly.

Nevertheless, the question whether a reflection is suitable for use as a variable in a starting set has to be decided by using the same criteria (1) and (2) as were formulated for selecting the origin-fixing reflections.

Summarizing what has been discussed for the choice of a good starting set:

(1) The starting set should consist of up to three reflections for origin definition, possibly one for specification of the enantiomorphous form and a number of variables. If the number of special reflections among the variables is n, the number of general reflections is m, a total of

$$t = 4^m 2^n \tag{5.54}$$

trials has to be calculated.

(2) All reflections included in the starting set should be chosen so as to satisfy the criteria (1) and (2).

(3) The number of variable reflections should be kept to a minimum, since t increases exponentially with m and n. For a medium-sized acentric structure with 2000–3000 reflections, $m = 2$ and $n = 2$ resulting in $t = 64$ trials should be a suitable choice. For centric structures where $m = 0$, $n = 2$ to 4, in addition to the origin-fixing reflections, is satisfactory in most cases.

Now the question is to discover which of the numerous trials is the correct solution of the phase problem. One way is to calculate an electron density map for each solution and to examine whether a structure can be recognized which is reasonable from chemical criteria. This is only possible if t is not too large, since such inspection of Fourier maps is very time-consuming. There are no general reliability criteria from which it can be determined unambiguously whether a given trial is the correct one or not. Several so-called "figures of merit" have been developed for use as criteria for the quality of a phase determination trial, but none of them are certain to indicate the correct solution. We shall mention here, among others, the R(E)-value introduced by Karle & Karle [Karle, J. and Karle, I. L., Acta Cryst. *21*, 849 (1966)]. R(E) is the relative average difference between the "observed" E-values $E_o(\mathbf{h})$ and the "calculated" E-values $E_c(\mathbf{h})$. The observed E-values are those obtained from the experimental data after (5.14). Calculated E-values can be derived from the application of the tangent formula. Note that the numerator and the denominator of the tangent formula can be interpreted as the imaginary and real part of $E(\mathbf{h})$. Therefore, if we write (5.43) as

$$\tan\varphi(\mathbf{h}) = \frac{E_i(\mathbf{h})}{E_r(\mathbf{h})}$$

we get

$$|E_c(\mathbf{h})|^2 = E_i(\mathbf{h})^2 + E_r(\mathbf{h})^2 \tag{5.55}$$

Then R(E) is defined by

$$R(E) = \frac{\Sigma_h ||E_o(\mathbf{h})| - |E_c(\mathbf{h})||}{\Sigma_h |E_o(\mathbf{h})|} \tag{5.56}$$

The trial having the lowest R(E)-value is most likely to be the correct solution. Experience has shown that R(E) values between 20 and 30% are promising, while R(E) values above 40% usually indicate an incorrect solution. However, the R(E) limits given above have to be used with great care, since they depend significantly on the number of reflections used in the phase determination process.

It has to be noted that only relatively few of the reflections measured will be used. Usually a minimum E-value E(min) is selected, and only those reflections with $E > E(\min)$ are included in the phase determination. In the early applications of direct methods, it was customary to set $E(\min) = 1.5$ for centric structures and $E(\min) = 1.2$ to 1.0 in the acentric case. Today, with more experience, it is recommended that the choice of E(min) is such that 10 to 15% of the whole data set is included in the phasing calculations, independently of whether the structure is centric or acentric. From our experience with relatively large acentric structures, we have found that even less than 10% of all measured reflections may be sufficient. For example, for the structure of 1, 2, 3, 4-tetra-O-benzoyl-β-D-xylopyranose, space group $P2_1$ [Luger, P., Kothe, G. and Paulsen, H., Angew. Chem. Int. Ed. *16*, 52 (1977)], which has 84 non-hydrogen atoms in the asymmetric unit, a successful phase determination was obtained using only 8% of the measured data set, corresponding to an $E(\min) = 1.68$.

Two reason can be given for this limitation of data set. On the one hand, the exclusion of reflections having small E-values reduces the possibility of estimating incorrect phases. On the other hand, it has been shown that the reduced set of high E-value reflections with their proper phases will generally be sufficient for the calculation of an electron density map which is a good enough approximation of the structure that it can be interpreted. To obtain this electron density map, it is customary to calculate a Fourier synthesis using E-values with their most probable phases (type 6 in (5.24)). E-values are preferred to F-values, since the resolution of an E-map is usually better than that of an F-map. However, E-maps more frequently show maxima which are not related to an actual atom position. These maxima, denoted as "ghosts" or "false peaks" must therefore be recognized and carefully eliminated from further calculations.

5.3.3 *Sign Determination for NITROS – An Example of the Centric Case*

From the considerations made in 3.3.3, and from the results of the E-value calculation (5.1.2), it is very likely that NITROS crystallizes in the centrosymmetric space group C2/m. We will therefore attempt the structure determination in this space group. We know that we have four molecules in the unit cell. Since C2/m requires eight-fold symmetry, the asymmetric unit must consist of one-half formula unit. As mentioned in 3.3.3, we expect that there is a mirror plane associated with the anion as well as with the cation, so that several atoms must be situated on this symmetry element (see Fig. 3.33).

A total of 1207 reflections have been measured for NITROS. We decided to include the 157 reflections with $E \geqq E(\text{min}) = 1.4$ in the phasing process, which is 13% of the total. The quantities M and **m** for this centered space group are

$$M = \begin{pmatrix} 1 & 1 & 0 \\ 0 & 0 & 1 \\ 0 & 0 & 0 \end{pmatrix} \qquad \mathbf{m} = \begin{pmatrix} 2 \\ 2 \end{pmatrix}$$

so that $\mathbf{h}(m)$ is given by

$$\mathbf{h}(m) = \begin{pmatrix} h+k \\ 1 \end{pmatrix} \bmod \begin{pmatrix} 2 \\ 2 \end{pmatrix}$$

referring however to the corresponding *primitive cell*. The origin is then defined by two reflections, for which the $\mathbf{h}_i(m)$ $(i = 1, 2)$ satisfy (5.50). Since we have a centric space group, no enantiomorph-fixing is necessary.

To get an idea of how to select an optimal starting set, let us proceed with a limited number of reflections. For this we have listed the 56 reflections with $E > 2.0$, together with all Σ_2-relations occurring among any three reflections of this set (Table 5.5, calculated by the program SINGEN of the X-RAY System). We must find a minimum number of starting reflections, for which a maximum number of signs can be derived using the Σ_2-relations. The starting set must include two reflections which constitute a valid origin choice.

Since we have a centered cell, the origin definition is not a trivial problem, but can be derived by the application of the formalism derived in 5.3.2. Since the vector $\mathbf{h}(m)$ has to be calculated from indices which refer to the primitive cell, it seems reasonable to transform the indices of a few-reflections with high E's to the primitive case. The matrix for the transformation

$$\mathbf{h}_c = S\,\mathbf{h}_p$$

is

$$S = \begin{pmatrix} 1 & 1 & 0 \\ 1 & -1 & 0 \\ 0 & 0 & -1 \end{pmatrix}$$

To get $\mathbf{h}_p$ from $\mathbf{h}_c$, we must do the inverse transformation

$$\mathbf{h}_p = S^{-1}\mathbf{h}_c$$

The inversion of S is done by using equation (1.6). We get

$$S^{-1} = 1/2 \begin{pmatrix} 1 & 1 & 0 \\ 1 & -1 & 0 \\ 0 & 0 & -2 \end{pmatrix}$$

(Check that result by calculating $S^{-1}S$)

With this matrix we have calculated the primitive indices, and then $\mathbf{h}(m)$ for the upper ten reflections of NITROS. The result is given in Table 5.6. It can be seen immediately

Table 5.5. 56 Reflections of NITROS having E > 2.0, with their Σ_2-Relations ($\mathbf{h}_j$ given by their current numbers).

			Σ_2-Relations ($\mathbf{h}_1+\mathbf{h}_2+\mathbf{h}_3=\mathbf{0}$)					
No.	h k l	E	$\mathbf{h}_1$	$\mathbf{h}_2$	$\mathbf{h}_3$	$\mathbf{h}_1$	$\mathbf{h}_2$	$\mathbf{h}_3$
1	−12 2 9	3.85	1	−1	−55	1	−17	−28
			1	−2	−39	1	−18	−42
			1	−5	−25	1	−22	−50
			1	−5	−38	1	−25	−26
			1	−6	−28	1	−30	−37
			1	−7	−51	1	−36	−54
			1	−9	−34	1	−44	−54
			1	−14	29	1	−49	−56
2	8 2 2	3.39	2	−2	−55	2	19	−27
			2	−3	56	2	25	28
			2	−4	29	2	28	−50
			2	−5	6	2	−28	38
			2	−5	17	2	30	−4
			2	6	−26	2	−30	43
			2	−6	22	2	30	−36
			2	−7	42	2	−30	48
			2	−9	32	2	32	−41
			2	−11	29	2	−34	−52
			2	−12	19	2	−37	54
			2	−17	22	2	45	−56
			2	−18	51			
3	8 2 7	3.32	3	−3	−55	3	−9	43
			3	−4	−26	3	−9	48
			3	−4	−5	3	−37	−52
			3	−5	−11	3	−41	43
4	9 3 2	3.26	4	−5	8	4	−19	35
			4	−7	30	4	19	−21
			4	−8	−26	4	−20	28
			4	−9	10	4	22	−45
			4	10	−41	4	−28	49
			4	−11	55	4	−42	48
			4	−14	39	4	−46	−52
			4	17	−56			
5	−1 5 5	3.08	5	6	−16	5	−18	34
			5	−8	−11	5	−23	−27
			5	−10	43	5	−26	−55
			5	−10	48	5	−31	51
			5	−12	−23	5	−37	46
			5	−14	57	5	−41	42
			5	−16	17			

Continued from Table 5.5

No.	h k l	E	Σ_2-Relations ($\mathbf{h}_1 + \mathbf{h}_2 + \mathbf{h}_3 = \mathbf{0}$)					
			$\mathbf{h}_1$	$\mathbf{h}_2$	$\mathbf{h}_3$	$\mathbf{h}_1$	$\mathbf{h}_2$	$\mathbf{h}_3$
6	−9 3 3	3.05	6	7	−9	6	−17	55
			6	−7	40	6	18	−31
			6	7	−41	6	−19	−23
			6	−8	−29	6	25	−39
			6	−10	30	6	29	−45
			6	11	−56	6	34	−51
			6	−16	26	6	46	−54
7	2 2 5	3.03	7	−7	−55	7	−25	−52
			7	−9	17	7	−28	34
			7	−11	−30	7	−29	−36
			7	−13	19	7	−29	−44
			7	−14	54	7	−38	−52
			7	−17	−40	7	−39	59
			7	22	−32			
8	−10 2 3	3.02	8	−8	−55	8	−31	37
			8	−10	42	8	−36	40
			8	−16	56	8	40	−44
			8	−17	29	8	46	−51
9	−7 5 8	2.93	9	−10	−11	9	−36	−45
			9	−13	−23	9	−39	−52
			9	−14	46	9	−41	−55
			9	−26	−42	9	−44	−45
			9	−33	−34			
10	−16 2 6	2.73	10	−10	−55	10	−26	−43
			10	−17	−30	10	29	−32
			10	−22	−36	10	−39	46
			10	−22	−44			
11	9 7 2	2.72	11	−19	35	11	−28	49
			11	19	−21	11	−37	51
			11	−20	28	11	−42	43
			11	22	−45	11	−46	−52
12	−7 3 5	2.65	12	−15	30	12	−24	40
			12	−21	29	12	−27	55
			12	23	−26			
13	−13 3 8	2.65	13	−15	−29	13	23	−41
			13	−21	−30	13	−27	−42
			13	−22	−24			
14	−11 3 9	2.64	14	−20	−22	14	−38	−56
			14	−28	−45	14	−41	−46

Continued from Table 5.5

			Σ_2-Relations ($\mathbf{h}_1+\mathbf{h}_2+\mathbf{h}_3=\mathbf{0}$)					
No.	h k l	E	$\mathbf{h}_1$	$\mathbf{h}_2$	$\mathbf{h}_3$	$\mathbf{h}_1$	$\mathbf{h}_2$	$\mathbf{h}_3$
15	−14 2 8	2.63	15	−15	−55	15	−19	−44
			15	−19	−36	15	−27	−30
16	−10 2 8	2.59	16	−16	−55			
17	−9 7 3	2.53	17	−19	−23	17	38	−39
			17	29	−45	17	46	−54
			17	−32	42			
18	−6 2 6	2.51	18	−18	−55	18	40	−50
			18	29	−37			
20	6 2 8	2.42	20	−20	−55	20	−37	−40
			20	−29	−50			
21	−6 2 5	2.42	21	−21	−55	21	−27	−29
			21	23	−56	21	40	−53
22	−17 5 1	2.40	22	28	−39			
23	6 2 0	2.39	23	−23	−55			
24	4 2 7	2.36	24	−24	−55	24	29	−53
			24	−27	−40			
25	−11 3 4	2.34	25	−29	−49	25	30	−47
			25	29	−57	25	−34	−42
			25	−30	−46	25	−38	55
26	−1 1 5	2.33	26	29	−56	26	−51	−52
			26	−37	46			
28	−3 5 6	2.32	28	−36	−46	28	−44	−46
			28	43	−47	28	−47	48
29	1 5 0	2.31	29	34	−46	29	38	−57
			29	−38	49			
30	−7 5 3		30	−32	35	30	38	−47
			30	34	−49	30	−39	54
			30	−38	46	30	−41	56
32	−15 3 6	2.25	32	−33	52	32	−48	−56
33	−2 2 7	2.25	33	−33	−55			
34	−5 7 1	2.24	34	−37	56	34	45	−54
35	−24 2 1	2.24	35	−35	−55			

Continued from Table 5.5

No.	h k l	E	Σ_2-Relations ($\mathbf{h}_1+\mathbf{h}_2+\mathbf{h}_3=\mathbf{0}$) $\mathbf{h}_1$	$\mathbf{h}_2$	$\mathbf{h}_3$	$\mathbf{h}_1$	$\mathbf{h}_2$	$\mathbf{h}_3$
36	1 3 5	2.21	36 36	−41 −44	45 55	36	−49	−52
37	−5 3 6	2.20	37	−39	43			
39	−20 0 7	2.19	39	−42	−51	39	−45	−49
42	−6 4 3	2.14	42	46	−57	42	−47	49
43	−15 3 1	2.13	43	−48	55			
44	1 7 5	2.09	44	−49	−52			
45	−8 2 3	2.09	45	−45	−55			
46	−4 2 1	2.08	46	−46	−55			
47	−18 2 7	2.07	47	−47	−55			
49	−12 2 4	2.06	49	−49	−55			

that the reflections 2, 9, and 10 cannot be chosen for origin definition because both their components in **h**(m) are zero. Among the remaining seven reflections, however, several choices of a valid reflection pair is possible (see bottom of Table 5.6).

The decision in favor of one of these pairs depends on their appearance in as many Σ_2-relations as possible. So we exclude reflections 3 and 8, since they are included in fewer relations than the others. For a choice among the other reflection pairs, no

Table 5.6. Index Transformation for the Upper Ten NITROS Reflections.

No.	$(h\ k\ l)_c$	$(h\ k\ l)_p$	**h**(m)	
1	−12 2 9	−5 −7 −9	(0, 1)	
2	8 2 2	5 3 −2	(0, 0)	no o.d.r.
3	8 2 7	5 3 −7	(0, 1)	
4	9 3 2	6 3 −2	(1, 0)	
5	−1 5 5	2 −3 −5	(1, 1)	
6	−9 3 3	−3 −6 −3	(1, 1)	
7	2 2 5	2 0 −5	(0, 1)	
8	−10 2 3	−4 −6 −3	(0, 1)	
9	−7 5 8	−1 −6 −8	(0, 0)	no o.d.r.
10	−16 2 6	−7 −9 −6	(0, 0)	no o.d.r.

Pairs of valid origin-defining reflections (o.d.r.) are 1,4; 1,5; 1,6; 3,4; 3,5; 3,6; 4,5; 4,6; 4,7; 4,8; 5,7; 5,8; 6,7; 6,8.

general recommendation can be given. The best selection is more a question of the experience or intuition of the crystallographer. We have decided to define the origin by the reflections no. 4 (9 3 2) and 7 (2 2 5). The determinant D[**h**(m)] is

$$\mathrm{D}[\mathbf{h}(\mathrm{m})] = \begin{vmatrix} 1 & 0 \\ 0 & 1 \end{vmatrix} = 1,$$

so we have made a correct choice. We assign the sign "+" to both reflections.

Let us now try to determine some signs from the table of Σ_2-relations (Table 5.5). First we have to explain the table more fully. Each Σ_2-relation is given by the number of the three contributing reflections. A negative sign before a number indicates that the original reflection indices are not used, but instead those of a symmetry-related reflection for which the structure factor is known (from space group symmetry) to have an opposite sign of that initially on the list. Let us consider, for example, the relation 4 −7 30. The original indices are 4: 9 3 2; 7: 2 2 5; 30: −7 5 3. The sum of these three reflections is not zero. This can be achieved, however, if, instead of 7, the reflection 7′: −2 2 −5 is used, and if instead of 30, the reflection 30′: −7 −5 3 is used. Then

$$4 + 7' + 30' = \begin{pmatrix} 9 \\ 3 \\ 2 \end{pmatrix} + \begin{pmatrix} -2 \\ 2 \\ -5 \end{pmatrix} + \begin{pmatrix} -7 \\ -5 \\ 3 \end{pmatrix} = \begin{pmatrix} 0 \\ 0 \\ 0 \end{pmatrix}$$

The structure factor of 30′ has the same sign as that of 30; however, that of 7′ is opposite to that of 7. So, if the sign of 30 is determined from that relation after (5.34), the change of sign for 7 must be observed. 4 and 7 were defined to be "+", then 7′ is "−". It follows that 30′ is also "−", hence 30 also has the "−" sign. In this way, all the relations can be used. The use of symmetry-related reflections is always recommended, since it significantly increases the number of relations obtained.

More signs can be derived from the two origin-fixing reflections and the additional reflection no. 30, determined above. This is done systematically be searching for all Σ_2-relations containing two known reflections. The third one is then determined and can be used immediately in a further search as a known phase. So we can derive 55, 11, after which all the relations are exhausted. Searching with another origin-defining set will give the same result. It is necessary, therefore, to include one or more variable reflections in the sign-determination process. We decide to add the reflections no. 1 and 3 as variables to the starting set (the reader may find a better choice). Although reflection no. 3 has only a small number of Σ_2-reflections, we include this reflection in the starting set since it seems to play a key role in connection with the signs determined so far. With no. 3 included, we get at once, nos. 26 and 5. No. 5 together with no. 1 gives 25 and 38. No. 5 together with 11 gives 8, etc.

This sign determination process is laborious when done by hand, since many more relations than are listed in Table 5.5 have to be checked (these are the relations between all 157 reflections included in the phasing process). So the complete sign determination is done on the computer, but a preliminary search by hand with a limited number of relations is a good way to get an idea of how to obtain a good starting set.

The computer calculation of sign determination is done with the program PHASE of the X-RAY System. The known reflections to be input are the two origin-fixing reflections no. 4 and no. 7 and the two variables no. 1 and no. 3. Since each of the variables can have the sign "+" and "−", we have to calculate four trials, one for each combination ++, +−, −+, and −−.

Table 5.7. Summary of Sign Determination Results for NITROS.

	1	2	3	4
Signs of 1 and 3	++	+−	−+	−−
No. of Signs Determined	157	157	157	157
Number of Signs to be +	157	72	77	80
Number of Signs to be −	0	85	80	77

The results are summarized in Tables 5.7 and 5.8. Table 5.7 shows that the number of starting reflections was sufficient, since the signs of all 157 reflections considered were determined. It can be seen, furthermore, that trial no. 1 is probably wrong, since it is very unlikely that all the signs are plus. A result of that kind, known as an "all plus

Table 5.8. Distribution of the Upper Ten Peaks in a Subsequent E-map Calculation.

Peak Height	x	y	z	Peak Height	x	y	z
	Trial 1				*Trial 2*		
1155	.450	.500	.997	950	.661	.500	.328
637	.903	.210	.157	710	.964	.212	.912
557	.198	.212	.413	675	.938	.209	.521
522	.526	.500	.393	492	.101	.205	.300
497	.876	.212	.764	490	.635	.500	.935
479	.179	.500	.351	420	.303	.500	.851
425	.541	.500	.831	351	.312	.500	.282
413	.224	.209	.806	343	.745	.211	.832
379	.739	.206	.378	331	.725	.500	.523
377	.436	.500	.802	288	.165	.421	.336
	Trial 3				*Trial 4*		
837	.450	.212	.718	1030	.013	.500	.196
807	.148	.500	.134	681	.390	.211	.963
713	.827	.500	.484	656	.711	.211	.607
458	.409	.207	.889	432	.950	.500	.990
395	.788	.500	.657	428	.692	.500	.545
384	.612	.203	.493	399	.751	.209	.437
349	.521	.209	.902	338	.666	.500	.152
335	.769	.211	.361	328	.084	.500	.384
305	.426	.205	.324	318	.318	.211	.778
296	.812	.500	.044	303	.511	.421	.188

catastrophe" is sometimes observed with the application of direct methods. It always indicates an incorrect trial.

A decision in favor of one of the remaining three trials is not possible from the sign determination results. Let us therefore inspect the peak distribution of the resulting E-maps (Table 5.8). Except in no. 1, which is already recognized as incorrect, we find for all trials three peaks with a height large enough to correspond to the expected three independent sulfur atoms. From considerations concerning the anion position relative to the mirror plane (see Fig. 3.33), if follows that two sulfur atoms, in addition to a nitrogen and oxygen atom, must be situated exactly on the mirror plane, i.e. must have a so-called "special position" (an atom is said to be in a special position if at least one of its coordinates cannot have an arbitrary magnitude). Only the upper three peaks of trial no. 3 have this property. An inspection of the bond lengths and angles (see Fig. 5.15) between the peaks of trial no. 3 shows that a resonable structural model is formed. Peaks no. 1, 2, and 3 are indentified as the three independent sulfur atoms, nos. 4, 5, and 6 as the nitrogens of the anion, and 10 as the oxygen. Only the nitrogen atom of the cation is missing from the top ten peaks. Peaks no. 7, 8, and 9 are "ghosts". No chemically reasonable model can be derived from the other trials.

So the results are that from the four trials of sign determination, that the sign combination "−" for reflection no. 1 and "+" for no. 3 was correct. From the E-map calculated with 157 reflections with $E > 1.4$, all but one atom of the expected structure could be located, so the structure is solved. The missing nitrogen atom of the cation will be located from a difference electron density map in the course of the structure refinement.

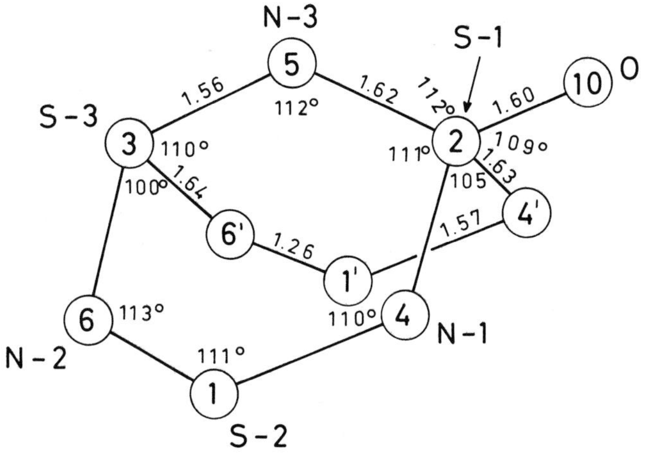

Fig. 5.15. Structural model resulting from E-map calculation, trial 3, for NITROS (peak numbers correspond to peak sequence given in Table 5.8).

5.3.4 *Phase Determination for SUCROS – An Example of the Acentric Case*

In contrast to the structure of NITROS, we have an *acentric* space group ($P2_1$) for SUCROS, so that phases have to be determined which vary over the whole range from 0 to 2π. Therefore, we have to apply the tangent formula (5.43). For the complete phase determination of SUCROS, we shall make use of the program "MULTAN", which is currently the most frequently used and most successful program for phase determination purposes. The first version was written in 1971 [Germain, G., Main, P. and Woolfson, M. M., Acta Cryst. *A27*, 368 (1971)] and updated versions are issued every two or three years. Because of the great success with the application of MULTAN, it is recommended that everyone doing single crystal analysis has this program available. Although we shall not describe all the details of MULTAN (which are provided in the program instructions), the principle of how it works and some of its main advantages will be noted here.

A rough flow diagram of the most important program routines of MULTAN is shown in Fig. 5.16. In the NORMAL part, the data-reduction and normalization is executed, so that E-values are calculated for the next stage of the program. The next section is the real MULTAN, which consists of three parts: the SIGMA-routine for the preparation of Σ_2-relations, a CONVERGE part for a selection of a starting set, and the FASTAN part which does the actual phasing calculation for all possible phase combinations from the starting set provided by CONVERGE. With the help of va-

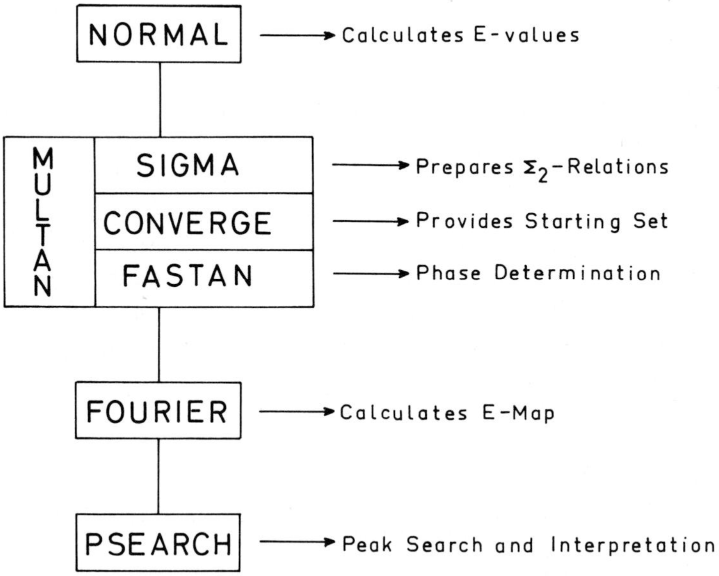

Fig. 5.16. Flow diagram of MULTAN.

rious figures of merit, the best phase determination trial is selected and used as input to the program FOURIER, which calculates an E-map. The E-map is usually not printed, but PSEARCH is executed, which searches for the highest peaks, calculates distances and angles, tries to construct one or more fragments of atoms presumably bonded, and prints a plot of the fragments found.

In comparison with most other programs for phase determination, MULTAN has the following advantages:

(1) All calculations from data reduction to the output of a structural model can be executed in one operation.

(2) The choice of starting set, including the proper selection of origin-defining reflections, can be left to the computer (CONVERGE part). The user has only to specify the number of reflections to be used in the starting set. An optimal choice is then done by the program.

(3) A phase determination is calculated for all combinations of starting values for the variables reflections, so that the user is not concerned with the number of trials executed.

(4) The program carries out its own recognition of the "best" phase set, so that an E-map of the most probable structure can be calculated immediately for possible interpretation.

In our application of MULTAN to SUCROS, we shall describe the actions of the user and demonstrate the way in which MULTAN works in practice.

The input to NORMAL consists of the unit cell and symmetry information and the reflection intensities. The number of reflections with highest E-values to be used must be specified. This is a very sensitive parameter and if MULTAN fails, it can be tried again with a change in this parameter. For SUCROS, we have a total of 1297 reflections. We include 180 reflections with highest E-values for the phase determination. This is a relative large number, approximately 15%. As mentioned in the last section, good results can often be obtained when only 10%, or even less, of the reflections are used in the phasing calculation.

Except for some organizational parameters, the most important input to MULTAN is the number of variable reflections to be added to the starting set. The choice of this parameter is, in principle, the only decision left to the user. Since he can discriminate between special and general reflections (he can also leave this decision to the program by specifying "any" reflections), he can predetermine the number of trials to be executed. What he has to do is to ensure that not too many trials are in the first run.

For the first run on SUCROS, we decided to use two variable reflections in the starting set, so we expect a total of 16 trials of phase determination (see 5.54). The input for the remaining program parts of MULTAN is trivial, so we can proceed to the results.

From the CONVERGE part, the reflections given in Table 5.9 have been found to be an optimal starting set.

We can check that the origin definition made by CONVERGE is a valid selection using the formalism derived in 5.3.2. The matrix of semi-invariants for space group $P2_1$

Table 5.9. Starting Set for SUCROS, Result of CONVERGE.

No.	Indices	E	Phase	Type
2	−1 7 1	2.70	$\pi/4$	Origin + Enantiomorph
8	−9 1 2	2.34	0	Origin
92	5 0 2	1.69	0	Origin
3	−5 2 6	2.67	$\pi/4,\ 3\pi/4,\ 5\pi/4,\ 7\pi/4$	Variable
6	−10 1 1	2.43	$\pi/4,\ 3\pi/4,\ 5\pi/4,\ 7\pi/4$	Variable

is the unit matrix, the semi-invariant modulus is

$$\mathbf{m} = \begin{pmatrix} 2 \\ 0 \\ 2 \end{pmatrix}$$

Then $\mathbf{h}(\mathrm{m})$ for the three reflections specified for origin-fixing reads

$$\mathbf{h}(\mathrm{m})_2 = \begin{pmatrix} 1 \\ 7 \\ 1 \end{pmatrix}; \quad \mathbf{h}(\mathrm{m})_8 = \begin{pmatrix} 1 \\ 1 \\ 0 \end{pmatrix}; \quad \mathbf{h}(\mathrm{m})_{92} = \begin{pmatrix} 1 \\ 0 \\ 0 \end{pmatrix}$$

Then we get

$$\mathrm{D}[\mathbf{h}(\mathrm{m})] = \begin{vmatrix} 1 & 7 & 1 \\ 1 & 1 & 0 \\ 1 & 0 & 0 \end{vmatrix} = -1$$

hence the choice made is correct.

Note that in cases where **m** includes one or more zero components, the correct choice of three origin-fixing reflections is not easy. The zero component causes the actual reflection index to be input to $\mathrm{D}[\mathbf{h}(\mathrm{m})]$, and it is then very difficult to get a determinant of ± 1.

A fixing of enantiomorph is provided by setting the phase of reflection no. 2 to $\pi/4$, so that all the specifications for this acentric space group are now made. The two variable reflections no. 3 and no. 6 are allowed to have the initial phases $\pi/4$, $3\pi/4$, $5\pi/4$, and $7\pi/4$. A summary of results for the 16 trials is given in Table 5.10.

Although the program uses different criteria for each trial, which are summarized in a so-called "combined figure of merit", the R (E) value already shows in our case, that trial no. 9 is probably the correct one, since R (E) = 17.1% is significantly smaller than for any other trials. The phase set determined by this trial is then used for an E-map calculation from which a fragment containing 25 peaks is constructed. A plot of this fragment is shown in Fig. 5.17. An analysis of distances and angles shows without doubt that this structural model is correct and the phase problem is therefore solved.

An inspection of the model when compared with the chemical formula of SUCROS (Fig. 5.17) shows that all atoms but one are present in their correct locations. Only

Table 5.10. FASTAN Calculations for SUCROS (phases in °).

No	2	8	92	3	6	R(E)	Combined Figure of Merit
1	45	0	0	45	45	30.0	0.77
2	45	0	0	135	45	27.7	1.46
3	45	0	0	225	45	28.3	0.71
4	45	0	0	315	45	28.5	1.19
5	45	0	0	45	135	24.9	1.15
6	45	0	0	135	135	27.1	0.99
7	45	0	0	225	135	27.5	1.27
8	45	0	0	315	135	30.8	0.65
9	45	0	0	45	225	17.1	2.93
10	45	0	0	135	225	28.5	1.19
11	45	0	0	225	225	23.6	1.51
12	45	0	0	315	225	29.4	1.49
13	45	0	0	45	315	27.0	1.30
14	45	0	0	135	315	30.7	1.38
15	45	0	0	225	315	27.5	0.76
16	45	0	0	315	315	30.2	1.09

the oxygen attached to C-3′ (peak no. 10) is in doubt. There are two peaks, nos 21 and 24, which might be the missing atom. Since we are not sure, we leave the determination of that atom to a difference synthesis following the first refinement.

Summarizing the result of the MULTAN calculation, we can state that a total of 180 reflections out of about 1300 ($\approx 15\%$) and two variables in the starting set, gave 16 trials which were sufficient for a successful structure determination. Of the 23 non-hydrogen atoms present in the molecule, 22 could be located amongst the upper 23 peaks of the corresponding E-map. Only one peak (no. 21) was a ghost. This was a very convincing result.

At this point, it should be mentioned that the first X-ray structure determination of sucrose was very difficult and took several years to complete. The first results were published in 1952 [Beevers, C.A., McDonald, T.R.R., Robertson, J.H. and Stern, F., Acta Cryst. *5*, 689 (1952)] with a hint from the authors that the first intensity data were collected in 1944. From 1952, it was more than 10 years before the first three-dimensional neutron diffraction analysis of sucrose was published [Brown, G.M. and Levi, H.A., Science *141*, 921 (1963)]. A further ten years passed until an improved X-ray investigation and the results of a further refinement of the 1963 neutron data of sucrose were published [Brown, G.M. and Levi, H.A., Acta Cryst. *B29*, 790 (1973); Hanson, J.C., Sieker, L.C. and Jensen, L.H., Acta Cryst. *B29*, 797 (1973)]. Today, sucrose is one of those compounds for which the most precise structural information is available.

This history of the sucrose structure is one of the best examples to demonstrate the

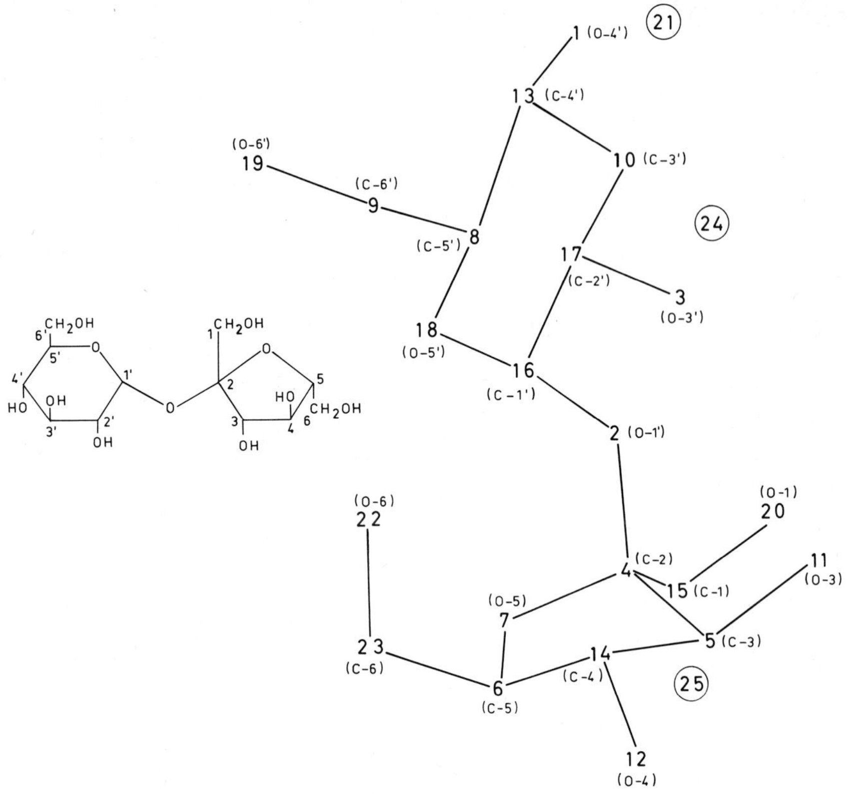

Fig. 5.17. Result of peak search and fragment construction for SUCROS (trial no. 9).

progress of structure analysis. In 1950 or so, it was an almost unsolvable problem, whereas today it could be treated without any difficulty and can be solved almost automatically by the computer in a minimum of time. The precision of atomic positions in 1952 was of the order of ± 0.08 Å, while in 1973 it was improved to ~ 0.001 Å.

6 Refinement

6.1 Theoretical Aspects

6.1.1 F_c-Calculation, Residual Index

The structural model determined so far can be expressed mathematically by a number of parameters. These are the positional coordinates and the temperature factors of each atom and one or more scale factors. Since, as pointed out in the last section, the parameters determined so far can only be regarded as preliminary, a method has to be derived which allows an improvement of the model.

First, however, we have to define a criterion which can be used to examine the accuracy of the present model. This is done by calculation of the so-called "residual or agreement index". Since this quantity is usually designated by the capital letter R, it is called the R-value. The R-value is the relative average difference between the calculated structure amplitudes $|F_c|$ (see 5.20) and the observed structure amplitudes $|F_o|$ (see 5.22) given by,

$$R = \frac{(1/r)\, \Sigma_h \big|\, |F_o(\mathbf{h})| - |F_c(\mathbf{h})| \,\big|}{(1/r)\, \Sigma_h |F_o(\mathbf{h})|}$$

where r is the total number of reflections. Hence,

$$R = \frac{\Sigma_h \big|\, |F_o(\mathbf{h})| - |F_c(\mathbf{h})| \,\big|}{\Sigma_h |F_o(\mathbf{h})|} \qquad (6.1)$$

Since the $|F_o|$'s are derived directly from the experiment and the $|F_c|$'s are calculated from the structural model, the R-value can be regarded as an indication of how well the model fits the real structure. Since the scaling provides that the sum of the $|F_o|$'s is equal to the sum of $|F_c|$'s, the R-value will always be smaller than 1.0, although the model might be completely wrong. Wilson (1950) has shown that the R-value of a structure which is oriented randomly in the unit cell, can be calculated theoretically [Wilson, A.J.C., Acta Cryst. *3*, 397 (1950)]. The result depends on whether the space group is centric or acentric;

$$\begin{aligned} R(\text{centric}) &= 2(\sqrt{2} - 1) = 0.828 \\ R(\text{acentric}) &= 2 - \sqrt{2} = 0.586 \end{aligned} \qquad (6.2)$$

So it is clear that an R-value calculated for an actual structure or structural fragment must be significantly smaller than the values given in (6.2) to indicate that the solution is at least partially correct. The first F_c-calculation for KAMTRA, based on the K^+ position only and the isotropic temperature factor obtained from the Wilson plot, led to an R-value of 0.465 (or 46.5%; note that frequently the R-value is given in %). F_c-calculations for NITROS and SUCROS, based on the models obtained from an E-map with phases assigned from direct methods, led to an R-value of 25.6% for NITROS and 29.2% for SUCROS. All three values are significantly lower than those for random structures, indicating that progress towards the correct structure has been made, but the R-values are not small enough to prove conclusively that a correct solution has been found.

6.1.2 *Theory of Least-Squares Refinement*

The problem of improving a given structure model needs some mathematical considerations, which are, however, not very complicated. Let us represent the parameters of a given model M′ by $x'_1, \ldots x'_n$ and note that this parameter set includes *all* parameters of the model contributing to the F_c's. So x'_j may be a positional, a thermal, or a scaling parameter, or possibly one of some special parameters which shall be discussed later.

The problem is to obtain a model $\bar{M}$ having the parameters

$$\bar{x}_j = x'_j + x_j \qquad j = 1, \ldots n \tag{6.3}$$

so that $\bar{M}$ is a "best" approximation of the experimental data. The problem expressed in terms of F's reads: Given a set of F_c's dependent on the parameters $x'_1, \ldots x'_n$

$$F'_c = F_c(x'_1, \ldots x'_n),$$

how do we find improvements for the x'_j, so that a set of F_c's derived from $\bar{x}_j = x'_j + x_j$ $(j = 1, \ldots n)$

$$\bar{F}_c = F_c(\bar{x}_1, \ldots, \bar{x}_n)$$

is a best approximation of the F_o's? The question immediately arises, which approximation is said to be the best, and how do we get this approximation? This problem is a typical "fitting problem" and it will be solved by a method which is almost 200 years old, the "least-squares method". It is based on the principle expressed first by Gauss and Lagrange (~1800), who stated that a set of theoretical values T_k $(k = 1, \ldots, r)$ is the best approximation for a set of observations L_k $(k = 1, \ldots, r)$ if the sum

$$\sum_{k=1}^{r} (T_k - L_k)^2 \tag{6.4}$$

is a minimum. If the T_k depend on the parameters $x_1, \ldots x_n$, the condition of (6.4) to be minimal can be used to calculate the most favorable parameter set.

In the application of that principle to structure analysis, the observations L_k are given by the observed structure amplitudes $|F_o(\mathbf{h})|$, the theoretical values T_k by the calculated structure amplitudes $|F_c(\mathbf{h})|$. Then the principle of least-squares for single crystal analysis is

$$Q = \Sigma_h (|F_o(\mathbf{h})| - |F_c(\mathbf{h})|)^2 = \text{minimum} \tag{6.5a}$$

Weighting factors $w(\mathbf{h})$ are frequently given to the various terms in that sum, to take into account the different precision of the $|F_o|$'s. Then we get

$$Q = \Sigma_h w(\mathbf{h}) (|F_o(\mathbf{h})| - |F_c(\mathbf{h})|)^2 = \text{minimum} \tag{6.5b}$$

We define a model $\bar{M}$ with parameters $\bar{x}_1, \ldots, \bar{x}_n$ to be the best approximation for the observations F_o if the F_c's calculated from that model,

$$F_c = F_c(\bar{x}_1, \ldots, \bar{x}_n)$$

satisfy (6.5).

Provided that a first approximation model M' exists with $F_c = F_c(x'_1, \ldots x'_n)$, we shall develop an algorithm for the calculation of improvements x_j ($j = 1, \ldots n$) by using (6.5b). Suppose that we have measured r reflections $\mathbf{h}_1, \ldots \mathbf{h}_r$, so that r $|F_o|$'s, $|F_o|_1 \ldots |F_o|_r$ and r $|F_c|$'s, $|F_c|_1 \ldots |F_c|_r$ are present. Let us denote by v_k the weighted difference between observed structure amplitude and that of the desired model

$$v_k = \sqrt{w_k}(|F_c|_k(\bar{x}_1, \ldots \bar{x}_n) - |F_o|_k) \qquad k = 1, \ldots r$$

Then (6.5b) can be expressed as

$$Q = \sum_{k=1}^{r} v_k^2 \doteq \text{minimum} \tag{6.5c}$$

For the correction x_j (see 6.3), we assume that

$$x_j \ll x'_j \qquad j = 1, \ldots n$$

so that we can expand $F_c(\bar{x}_1, \ldots \bar{x}_n)$ into a Taylor series and neglect all but the linear terms:

$$\begin{aligned} |F_c|(\bar{x}_1, \ldots \bar{x}_n)_k &= |F_c|(x'_1 + x_1, \ldots x'_n + x_n)_k \\ &= |F_c|(x'_1, \ldots x'_n)_k + \alpha_{k1} x_1 + \ldots \alpha_{kn} x_n \qquad k = 1, \ldots r \end{aligned} \tag{6.6}$$

with

$$\alpha_{kj} = \frac{\partial |F_c|(x'_1, \ldots, x'_n)_k}{\partial x_j} \qquad \begin{matrix} k = 1, \ldots, r \\ j = 1, \ldots, n \end{matrix} \tag{6.7}$$

Note that (6.6) is true for every reflection $\mathbf{h}$, so that k varies from 1 to r. (6.7) holds for every reflection (subscript k) and for every parameter (subscript j). Setting

$$|F_o|_k - |F_c|(x'_1, \ldots x'_n)_k = l_k \qquad k = 1, \ldots, r$$

we get

$$v_k = \sqrt{w_k}\,(\alpha_{k1} x_1 + \ldots \alpha_{kn} x_n - l_k) \qquad k = 1, \ldots r \tag{6.8a}$$

For convencience, we proceed by using a matrix notation. We introduce an $r \times n$

matrix

$$A = (a_{kj}), \quad \text{with } a_{kj} = \sqrt{w_k}\,\alpha_{kj}$$

two r-dimensional columns

$$V = \begin{pmatrix} v_1 \\ \vdots \\ v_r \end{pmatrix} \qquad L = \begin{pmatrix} l_1 \\ \vdots \\ l_r \end{pmatrix}$$

and an n-dimensional column

$$X = \begin{pmatrix} x_1 \\ \vdots \\ x_n \end{pmatrix}$$

Then we can express (6.8a) by the matrix equation

$$V = AX - L \tag{6.8b}$$

Q (see 6.5c) is a function of the x_j, so it follows from the condition $Q = \min$ that all partial derivatives must be zero:

$$\frac{\partial Q}{\partial x_1} = \frac{\partial Q}{\partial x_2} = \ldots = \frac{\partial Q}{\partial x_n} = 0$$

With (6.5c) we get

$$0 = \frac{\partial Q}{\partial x_j} = \frac{1}{2}\frac{\partial Q}{\partial x_j} = v_1 \frac{\partial v_1}{\partial x_j} + \ldots v_r \frac{\partial v_r}{\partial x_j} \qquad j = 1, \ldots, n$$

From (6.8a) it follows that

$$\frac{\partial v_k}{\partial x_j} = a_{kj} \qquad k = 1, \ldots, r; \qquad j = 1, \ldots, n$$

Thus we get

$$\begin{array}{l} a_{11} v_1 + a_{21} v_2 + \ \ldots \ a_{r1} v_r = 0 \\ \vdots \qquad\qquad\qquad\qquad\quad \vdots \qquad \vdots \\ a_{1n} v_1 + a_{2n} v_2 + \ \ldots \ a_{rn} v_r = 0 \end{array}$$

or

$$A'V = 0$$

with A' being the transposed matrix of A. Using (6.8b), it follows that

$$A'(AX - L) = 0$$

Setting

$$N = A'A$$
$$C = A'L$$

we obtain finally a system of linear equations of order n

$$NX = C \tag{6.9}$$

which can be solved to get the improvements $x_1, \ldots, x_n$ (Note that A' is of order $n \times r$, so $N = A'A$ is a square matrix of order n, $A'L$ is of order $n \times 1$, see 1.1.2).

Since we have restricted the Taylor expansion of $|F_c|(\bar{x}_1, \ldots \bar{x}_n)$ to linear terms only, the improvements $x_1, \ldots, x_n$ obtained as the solution from (6.9) will not exactly satisfy (6.3), but we shall get a parameter set

$$x_j'' = x_j' + x_j \qquad j = 1, \ldots n \tag{6.10}$$

where the x_j'' are a more refined approximation of the model than the x_j'.

If necessary, the calculation to obtain improvements x_j can be repeated by using the x_j'' as input parameters, so that this procedure can be repeated until convergence of refinement is obtained.

An additional great advantage of the least-squares procedure is that an estimation of the standard deviations of the parameters can be easily obtained. It follows from (6.9) and the definition of C that the x_j are obtained by

$$X = N^{-1}C = N^{-1}A'L$$

Introducing a further matrix $B = N^{-1}A'$, we get

$$X = BL$$

with B being of type $n \times r$. If b_{jk} are the elements of B, the j th variable x_j is then given by

$$x_j = b_{j1}l_1 + \ldots b_{jr}l_r \qquad j = 1, \ldots, n$$

For the calculation of the standard deviation $\sigma^2(x_j)$, we use the law of propagation of errors. If, for instance, the quantities u, v, w are related by $w = au + bv$, it follows that $\sigma^2(w) = \sigma^2(au + bv) = a^2\sigma^2(u) + b^2\sigma^2(v)$. Hence

$$\sigma^2(x_j) = b_{j1}^2\sigma^2(l_1) + \ldots b_{jr}^2\sigma^2(l_r)$$

l_k was defined as

$$l_k = |F_o|_k - |F_c|_k \qquad k = 1, \ldots, r$$

We approximate the standard deviation of *all* l_k's by a common $\sigma(l)$, resulting in (see 4., equation (4.7c)),

$$\sigma^2(l) = \frac{1}{r-n}\sum_{k=1}^{r} w_k(|F_o|_k - |F_c|_k)^2$$

(Note that the degree of freedom is now given by $r - n$, the difference between the number of observations r and the number of parameters n to be determined.)

It follows for $\sigma^2(x_j)$ that

$$\sigma^2(x_j) = \frac{1}{r-n}\sum_{k=1}^{r} w_k(|F_o|_k - |F_c|_k)^2 \sum_{k=1}^{r} b_{jk}^2$$

For the sum over the b_{jk}, an expression more suitable for numerical calculation can be derived. If the matrix product BB' is calculated, we get a square matrix

$$D = BB'$$

with its diagonal elements d_{jj} equal to

$$d_{jj} = \sum_{k=1}^{r} b_{jk} b'_{kj} = \sum_{k=1}^{r} b_{jk}^2 \qquad j = 1, \ldots, n$$

On the other hand BB' is given by

$$BB' = N^{-1}A'(N^{-1}A')' = N^{-1}A'(A')'(N^{-1})'$$

Since $N = A'A$ is symmetrical with respect to its principal diagonal, $N = N'$ and also $N^{-1} = (N^{-1})'$ holds. Moreover, $(A')' = A$, so we get

$$BB' = N^{-1}NN^{-1} = N^{-1}$$

It follows that

$$\sigma^2(x_j) = \frac{d_{jj}}{r-n} \sum_{k=1}^{r} w_k(|F_o|_k - |F_c|_k)^2 \qquad j = 1, \ldots, n \tag{6.11}$$

with the d_{jj} being the elements of the principal diagonal of N^{-1}. Note that if for w_k the proper statistical weight $1/\sigma^2(F_{ok})$ is taken, it can be shown that for $r \to \infty$, the sum

$$\left(\sum_{k=1}^{r} w_k(|F_o|_k - |F_c|_k)^2\right)/(r-n)$$

converges towards 1. Then

$$\sigma(x_j) = \sqrt{d_{jj}} \qquad j = 1, \ldots, n$$

For real structures, a more or less good approximation is therefore given by

$$\sigma(x_j) \approx \sqrt{d_{jj}} \tag{6.12}$$

All the information needed for the numerical calculation of a least-squares cycle can be obtained from the experimental data and a given approximation of the structural model. Note that the application of the least-squares method requires the existence of a good model (the mathematical reason is given by the condition $x_j \ll x'_j$, which permits the neglect of the non-linear terms in the Taylor expansion), so that one of the methods of phase determination described in section V is a necessary prerequisite. The quantities needed are the matrix

$$N = A'A$$

and the column

$$C = A'L$$

Then the problem is reduced to obtaining A and L. The elements of A (see equation

(6.7)) are given by the partial derivatives of $|F_c|$ with respect to the variables x_j and a weighting scheme to be chosen by the user. Since F_c can be expressed analytically in terms of the x_j, an analytical expression can be derived for the a_{kj}. Input of the given model results in the actual numerical values of all matrix elements. The components l_k of L are easily obtained from the $|F_o|$'s and the $|F_c|$ values calculated from the given model. From A and L, the calculation of N and C can be done, Then, the inversion of N to get N^{-1} is necessary (in practice, an extensive numerical calculation), but finally, from N^{-1} all desired quantities can be obtained, that is, the x_j as well as the $\sigma(x_j)$.

Some additional remarks on the least-squares method should be added:

(1) The method described above is said to be a refinement based on $|F|$, since the quantity

$$Q = \Sigma_h w(\mathbf{h})(|F_o(\mathbf{h})| - |F_c(\mathbf{h})|)^2$$

is minimized. There are good arguments to minimize on $|F|^2$ instead, using

$$Q' = \Sigma_h w(\mathbf{h})(|F_o(\mathbf{h})|^2 - |F_c(\mathbf{h})|^2)^2$$

since the squares of the structure amplitudes are more directly associated with the experimental data. In this case, l_k has to be replaced by

$$l'_k = |F_o|_k^2 - |F_c|_k^2$$

and the partial derivatives are given by

$$\alpha'_{kj} = \frac{2|F_c|_k \partial |F_c|_k}{\partial x_j}$$

Using these modifications, the procedure will otherwise be the same as that for a refinement based on $|F|$. For ordinary structure determinations, there will be no great difference between the results obtained from the two methods of refinement. However, if very precise electron density details are to be determined, the $|F|^2$ refinement is preferable.

(2) With regard to the weights used during refinement, we noted that for statistical reasons, the proper weights are given by $1/\sigma^2(|F_o|)$ [or, if F^2-refinement was chosen, $1/\sigma^2(|F_o|^2)$]. Nevertheless, it is customary to experiment with a variety of more or less sophisticated weighting schemes. It should be mentioned, however, that even a trivial weighting scheme in which all weights equal one can sometimes be used to give satisfactory results, unless high accuracy is sought. From the large number of weighting schemes which can be applied, we shall describe only one method, which appears to be one of the most reasonable, in addition to the $1/\sigma^2$ weighting scheme. For this method, the range covered by $|F_o|$ is sub-divided into a number of intervals, each containing approximately the same number of reflections. A mean $|F_o|$ value can be assigned to each interval. Then the average $|F_o| - |F_c|$ difference is calculated for each interval

$$\Delta F(i) = \langle ||F_o| - |F_c|| \rangle_i$$

If $\Delta F(i)$ is plotted versus $|F_o|$ (see Fig. 6.1), the distribution of points can usually be

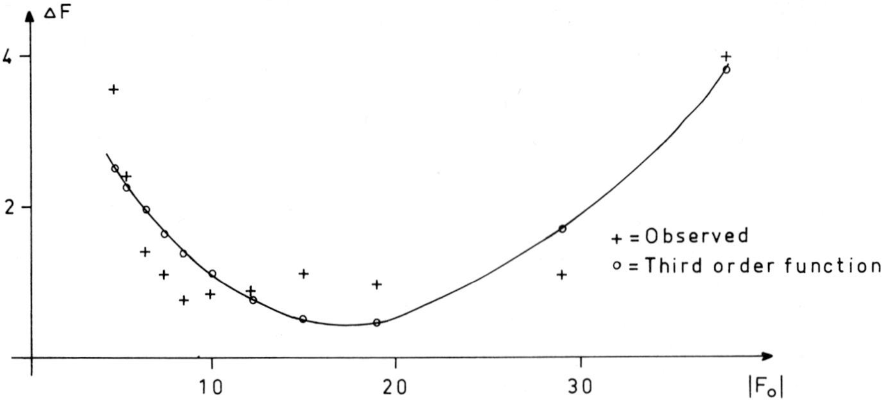

Fig. 6.1. Plot of ΔF versus $|F_0|$ for weighting purposes.

approximated by an analytical function of second or third order

$$\Delta F(i) = p(|F_o|) = a_0 + a_1|F_o| + a_2|F_o|^2 + a_3|F_o|^3 \tag{6.13}$$

The objective of the weighting scheme is to make ΔF independent of $|F_o|$, so w is chosen so that

$$w(\mathbf{h}) = 1/p(|F_o|)$$

(3) It should also be mentioned that the principles of least-squares are applied to the refinement of lattice constants. In this case, the observations L_k are the measured Bragg angles θ_{ok}, $k = 1, \ldots, r$, of a number r of reflections. The theoretical values T_k are the theoretical θ_{ck} values derived from Bragg's law

$$\theta_{ck} = \arcsin[(\lambda/2)d^*]$$

where θ_{ck} is a function of the lattice constants

$$\theta_{ck} = \theta_{ck}(a, b, c, \alpha, \beta, \gamma)$$

By minimizing

$$\sum_{k=1}^{r} (\theta_{ok} - \theta_{ck})^2$$

a refinement of the lattice constants can be obtained. In this case, the derivatives of θ_{ck} with respect to the lattice constants have to be calculated to get the matrix A. A formalism similar to that derived above can then be applied to obtain refined lattice constants and their estimated standard deviations. Computer programs for the application of this procedure are usually included in the software provided with a modern automatic diffractometer. A least-squares refinement was applied in 4.2.2 to the lattice constant determination for KAMTRA.

6.2 Practising Least-Squares Methods

6.2.1 *Aspects of Numerical Calculations*

The application of the least-squares technique requires the execution of extensive numerical calculations. It has to be noted that, for example, the order of the square matrix N introduced in 6.1.2 is given by the number of parameters. Even for a medium-sized structure, 100 or more parameters may be present, so it is evident that large order matrix operations have to be executed. Therefore, least-squares refinement cannot be applied unless a suitable computer program is available. However, even for a large computer, the refinement calculation is not a trivial problem. The first problem is that of storage. If n is the number of parameters to be refined, the matrix N contains n^2 elements. However, since N is symmetric, only the elements on one side of the principal diagonal and on the diagonal itself have to be calculated and stored. Thus the number of elements reduces to approximately $n(n+1)/2$; nevertheless, for $n = 100$, 5000, for $n = 150$, 11250, and for $n = 20$, 20000 words of core memory are required for the matrix elements. Even for large computers, such large amount of storage is not always available and restrictions have to be introduced. One way to save storage is to reduce the number of parameters to be refined in a cycle. For instance, one-half of the atoms may be refined in a first cycle, and the second half in a second cycle; or, all positional parameters can be refined in the first cycle and all thermal parameters in the second cycle. The parameters which are not refined contribute only to the F_c's, but are not taken into consideration for the refinement matrix.

A reduction in the number of refined parameters saves core very effectively. Suppose we have 50 atoms and wish to refine three positional and one thermal parameter. Then we have $50 \times 4 = 200$ parameters. The storage needed for a full-matrix refinement is 20000 words. For two cycles with refinement of only 25 atoms in each cycle, we have 100 parameters and only 5000 words of core are necessary. Since the computing time does not increase linearly with the number of parameters, two cycles with half the number of parameters refined in each may require less time than one cycle of refinement including all parameters. So from the view of computer time, the sub-division of parameters is no drawback. However, a more rapid convergence is obtained if all parameters are refined together. Thus a partial refinement should only be made if inadequate computer storage makes it necessary.

Another was to save considerable amounts of core is by the so-called "blocking technique". When applying the usual full-matrix refinement procedure, the generation and storage of one-half of the complete $n \times n$ matrix N is needed. When using the blocking technique, the fact that the matrix elements contribute to the result with different weights is used. It follows from statistical considerations that the elements of the inverse matrix N^{-1} are associated with the correlation of the corresponding parameters. That means, if m_{ij} is an element of N^{-1}, its magnitude is a measure for the correlation between the parameters x_i and x_j. Then it follows that the diagonal ele-

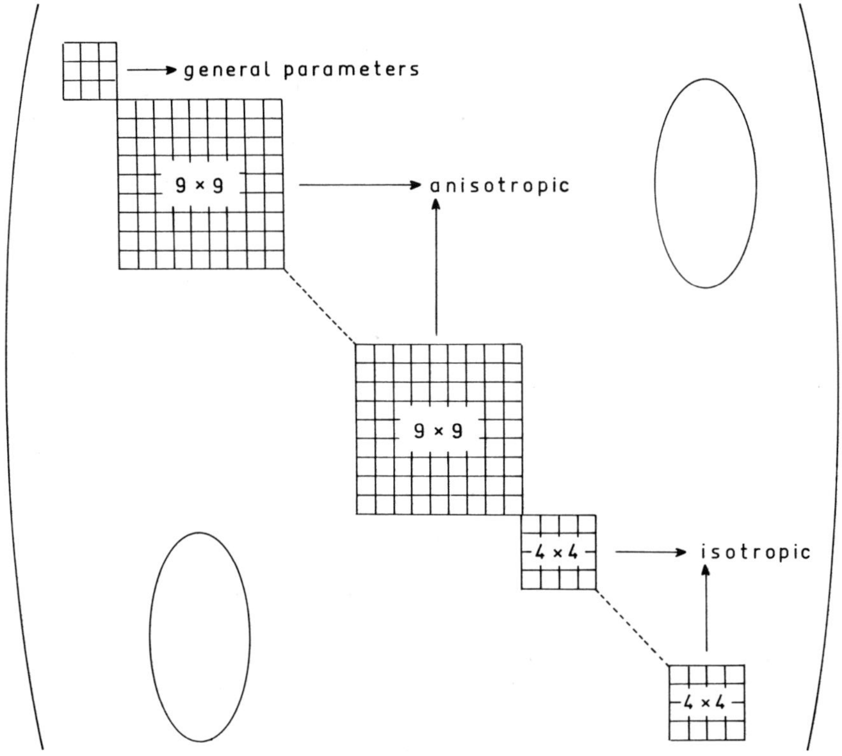

Fig. 6.2. Sub-division of the full matrix into blocks around the principal diagonal.

ments m_{ii} will mainly determine the matrix. Further major contributors will be the elements associated with two parameters of the same atom, while elements corresponding to parameters of different atoms will be of minor weight unless there is a special relation between these atoms.

To save storage, special blocks of the matrix elements which include the elements concerned with one atom are used. It is then ensured that the elements of major weight, as discussed above, are included in the calculation. All other elements are neglected, so the matrix consists only of blocks which are situated symmetrically around the principal diagonal. The order of each sub-matrix block is equal to the number of parameters for one atom. A typical blocking scheme is illustrated in Fig. 6.2. The first block consists of the elements concerned with the general parameters, such as scale factors, etc. For each non-hydrogen atom refined anisotropically, there is a square block of order 9, and for the hydrogens, which are usually refined with isotropic temperature factors, blocks of order 4 are used. A refinement of that type is called a "block diagonal refinement". Using this technique, an appreciable amount of core storage and computer time is saved for the calculation of each cycle. However, the block-diagonal method converges less rapidly than the full-matrix technique, so that the number of cycles to be calculated to convergence can be very much larger. Thus the time saved within one cycle is lost by the need for more cycles.

For medium-sized structures, a modification of the block-diagonal technique can be applied if relatively large core is available. This is done by including the parameters of more than one atom in one block, so that more non-diagonal elements are taken into consideration. If a group of atoms belongs to a special fragment of one molecule, such as a benzene ring, the parameters between these atoms will be more correlated than those between different molecular groups. It is then useful to use blocks of such molecular sub-groups. An improved convergence will be obtained at the cost of more core. Despite the convergence problem the block diagonal technique is an efficient method. Moreover, its application is essential if only minicomputers are available. This is usually the case if a structure determination and refinement program package is intended to run on the computer of a diffractometer system.

Among the computer programs available for least-squares refinement calculations, the most widely used is ORFLS, written by Busing, Martin and Levy, the first version of which was issued in 1959 [Busing, W. R. and Levi, H. A., ORNL report *59-4-37* (1959)]. Almost all programs in use today have been developed from the basis of ORFLS, such as the least-squares program incorporated in the present versions of X-RAY, called CRYLSQ, written by F. A. Kundell. CRYLSQ provides the user with the facility to use the program either in the full-matrix mode, in the pure block-diagonal mode, or to make use of blocks of arbitrary size. In the absence of specifications from the user, blocking is made automatically within the limits of storage. Since we shall execute all the refinement calculations of our three test structures with the CRYLSQ program, we shall give a description of how to use a refinement program when dealing with the examples (see 6.4).

6.2.2 *Execution of a Complete Refinement Process*

Before going to the details of numerical calculations, we shall give a survey of a complete refinement process for one structure. If a first structural model is obtained from a successful phase determination (which might be partially incomplete, as sometimes happens when the heavy-atom method is applied), the refinement of a scale factor, the positional parameters (i.e. the fractional coordinates x, y, z) and an isotropic temperature factor for each atom has to be calculated. The initial values for the scale and temperature factors can be taken from the results of a Wilson plot. If non-hydrogen atoms are still missing, a difference electron density map should be calculated after the first refinement, in order to locate the remaining atoms. If the initial model was very incomplete, all missing atoms may not be derived from the first difference map, and it might be necessary to repeat the process.

With all non-hydrogen atoms included in the model, the R-value should be significantly below 20%. If convergence is obtained with all atoms present with isotropic temperature factors, the R-value should be about 10–15%. A failure to solve the structure at this stage can be recognized in two ways:

(1) The highest maxima found in the difference map do not correspond to reasonable atomic positions;

(2) The R-value does not decrease significantly if additional atoms are included in the refinement.

If either of these are observed, either the original structural model was totally or partially incorrect or the known component of the complete structure is too small. If it is possible to complete the structure and reach an R-value of 12% or so, anisotropic temperature factors can be assigned to all the non-hydrogen atoms, using for the initial values those obtained from the isotropic refinement.

After refinement with anisotropic thermal parameters, the R-value should fall below 10%. At this stage, any hydrogen atoms present can be located from a difference synthesis and added to a further refinement cycle. Since the electron density of a hydrogen atom is significantly smaller than that of other atoms, there is little chance of observing the hydrogens at an earlier stage of the structure determination. It is essential that the non-hydrogen atoms are fitted as well as possible before the hydrogens are located. The locations of the hydrogens are the least precise of all the structural results, and for heavy-atom structures, hydrogen positions may be undetermined. For most organic compounds, generally all hydrogens can be located. If not, there may be errors in the intensity measurement, the sources of which are discussed in the next section. Since the hydrogen positions are about an order of magnitude less precise than those of other atoms, it is customary to include the hydrogens with isotropic temperature factors in the refinement process. Convergence can be assumed if the parameter shifts are in the range of the standard deviations. The final R-value should then be about 2–5%.

We can summarize the four steps of the refinement of a crystal structure as follows:

Step 1: Completion of the structural model with isotropic temperature factors.

Step 2: Isotropic refinement of the model ($R \approx 10$–15%).

Step 3: Anisotropic refinement and location of hydrogens ($R < 10\%$).

Step 4: Refinement of all parameters (in mixed temperature factor mode) to convergence ($R \approx 2$–5%).

The decrease of the R-value in the course of a refinement process is accompanied by an increase in accuracy for the structural parameters. At the end of step 2, the atomic positions should have a precision of a few hundredths of an Angstrom. After the execution of step 4, the non-hydrogen atom positions are usually precise to ± 0.003 to ± 0.006 Å; hydrogen positions are a factor of ten less precise.

6.2.3 *Corrections to be Applied During Refinement*

In the final stages of refinement, discrepancies in the results are sometimes observed which cannot be rationalized as experimental errors in the intensity measurements. It is customary to inspect the agreement between the observed and calculated structure factors for three effects, which might possibly be sources of errors:

(a) absorption,
(b) secondary extinction,
(c) anomalous dispersion.

The problem of absorption has already been discussed in 2.2.2. The question whether or not an absorption correction should be made was considered in 5.1.1. Although from a physical view, an absorption correction, if necessary, should be applied in the earliest stage possible, there are some practical reasons for deferring it to the refinement stage. The main input to any absorption program consists of a precise description of the crystal shape, which is usually done by specifying all crystal faces. This, however, is sometimes difficult because the crystal faces are hard to recognize. Errors in measuring the crystal shape are frequently a source of errors in the absorption correction, which cannot be recognized if the absorption correction is calculated at data reduction time, since no criteria exist to check the confidence of the data set. For structures with a moderate linear absorption coefficient, it is therefore a good practice to apply the absorption correction at some early refinement stage, for example, at the end of isotropic refinement. The user can then use the R-value criterion after an execution of a further refinement cycle with the corrected intensities, to check whether the absorption correction made any significant improvement. It is, however, important to note that in cases where more than the symmetry-independent reflections have been measured, absorption corrections have to be applied at the data reduction stage, so that the averaging of symmetry-related intensities is done with the *corrected* data set.

Another source of error in the intensities is due to extinction. This effect, first treated by Darwin [Darwin, C.G., Phil. Mag. *27*, 315, 675 (1914); Darwin, C.G., ibid. *43*, 800 (1922)] occurs in two different ways, the so-called "primary" and "secondary" extinc-

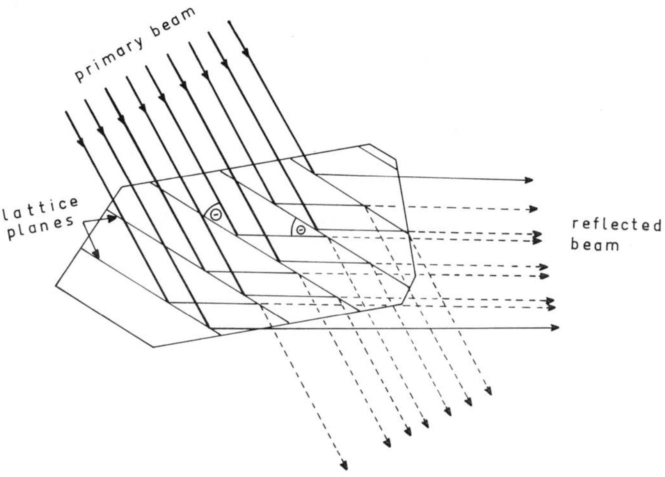

Fig. 6.3. Illustration of primary extinction.

tion. Primary extinction is a weakening of intensity caused by a multiple reflection process. Fig. 6.3 shows, that in an ideally perfect crystal a reflected ray strikes a neighbouring lattice plane at the proper angle for reflection, so that a further reflection occurs. Since there is a phase change of $\pi/2$ for each reflection, the doubly-reflected ray has a phase difference of π relative to the primary beam. This ray therefore not only does not contribute to the reflected beam, but also causes an additional decrease in the intensity of the incident beam. It follows that primary extinction results in an attenuation of both incident and reflected beams.

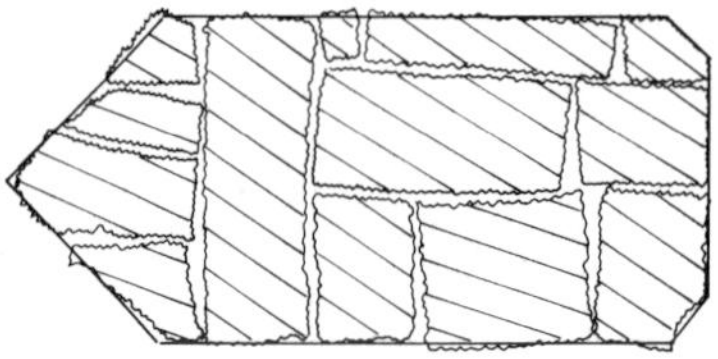

Fig. 6.4. Mosaic structure of a real crystal.

Most real crystals do not contain extended regions with perfect planes as illustrated in Fig. 6.3. Instead, they are sub-divided in a large number of mosaic blocks, all slightly disoriented over a certain angular range (Fig. 6.4). In a mosaic crystal, multiple reflections are less probable than in an ideally perfect crystal, and if the mosaic blocks are small enough, primary extinction can be neglected. The influence of primary extinction on the reflected intensity can be expressed analytically by

$$I \sim |F|^n \tag{6.14}$$

with $n < 2$. A crystal for which I is proportional to $|F|$ rather than $|F|^2$ is said to be ideally perfect, in contrast to an ideally imperfect crystal, for which $n = 2$. Since crystals used for single crystal investigations usually approach the ideally imperfect case, primary extinction is generally not considered.

In contrast, secondary extinction plays a role in all practical single crystal analyses. If a crystal is in a position to cause a reflection of relatively large intensity, the lattice planes firstly encountered by the primary beam will reflect away a significant fraction of the primary intensity. Hence the deeper planes receive less primary intensity and will therefore reflect less than the upper ones. The requirement that the complete crystal volume is exposed to the full incident radiation does not hold, and a weakening of reflected intensity results. This effect of shielding the inner lattice planes by reflection of a fraction of the primary intensity by the outer planes is called secondary extinction. This type of extinction is observed mainly in high intensity reflections of low $\sin\theta/\lambda$ value, and increases with the size and perfectness of a crystal. In a crystal with an almost randomly oriented mosaic structure, the number of lattice planes in the reflection position at the same time is small, so this shielding effect is also small.

Reflections which are affected by secondary extinction can easily be recognized at the refinement stage. Usually these are reflections with low indices and high intensity, for which $|F_o| < |F_c|$. In most organic structures, only a few reflections, less than ten, are affected by secondary extinction. Since no satisfactory analytical way of considering secondary extinction was known until recently, this problem was solved by simply excluding the most seriously affected reflections from further refinement. For several inorganic structures and in the case of neutron diffraction the intensity data may be affected strongly from secondary extinction so that a more effective solution of the extinction problem should be applied.

Larson has given a method for the inclusion of secondary extinction effects in least-squares calculations [Larson, A.C., Acta Cryst. *23*, 664 (1967)], which is based on investigations by Zachariasen [Zachariasen, W.H., Acta Cryst. *16*, 1139 (1963)]. A so-called "isotropic extinction coefficient", g, is introduced which depends only on the crystal specimen used for intensity measurement. A correction for the calculated structure amplitude can then be given by

$$|F_c|_{cor} = t|F_c|(1 + 2\Delta(\theta)\, g\bar{T}|F_c|^2)^{-1/4} \qquad (6.15)$$

where t is the scale factor, $\Delta(\theta)$ is a quantity which can be calculated from the experimental conditions, and $\bar{T}$ the average path length through the crystal [for comparison, see equation (5.6)]

$$\bar{T} = (1/A) \int [p(T) + q(T)]\, e^{-\mu[p(T) + q(T)]}\, dT \qquad (6.16)$$

Since the partial derivatives of $|F_c|_{cor}$ with respect to the structural parameters have been calculated by Larson, the extinction coefficient, g, can be included in the refinement procedure. It is only necessary to have the $\bar{T}$'s for each reflection, which can be calculated from an absorption correction program. In several of the newer least-squares programs, an option to refine an extinction parameter by the method of Larson is included, for instance in the CRYLSQ-link of the X-RAY system.

It should be mentioned that also an anisotropic extinction factor may be used [see Coppens, P. and Hamilton, W.C., Acta Cryst. *A26*, 71 (1970)]. This is rarely justified in X-ray work, but can be important in neutron diffraction.

We shall demonstrate the influence of secondary extinction on a severely affected crystal by the example of the crystal structure of 2, 6-anhydro-β-D-fructofuranose (Dreissig and Luger, see ref. in 5.3.2). In the course of refinement, several ΔF values ($\Delta F = |F_o| - |F_c|$) of strong reflections had $\Delta F \ll 0$, indicating the effect of secondary extinction. In the first attempt to solve this problem, the ten strongest reflections were eliminated from the calculations. An R-value of 4.9% and standard deviations for non-hydrogen bond lengths ranging from 0.005 to 0.008 Å were obtained. Since the structure had two molecules in the asymmetric unit, a comparison of equivalent bond lengths was possible. It was found that, in spite of the relatively good R-value, several bond lengths differed significantly. Ten out of 22 differences were greater than 3σ. Further refinements were then done with the inclusion of an extinction parameter.

After convergence, we obtained a g-value of $g = 2.4 \times 10^{-4}$. The R-value decreased to 2.8%. The extinction correction after (6.15) exceeded 10% of the original $|F_c|$ for 49 reflections, and even 50% for three reflections (the total number of reflections was 1481, including 61 less-thans). The standard deviations of the bond lengths decreased to 0.003 Å for the non-hydrogen atoms, and now only three pairs of equivalent bond lengths differed by more than 3σ, although the significance limit was much lower.

This example shows that secondary extinction may have a significant influence on the structural results. Therefore, if a large number of strong reflections are present with $|F_c| \gg |F_o|$, an extinction correction parameter should be included in the refinement.

Another correction which is introduced at the refinement stage is that for "anomalous dispersion". In 3.2.1, we introduced the atomic scattering factors as the Fourier transforms of the atomic electron densities. These scattering factors were real functions dependent only on $s = \sin\theta/\lambda$, so that all reflections with the same absolute value $|\mathbf{h}|$ have the same f. This simple representation of the scattering factors is only valid if the wavelength λ of the incident beam is significantly different from the wavelength λ_k of the K-absorption edge for an atom of the scattering material. If, however, $\lambda \approx \lambda_k$, the scattering process shows an unusual behaviour caused by an anomalous phase-shift of the scattered wave. This effect, called "anomalous dispersion", can be expressed analytically by replacing the real atomic scattering factor f by a complex quantity f_A which is obtained from f, by inclusion of a real and an imaginary correction term,

$$f_A = f + \Delta f' + i\Delta f'' \tag{6.17a}$$

The effect of anomalous dispersion increases with the wavelength of the incident radiation, and is usually significant only for those elements of the scattering material having an atomic number close to that of the target material of the X-ray tube.

For most atoms, the corrections $\Delta f'$ and $\Delta f''$ are tabulated in the International Tables, Vol. III, for Cr-, Cu- and Mo$K\alpha$ radiation. They are insensitive to s and in practice it is sufficient for most cases to use the values for $s = 0$ when such corrections are necessary.

In practical X-ray work, anomalous dispersion is negligible for Cu$K\alpha$ and Mo$K\alpha$ radiation for atoms having an atomic number less than 20. For very accurate analyses, or for the determination of absolute configurations, anomalous scattering factors are sometimes used for sulfur or chlorine atoms, but for organic structures generally, the effect is usually ignored. If heavier atoms are present, the magnitudes of $\Delta f'$ and $\Delta f''$ indicate whether is is worthwhile to correct the scattering factors for anomalous dispersion.

The effect of anomalous dispersion has important consequences on the structure factor expression, which can be utilized for some special purposes. The first is that Friedel's law no longer holds. From expression (3.11) for the structure factor

$$F(\mathbf{h}) = \sum_{j=1}^{N} f_j \, e^{2\pi i \mathbf{h}\mathbf{r}_j} \tag{3.11}$$

it follows immediately that $F^*(\mathbf{h}) = F(-\mathbf{h})$ and thus,

$$I(-\mathbf{h}) = I(\mathbf{h})$$

This last equation, which was expressed in (3.21) as Friedel's law, has to be modified if f is replaced by f_A. Since f_A is complex, let us write

$$f_A = |f_A|\, e^{2\pi i\alpha} \qquad 0 \leq \alpha < 2\pi \tag{6.17b}$$

Then we get the expression for the structure factor, corrected for anomalous dispersion, which is

$$F_A(\mathbf{h}) = \sum_{j=1}^{N} |f_{Aj}|\, e^{2\pi i\alpha_j}\, e^{2\pi i\mathbf{hr}_j}$$

or

$$F_A(\mathbf{h}) = \sum_{j=1}^{N} |f_{Aj}|\, e^{2\pi i(\alpha_j + \mathbf{hr}_j)} \tag{6.18}$$

For $F_A(-\mathbf{h})$ we get

$$F_A(-\mathbf{h}) = \sum_{j=1}^{N} |f_{Aj}|\ {}^{2\pi i(\alpha_j - \mathbf{hr}_j)}$$

and then

$$F_A(\mathbf{h})\, F_A^*(\mathbf{h}) \qquad = \Big(\sum_{j=1}^{N} |f_{Aj}|\, e^{2\pi i(\alpha_j + \mathbf{hr}_j)}\Big)\Big(\sum_{j=1}^{N} |f_{Aj}|\, e^{-2\pi i(\alpha_j + \mathbf{hr}_j)}\Big)$$

$$F_A(-\mathbf{h})\, F_A^*(-\mathbf{h}) = \Big(\sum_{j=1}^{N} |f_{Aj}|\, e^{2\pi i(\alpha_j - \mathbf{hr}_j)}\Big)\Big(\sum_{j=1}^{N} |f_{Aj}|\, e^{-2\pi i(\alpha_j - \mathbf{hr}_j)}\Big)$$

Since these two expressions are different generally and since $I(\mathbf{h}) \sim F(\mathbf{h})\, F^*(\mathbf{h})$, it follows that Friedel's law is no longer valid in the presence of anomalous dispersion. However, for centric structures, we get for the structure factor expression

$$F_A(\mathbf{h}) = \sum_{j=1}^{N/2} |f_{Aj}|\, e^{2\pi i\alpha_j}\big(e^{2\pi i\mathbf{hr}_j} + e^{-2\pi i\mathbf{hr}_j}\big)$$

$$F_A(\mathbf{h}) = \sum_{j=1}^{N/2} 2|f_{Aj}|\, e^{2\pi i\alpha_j} \cos 2\pi\mathbf{hr}_j$$

Since the cosine is an even function, we get

$$F_A(\mathbf{h}) = F_A(-\mathbf{h})$$

and therefore

$$I(\mathbf{h}) = I(-\mathbf{h})$$

Therefore, because of the anomalous dispersion, the structure factor of a centric structure is no longer real, but is a complex quantity, as for the acentric case. However, since $F_A(\mathbf{h}) = F_A(-\mathbf{h})$, Friedel's law holds.

We can now summarize: If anomalous dispersion has to be considered, the structure factor is always a complex quantity; Friedel's law is still valid for centric structures but not in the acentric case. An important and interesting application can be made from this last property. In 5.3.2, we discussed the problem of enantiomorphic structures, and it was found that a decision between a left-handed and a right-handed structure was impossible by X-ray methods, provided that Friedel's law was valid. When Friedel's law is no longer true, the effect of anomalous dispersion can be used to derive the correct absolute configuration. Denoting the atom position vectors of the left-handed (L) and right-handed (R) structure by $\mathbf{r}_j(L)$ and $\mathbf{r}_j(R)$ ($j = 1, \ldots, N$), we get the relation given be equation (5.52),

$$\mathbf{r}_j(R) = -\mathbf{r}_j(L) \qquad j = 1, \ldots, N$$

By using f_A from (6.17b) instead of f, we get for the structure factors

$$F_{AR}(\mathbf{h}) = \sum_{j=1}^{N} |f_{Aj}|\, e^{2\pi i[\alpha_j + \mathbf{h}\mathbf{r}_j(R)]} = \sum_{j=1}^{N} |f_j|\, e^{2\pi i[\alpha_j - \mathbf{h}\mathbf{r}_j(L)]} = F_{AL}(-\mathbf{h})$$

Since Friedel's law does not hold, we get

$$F_{AR}(\mathbf{h})\, F_{AR}^{*}(\mathbf{h}) = F_{AL}(-\mathbf{h})\, F_{AL}^{*}(-\mathbf{h}) \neq F_{AL}(\mathbf{h})\, F_{AL}^{*}(\mathbf{h})$$

It follows that the intensities of right- and left-handed structures are not equal, and the differences can be used for a decision in favor of the correct absolute configuration.

It has to be noted that the question of absolute configuration frequently arises for light atom organic compounds which show only small anomalous dispersion. Nevertheless, there are several examples in which the relatively small dispersion effect of oxygen with CuKα radiation was sufficient to determine the absolute configuration. However, this work requires exceptionally precise intensity measurements. The first successful application of anomalous dispersion for determining absolute configuration was described by Bijvoet and co-workers [Bijvoet, J.M., Peerdeman, A.F. and Van Bommel, A.J., Nature *168*, 271 (1951)]. From their famous investigation on a NaRb tartrate, the absolute configuration of tartaric acid was derived experimentally for the first time.

6.3 Analysis and Representation of Results

6.3.1 *Geometrical Data*

In all stages of refinement and especially at the end of the refinement, the results should be checked carefully by calculation of geometrical molecular data, which are a great help in judging the validity of the model. These geometrical results are generally the reason for the structure analysis, and they are, finally, the basis of structural discussion of the chemical or physical properties of the structure.

The quantities most frequently calculated are interatomic distances and valence and torsion angles. Vector algebra can be used to derive the necessary formulae. If $\mathbf{r}_1$ and $\mathbf{r}_2$ are the position vectors of two atoms A_1 and A_2, the distance d between A_1 and A_2 is given by the scalar product

$$d^2 = (\mathbf{r}_1 - \mathbf{r}_2)^2 \tag{6.19}$$

In crystallography, we cannot generally assume an orthonormal system, so all scalar products between the base vectors have to be included in the calculation (see 1.1.4).

Three different types of atomic distances are used in structure analysis:

(1) If A_1 and A_2 are bonded atoms, their distance is said to be a *bond distance* or a *bond length*

(2) If A_1 and A_2 are non-bonded atoms within *one* molecule, their distance is said to be an *intramolecular contact distance.*

(3) If A_1 and A_2 are atoms of different molecules within the crystal lattice, their distance is called an *intermolecular contact distance.*

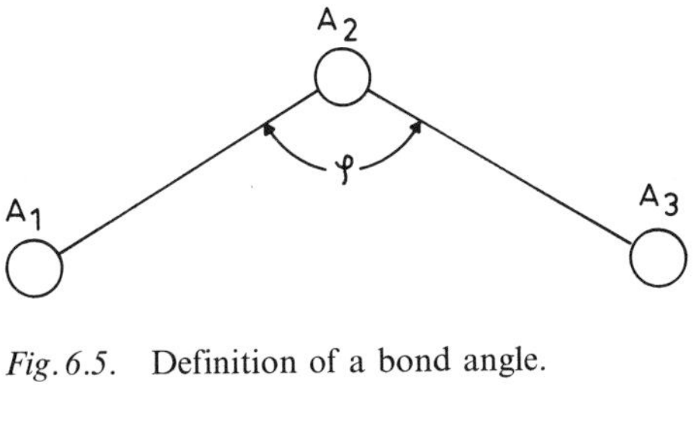

Fig. 6.5. Definition of a bond angle.

Angles are usually calculated only between bonded atoms. If $\mathbf{r}_1$, $\mathbf{r}_2$, $\mathbf{r}_3$ are the position vectors of atoms A_1, A_2, and A_3, with A_1 and A_3 bonded to A_2, the bond angle φ (see Fig. 6.5) is given by

$$\cos\varphi = \frac{(\mathbf{r}_1 - \mathbf{r}_2)(\mathbf{r}_3 - \mathbf{r}_2)}{|\mathbf{r}_1 - \mathbf{r}_2||\mathbf{r}_3 - \mathbf{r}_2|} \tag{6.20}$$

The result of a bond lengths and angles calculation provides the user with useful information. However, some stereochemical aspects cannot be discussed with these data alone. If we have, for instance, a cyclohexane or pyranosyl fragment in a molecule (as in sucrose), it is difficult to decide from bond lengths and valence angles whether the substituents are linked axially or equatorially.

In Fig. 6.6a and b, the molecular models of α- and β-D-glucose are represented. They differ only by the position of the OH-group at C-1 which is linked axially for the α-compound and equatorially for the β-compound. If both structures are described by their bond lengths and valence angles, it is difficult to distinguish between the α- and β-forms. This can be done immediately if the torsional angle (see definition in 1.1.6) O-1 – C-1 – C-2 – H-2 is calculated. If it has the magnitude 180°, we have the α-form, if it is +60°, the β-form is present.

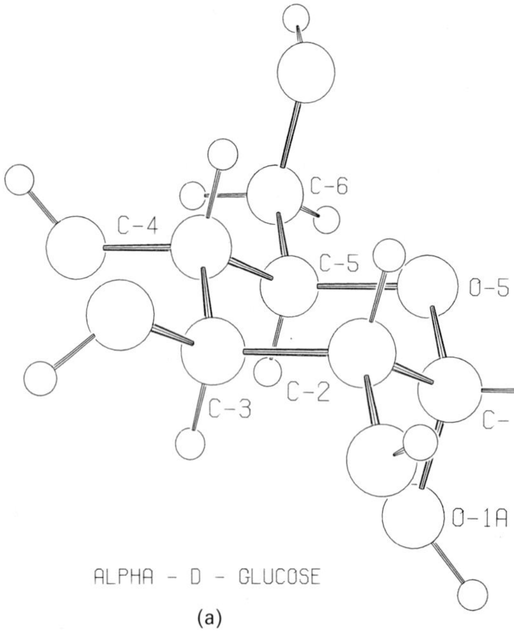

(a)

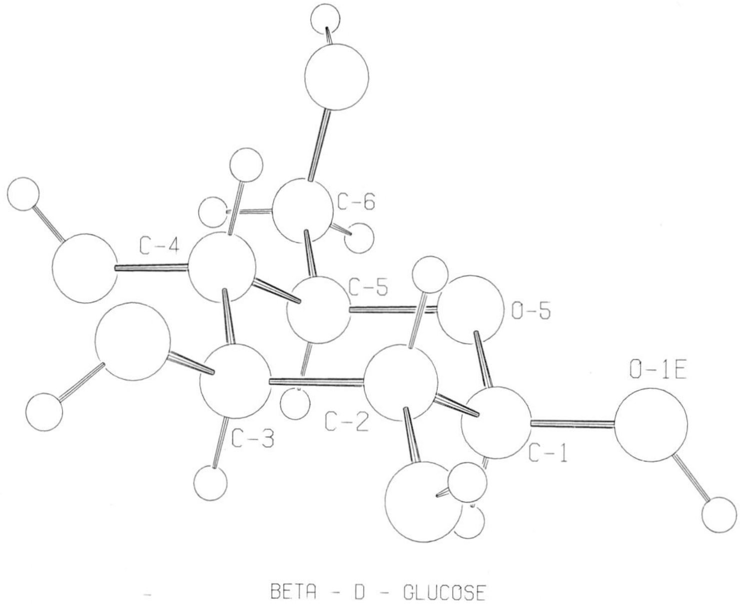

(b)

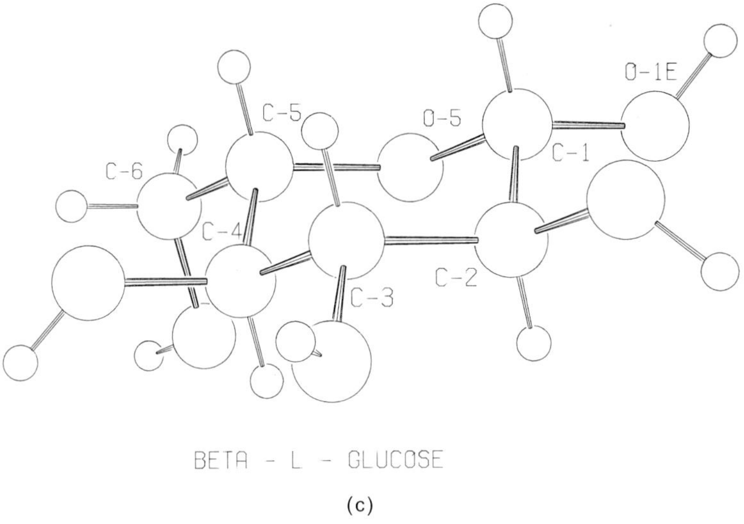

Fig. 6.6. Different forms of glucose:
(a) α-D-glucose; (b) β-D-glucose; (c) β-L-glucose.

For describing molecular configurations and conformations, the calculation of torsional angles is therefore necessary. The sign of a torsional angle (defined in 1.1.6) can be used to decide between D- and L-forms of optically active compounds. For instance, β-D-glucose and the corresponding L-form (Fig. 6.6b and c) are equal in their bond lengths and valence angles, and in the magnitudes of their torsional angles. A decision between the two forms is possible only from the signs of the torsional angles (see Table 6.1).

Table 6.1. Signs of Torsional Angles in the Pyranosyl Ring of D- and L-Glucose(Fig. 6.6 b,c).

Sequence	Sign (D-form)	Sign (L-form)
C-1 – C-2 – C-3 – C-4	−	+
C-2 – C-3 – C-4 – C-5	+	−
C-3 – C-4 – C-5 – O-5	−	+
C-4 – C-5 – O-5 – C-1	+	−
C-5 – O-5 – C-1 – C-2	−	+
O-5 – C-1 – C-2 – C-3	+	−

For molecules containing sub-groups which for chemical reasons might be planar (such as phenyl groups, etc.), it is also customary to calculate a so-called "*least-squares plane*". Using a formalism similar to that described in 6.1.2, a plane is calculated which best fits the atoms which are assumed to be planar. The mean-square deviation of atoms contributing to that plane is a good measure for the planarity of a group. Phenyl rings, for example. are usually planar within 0.005 to 0.01 Å. From the least-squares-plane equations, the angles between different planar groups can be calculated to give further useful information about the shape of the molecules.

Computer programs are available for all these geometrical calculations. In the X-RAY system, distances, angles, and torsional angles (including their standard deviations derived from the errors of atomic positions) are calculated in a sub-program called BONDLA. Least-squares planes (and lines) can be obtained by the so-called LSQPL sub-program.

6.3.2 *Graphical Representations*

The most illustrative way of representing structural results is given by graphical methods, which involves, however, always the problem that a three-dimensional model must be represented two-dimensionally. Although it is a question of individual taste which kind of representation is preferred, some types of graphical representations, being in parts supported by computer programs, are generally in use. Of these, we shall describe three methods

(1) The method which has been used from the earliest days of X-ray analysis is the drawing of contours obtained from the electron density map. At given intervals (0.5 or 1.0 e/Å^3, or so) points of equal electron density are contoured on each page-to-page level of the Fourier map and then a projection is drawn. An example of this type of drawing is given in Fig. 6.7, showing the projection of one unit cell of the p-Cl-phenyl-diphenylphosphin oxide structure onto the **a**-**b** plane [Dreissig, W., Dissertation Thesis, Freie Universität Berlin (1969)]. An advantage of this type of representation is that a good three-dimensional impression usually appears and that the shape of the atomic density maxima is generally visible. It has the disadvantage that these drawings must generally be made by hand, since computer programs which carry out the contouring and give the output on a plotter or a line printer, are rarely available. For this reason, this representation is seldom used these days, except for special studies where the objective is to measure very accurately the details of the valence electron distributions.

(2) A very simple but fast method is the projection of atomic positions and their corresponding bond lengths on a selected plane. Computer programs for the execution of this type of drawing exist (e.g. the sub-program PROJCT of the X-RAY system, which runs only for a few seconds and gives a fairly good output on the line printer), and even if they are not available, such ball and stick drawings onto the unit cell planes can be made by hand with relatively little effort. The draw-back of this method is that a three-dimensional impression of the structure is lost. Since these projections are easily obtained, however, they are frequently used for a first view of the structure. An example of such a projection obtained on the line printer from the program PROJCT is illustrated in Fig. 6.8., representing two unit cells of methyl-2,4-bis(N-acetyl-N-

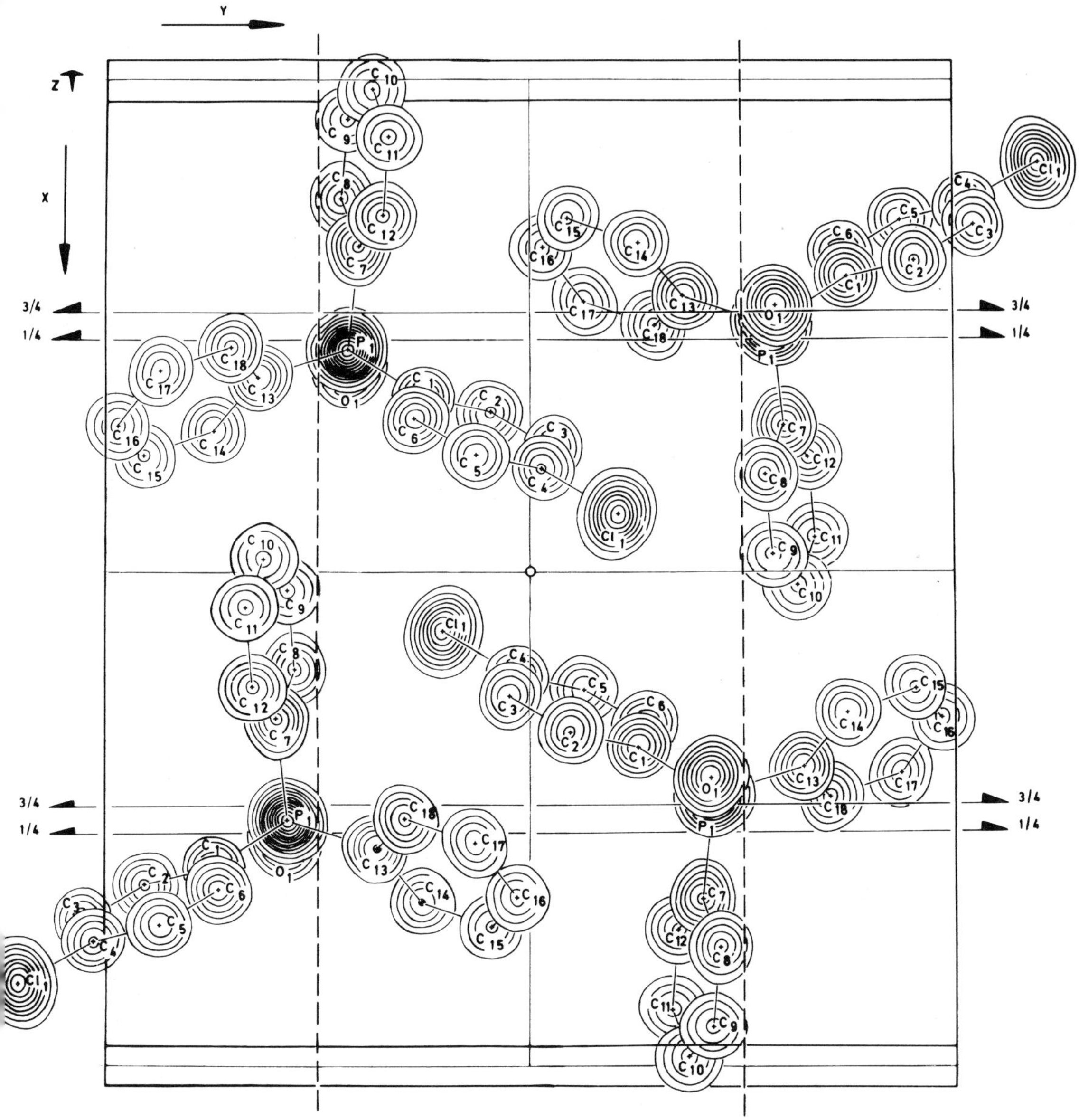

Fig. 6.7. Contoured Fourier map of one unit cell of p-Cl-phenyl-diphenylphosphine oxide.

benzoylamino)-3,6-di-O-benzoyl-2,4-dideoxy-α-D-idopyranoside. [Luger, P. and Paulsen, H., Acta Cryst. *B34*, 1254 (1978)]. Although it is a very simple drawing the orientation of the large substituents can be visualized and a first impression of the crystal packing can be obtained. For more complex molecules, this method has serious limitations.

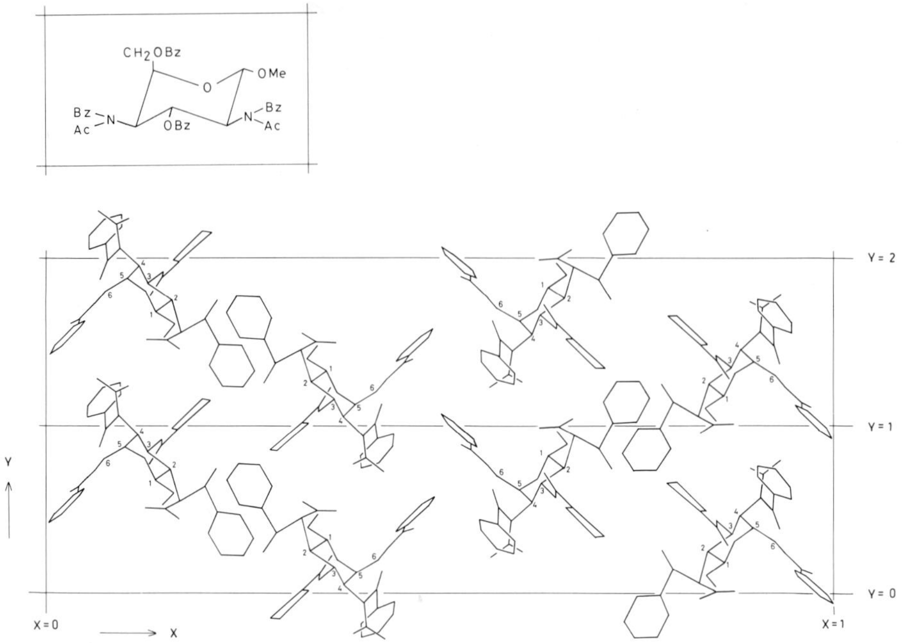

Fig. 6.8. Example of a simple structure projection, showing two units cells of methyl-2,4-bis(N-acetyl-N-benzoylamino)-3,6-di-O-benzoyl-2,4-dideoxy-α-D-idopyranoside.

(3) A very impressive way of representing crystal structures was introduced in 1965 by the development of the first version of the ORTEP plotter program [Johnson, C.K., ORTEP, Report ORNL-3794, Oak Ridge National Laboratory, Oak Ridge, Tennessee (1965)]. This program provides the drawing of a structural model on a plotter with the atoms represented by their thermal ellipsoids and the bonds drawn stick-like. These drawings give a spectacular impression of the structure. A pair of stereoscopic drawings can also be plotted, thereby providing for the first time a representation which has full three-dimensional quality. A number of further options enables the user to plot the complete unit cell or arbitrary sections of the crystal lattice in any projection direction. Several types of atom and bond representations are also provided. Fig. 6.9 and Fig. 6.10 show two examples of ORTEP applications. In Fig. 6.9, a stereoscopic pair of drawings of the molecular model of trans-cyclohexane dicarboxylic acid (1.4) is represented. Fig. 6.10 shows a section of the crystal lattice of a germanium hemi-porphyrazin structure [Hecht, H.J. and Luger, P., Acta Cryst. *B30*, 2843 (1974)].

ORTEP is now the most widely distributed and most frequently used plotter program for the representation of crystal structures.

Although an ORTEP run needs little calculation time in practical use, the time needed by the plotter is much longer. Unless ample plotting time is available, it may take more than a day to get one plot. Several plots are usually necessary to obtain an

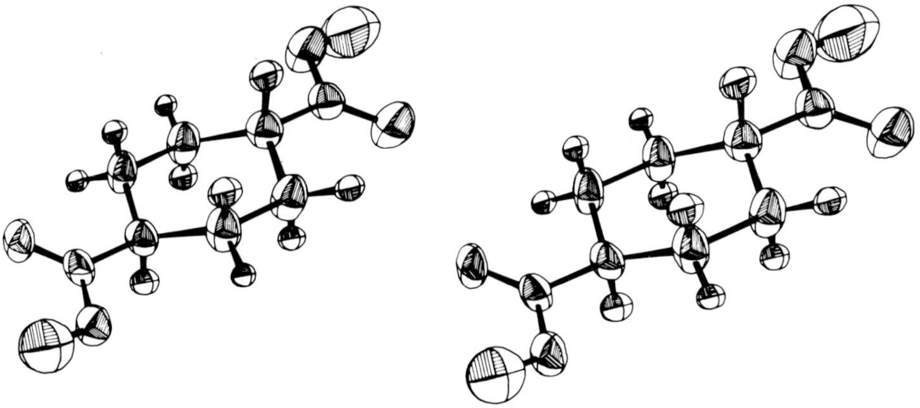

Fig. 6.9. Stereoscopic pair of drawings of the molecular model of trans-cyclohexane dicarboxylic acid (1,4) drawn by ORTEP.

optimal representation, and the generation of a good ORTEP diagram can take a relatively long time. More recently, modifications of ORTEP have been developed which give the diagram on a rapid printer, or on a TV display console. ORTEP diagrams can then be obtained in a few minutes. The user can produce several plots in rapid succession, and then decide which to reproduce on the slower plotter. An example of a molecular model generated on a Tectronix screen attached to a PDP-15/40 computer [ORTEP, locally modified by W. Dreissig, Freie Universität Berlin (1978)] will be shown in 6.4.3, for the SUCROSE molecule. The combinination of representing an ORTEP model optinally on a TV display console and a plotter seems to be the most effective method of obtaining the best structural models in the fastest way.

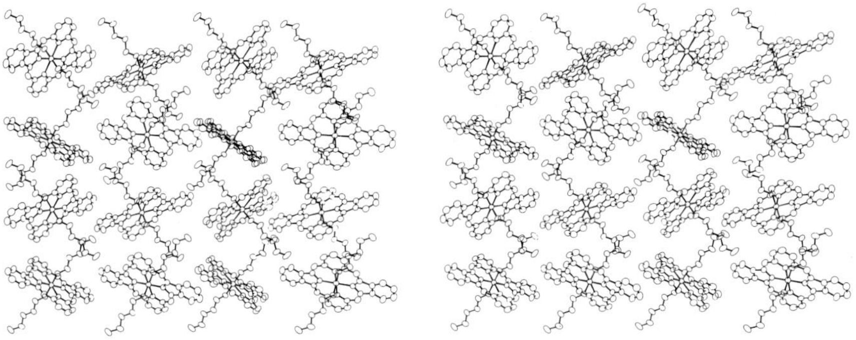

Fig. 6.10. Example of a stereoscopic lattice representation, drawn by ORTEP.

6.4 Applications to the Test Structures

We shall now complete the structure determinations of our three test structures. In all three cases, we will describe the completion and refinement of the model and give a brief description of the most important results.

6.4.1 *Completion and Refinement of the KAMTRA Structure*

Starting with the KAMTRA structure, for which we have only the potassium position, derived by application of the heavy-atom method, we start a first refinement using only the K^+ position with a temperature factor and a preliminary F_{rel} scale factor obtained from a Wilson plot ($B = 1.0$ Å^2, $c = 1/t = 0.207$). The phases derived from this model are then used in a subsequent ΔF-synthesis, from which we should get more atomic positions. The R-value after one cycle of refinement was $R = 46.5\%$, significantly below the theoretical value for random structures (see equation (6.2)). The peak distribution obtained from the difference map was not interpretable from the chemical point of view. The reason for this is that the K^+ position has a y-coordinate of 0.034, which differs only slightly from zero. If this value were exactly zero, the difference map would show the superposition of the structure and its reflected image relative to the x-z plane. This can be understood from the symmetry of $|F(hkl)|$ in the orthorhombic system.

Let us denote the KAMTRA structure by $\varrho = \varrho(x, y, z)$. Then the corresponding structure reflected on the x-z plane is given by $\varrho' = \varrho(x, -y, z)$. If $F(\mathbf{h}) = \mathbf{F}(\varrho)$ and $F'(\mathbf{h}) = \mathbf{F}(\varrho')$, we get

$$F(hkl) = \sum_{j=1}^{N} f_j \, e^{2\pi i(hx_j + ky_j + lz_j)}$$

and

$$F'(hkl) = \sum_{j=1}^{N} f_j \, e^{2\pi i(hx_j - ky_j + lz_j)} = F(h-kl)$$

Since for the orthorhombic system, $|F(hkl)| = |F(h-kl)|$, we get $|F'(hkl)| = |F(hkl)|$. It follows that the reflection intensities of ϱ and ϱ' are equal and that the two models are indistinguishable. Both structures are illustrated schematically in Fig. 6.11. If the potassium ion is situated at $y = 0$, its position remains unchanged when reflected on the x-z plane. So it could belong to either of the two reflected images of the structure (Fig. 6.11a). The difference map must then show the superposition of both images as illustrated in Fig. 6.11b, which is impossible to interpret.

We can avoid this unfavorable situation only if we can find a structural fragment which is unsymmetrical about the x-z plane. We inspect the difference map for a peak in a general position for which the difference vector of the K^+ position appears in the Patterson map. As an additional argument, we can use the known chemical property that K-O contact vectors generally have a length of about 2.8 Å. So, if we can find a large peak in the difference synthesis for which the peak-potassium vector is present in

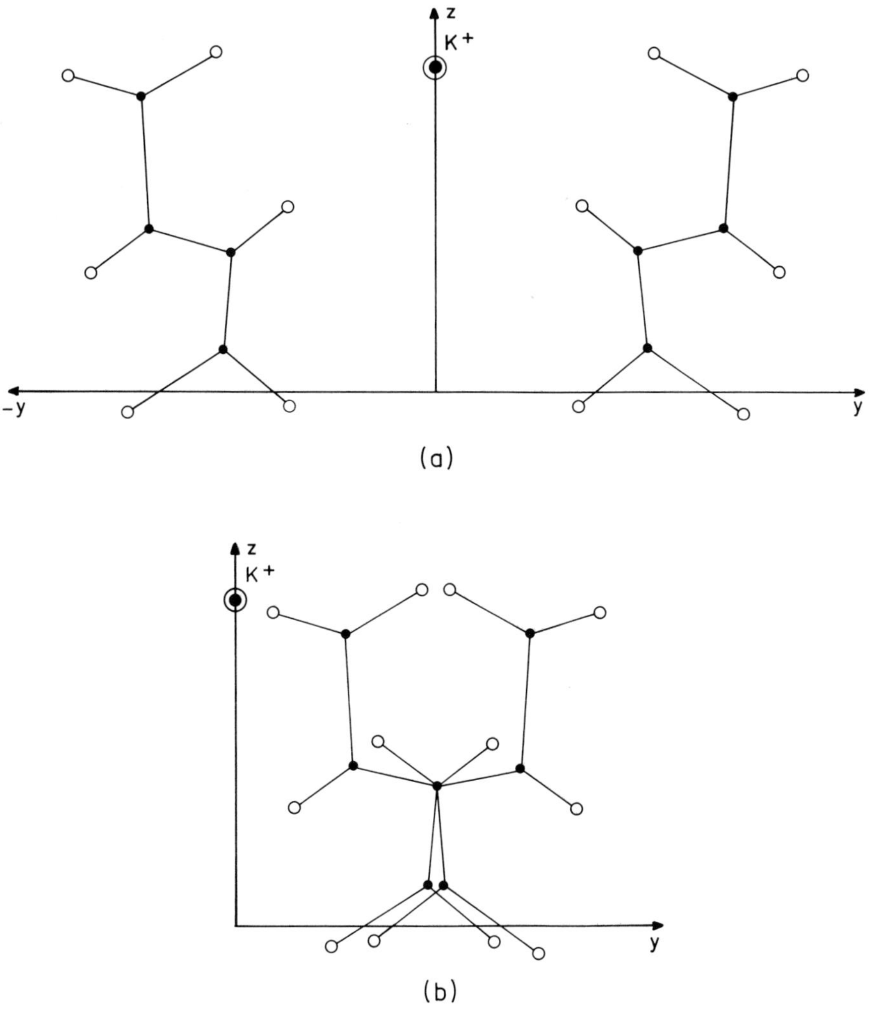

Fig. 6.11. Schematic drawings of (a) mirror-related images of the KAMTRA structure, and (b) their superposition.

the Patterson map with a length of 2.8 Å, it is very probable that this peak represents the position of an oxygen atom. We find that peak no. 3 has the desired property (the numeration chosen refers to the peak height). Designating this peak as O-1, we start a second run of refinement and subsequent difference synthesis with K^+ and O-1 used as input (see Fig. 6.12 in which this and the following steps are illustrated). The R-value only decreased to 44.5%, but a fragment is present in the difference map corresponding to the expected geometry of a carboxyl group. Using this group as input in the third run, the R-value dropped to 39.1% and a relatively large fragment can be recognized, having in part the geometry expected for the tartrate anion. This fragment, which consists of seven atoms, has, unfortunately, no connection to be carboxyl group obtained from the previous run. But for the moment, we ignore that and start a fourth run. Since we are uncertain at this moment about the chemical identity of the peaks in the large fragment, they are all treated as carbon atoms. After this run, the R-value was reduced to 21.4% and the connection between the two fragments was found. The C-1 position, found after the second run, was a "ghost", and the true position of this atom is now given by the highest peak (Z1, see Fig. 6.12). With this last atom position, we have the complete model of the tartrate anion as illustrated at the bottom of Fig. 6.12. Since this model corresponds to that expected from chemical considerations, there is no doubt that step 1 of the refinement process, as defined in 6.2.2 is finished.

We now carry out three cycles of isotropic refinement of all atoms (having assigned the correct chemical identity to each peak). The R-value again decreases significantly

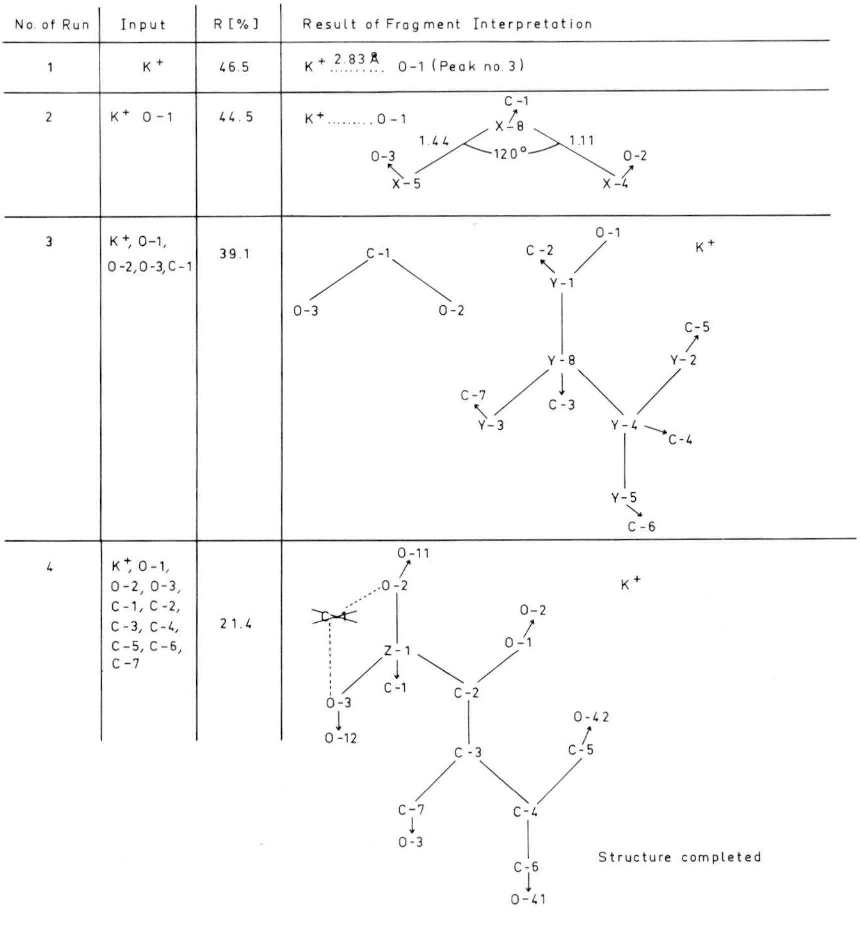

Fig. 6.12. Completion and refinement steps for the tartrate anion. The enumeration of the peaks X – i, Y – i, Z – i(i = 1,2,...) found in the subsequent difference syntheses refers to the peak heights.

to 5.7%. It should be noted that for organic structures without a heavy atom, the R-value is usually larger at this stage, being in the range of 10–15%.

Two further cycles of refinement with anisotropic thermal parameters gives an R-value of 4.0%. From the corresponding difference synthesis, we obtain all five hydrogen positions unambiguously. These are included in two more cycles of refinement with isotropic temperature factors. After this final calculation, all parameters have converged. The R-value is 2.7%, the standard deviations are $5–6 \times 10^{-3}$ Å for C-C and C-O bond lengths and 0.3° for angles between atoms of this type. The standard deviations of bond lengths and angles involving hydrogens have larger values by approximately one order of magnitude.

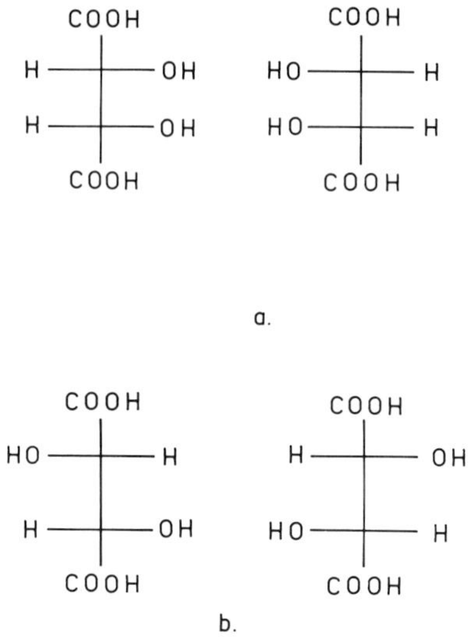

Fig. 6.13. Fischer projections of the different forms of tartaric acid:
(a) the two identical models of the meso form. One form can be obtained from the other by rotation about 180°;
(b) the two optically active forms; (−)-(2S.3S)-form on the left, (+)-(2R.3R)-form on the right, both related by a reflection.

The structure determination of KAMTRA is completed, and an analysis and discussion of the results can be made. This structure has been known for 20 years, since in 1958 it was described by van Bommel and Bijvoet [van Bommel, A.J. and Bijvoet, J.M., Acta Cryst. *11*, 61 (1958)]. Nevertheless, there are a few interesting features which can be derived from our results. It is well known that tartaric acid exists in an optically inactive meso from (Fig. 6.13a) and in the two enantiomorphic (+)–(2R. 3R) and (−)–(2S. 3S)–forms (Fig. 6.13b). From the model of the tartrate anion (Fig. 6.14a) derived from this X-ray analysis, we can decide whether KAMTRA is the salt of the meso form or of one of the optically active forms. A comparison of the Fischer projections (note the conventions of the Fischer projection: bonds drawn horizontally point from the plane of the paper upward into the direction of the viewer; vertical bonds are directed below the plane of the paper) with our model, shows clearly that we have a salt of the optically active form. However, we cannot determine whether we have the + or the − form, since this question cannot be decided unless anomalous dispersion is utilized (see 6.2.3). Although our model is identical to the (+)−(2R. 3R) – form, as shown in Fig. 6.14a, this result is coincidental. It is a consequence of the choice of the second atom (designated O-1 above). If the reflected image of that peak were taken, the enantiomorphic structure would have resulted.

The bond lengths and valence angles are given in Fig. 6.14b, with their standard deviations. These results may be used as the basis for a chemical discussion. Fig. 6.14c gives an illustration of the KAMTRA lattice.

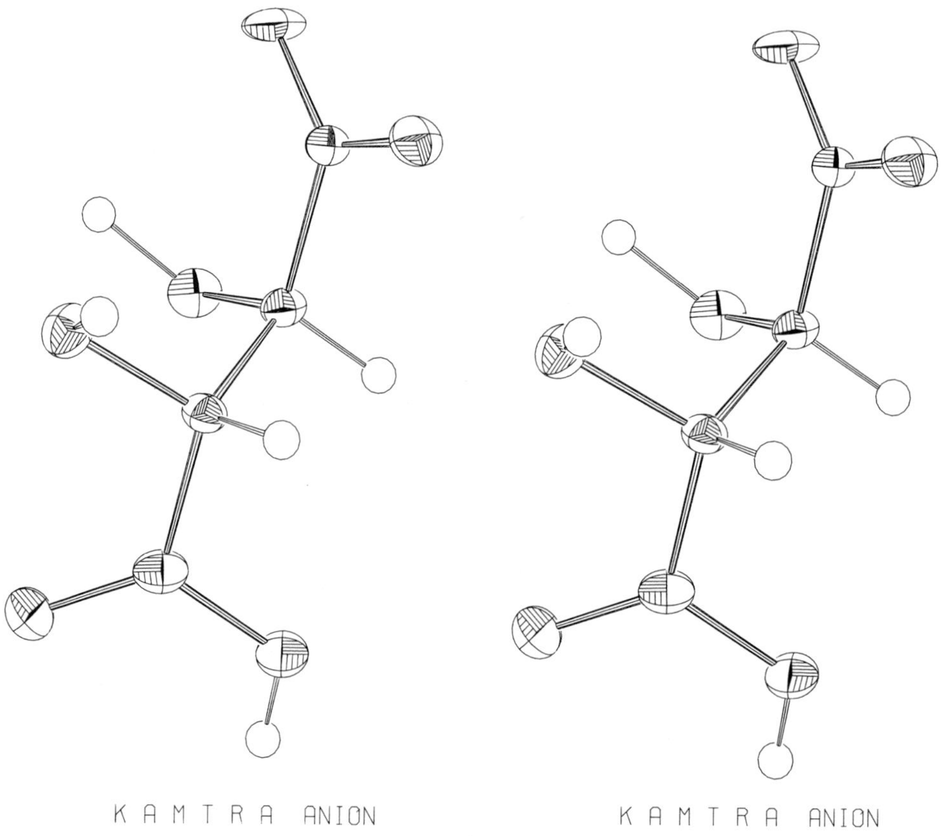

Fig. 6.14. (a) The tartrate anion found in the KAMTRA structure analysis;

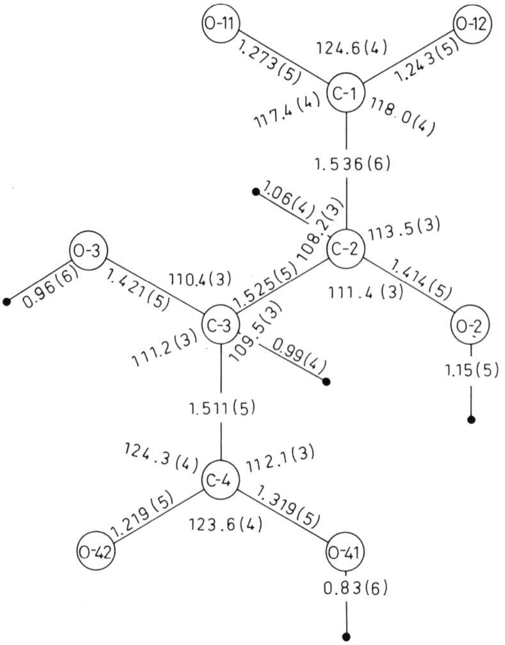

(b) bond lengths (Å) and valence angles in the tartrate anion (estimated standard deviations are given in parentheses);

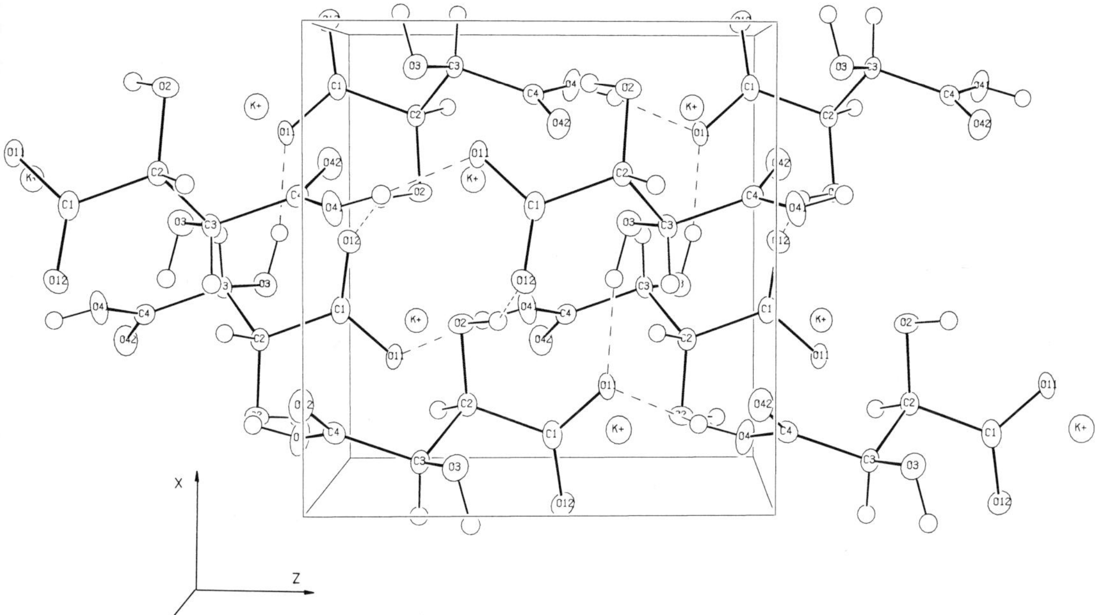

(c) crystal lattice of KAMTRA.

6.4.2 *NITROS Refinement with Extinction Correction*

As mentioned in 6.1.1, a first F_c-calculation using the NITROS structure model obtained from direct methods gave an R-value of 25.6%. From this it was clear that the model was correct, in principle. A difference Fourier synthesis was calculated to complete the structure by adding the missing nitrogen atom of the cation. One peak having a relative height four times the next peak in sequence was found on the mirror plane (x = 0.144, y = 0.0, z = 0.832), as was expected from symmetry. It was clear that this was the missing nitrogen position, so the structure could be regarded as completed, except for the hydrogen positions.

Next, a three-cycle isotropic refinement was carried out. Note that for the atoms situated in special positions (S-1, N-3, S-3, O, and the cation nitrogen N-4), the y-coordinates must be fixed and not refined. Usually the user of a refinement program has to specify the status of a parameter (refine or invariant), but in the CRYSLQ-link of the X-RAY system, special position parameters are recognized and are set invariant automatically by the program.

So, in addition to the scale factors and the atomic parameters (positional and thermal), the following quantities are needed as input to the least-squares program,

(a) number of cycles (we chose three),
(b) temperature factor mode (isotropic),
(c) refinement type (full-matrix mode, based on F). In this relatively small structure, we have only 11 independent atoms, so the full-matrix mode can be used throughout the refinement.
(d) weighting scheme (we chose the $1/\sigma$ weighting scheme)
(e) a number of output options, which will not be described in detail here.

From this input we get a three-cycle refinement calculation which dropped the R-value from 21.2% (before the first cycle), to 16.8% (after the first cycle), 16.4% (after the second cycle), and to 16.4% after the third cycle. Since there is no change in R-value from the second to the third cycle, we can regard the isotropic refinement (step 2, see 6.2.2) as completed. The standard deviations of bond lengths and angles, which should be calculated after each refinement, are close to 0.01 Å and 0.6°. Since the R-value is relatively large at this stage, we carry out a two-cycle anisotropic refinement, after which we request an output of those reflections for which $\Delta F = |F_o| - |F_c|$ is larger than a given threshold, S (we chose S = 20). After these two cycles of anisotropic refinement, the R-value decreases via 7.5% to 7.1% and the reflection list given in Table 6.2 shows seven strong reflections for which $|F_c|$ is considerably larger than $|F_o|$.

From the discussion in 6.2.3, we assume that these reflections are strongly affected by secondary extinction. We decide to apply an extinction correction, rather than exclude these reflections from further refinement. To get the $\bar{T}$'s (see equation (6.16)), we carry out an absorption correction, although this would not be necessary from the magnitude of μ ($\mu = 11.8\ cm^{-1}$ for MoKα). We then add a further three-cycle refine-

Table 6.2. Seven Reflections of NITROS Which are Affected by Secondary Extinction.

	Before Extinction Correction				After Extinction Correction			
h k l	$\|F_o\|$	$\|F_c\|$	ΔF	$\|\Delta F/F_o\|$	$\|F_o\|$	$\|F_c\|$	ΔF	$\|\Delta F/F_o\|$
-2 0 1	61.1	111.9	−50.8	0.83	54.1	73.5	−19.3	0.35
2 0 1	75.7	112.3	−36.6	0.48	67.0	70.3	−3.2	0.05
4 0 0	68.3	111.2	−42.9	0.63	60.5	79.9	−19.4	0.32
4 0 1	82.1	104.7	−22.6	0.38	72.7	75.2	−2.5	0.03
1 1 0	64.8	111.9	−47.8	0.74	57.4	69.5	−12.1	0.21
3 1 0	74.1	112.3	−38.2	0.52	65.6	75.5	−9.8	0.15
2 2 0	104.8	150.6	−45.8	0.44	95.4	101.6	−6.2	0.06

ment which includes the extinction coefficient, g (see equation (6.15)), for which the starting value is set to zero. After this refinement, the R-value decreases to 5.6%, the standard deviations which were 0.005 Å and 0.2° for bond lengths and angles before extinction correction are reduced to 0.003 Å and 0.1–0.2°.

At this stage, we seek to determine the hydrogen atom positions of the NH_4^+ group. However, it should be noted that groups, such as NH_4^+, CH_3, etc., are known to have large thermal motion, so that these hydrogens may be difficult to locate. Since in this structure, we have a relatively large sulfur atom contribution to the scattering power, the hydrogens with one electron only are of minor significance. Inspection of the difference map in the vicinity of N-4 (Fig. 6.15) shows indications of two maxima on the mirror plane, and one above and one below, as required from symmetry. Assuming that these maxima correspond to the hydrogens, we include them in the further refinement with isotropic temperature factors. Convergence is obtained after two more cycles of refinement at an R-value of 5.3%. The final standard deviations for S-N and S-O bonds and angles are 0.002–0.003 Å and 0.1–0.2°. The discrepancies between $|F_o|$

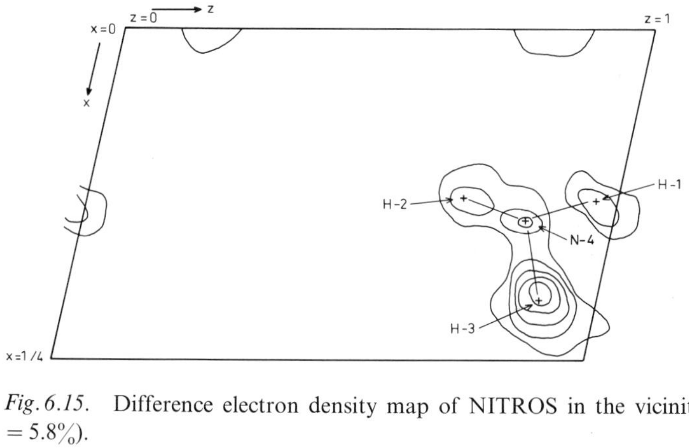

Fig. 6.15. Difference electron density map of NITROS in the vicinity of N-4 (calculated at R = 5.8%).

and $|F_c|$ of the seven reflections listed in Table 6.2 are now smaller. This can best be seen by comparing the two $\Delta F/F$ columns. Before extinction correction, all ratios were above 25%, four were even above 50%. After extinction correction, none exceeded 50%, only two values were above 25%, and three were even below 10%.

Fig. 6.16 shows as the final result of the structure analysis of NITROS, one unit cell of the crystal lattice. The most important result is the confirmation of a cage form for the anion. Moreover, the small intramolecular distance for S-2 – S-2 of 2.632 Å, which is significantly smaller than the S-S van der Waals distance (3.4–3.6 Å), has prompted chemical discussion as to whether this is a partial S-S bond. The hydrogens of the NH_4^+ cation, which could be refined to reasonable positions, make connections between the anion and the cation via N-H...O and N-H...N hydrogen bonds as illustrated in Fig. 6.16. A detailed discussion of this structure has been given elsewhere [Luger, P., Bradaczek, H. and Steudel, R., Chem. Ber. *109*, 3441 (1976)].

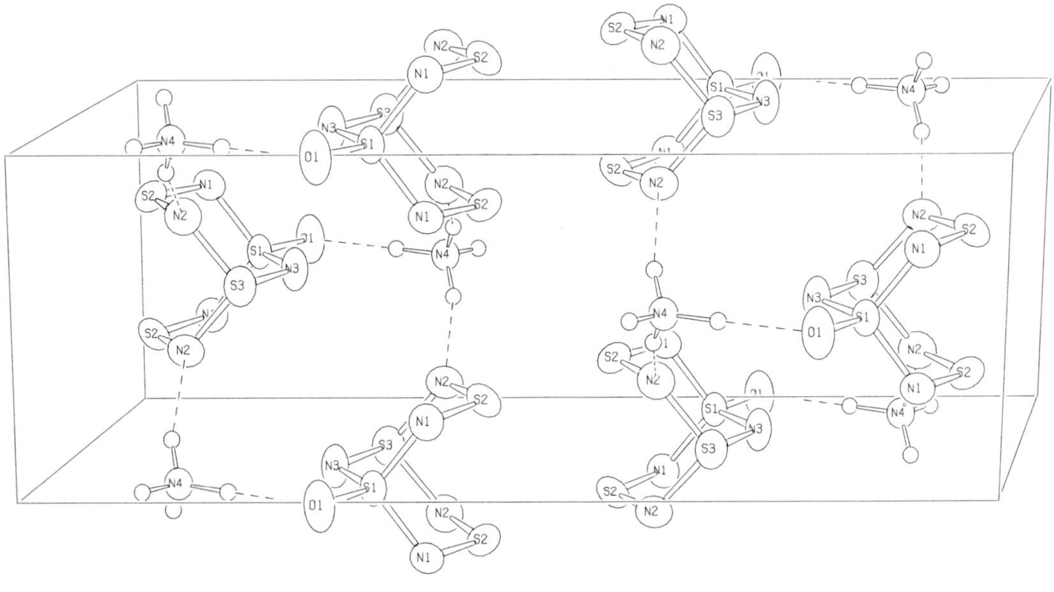

Fig. 6.16. Unit cell of the NITROS lattice. Dashed lines indicate hydrogen bonds.

6.4.3 *SUCROS – Refinement and the Problem of Enantiomorphy*

The R-value of 29.2% calculated from the MULTAN structure shows that this model is correct. From a subsequent difference synthesis, we find one significantly large peak having a chemically reasonable position for O-3′ (as shown by a BONDLA calculation), so we have found the missing oxygen atom of the OH-group at C-3′. Now with all non-hydrogen atoms included in the model, we calculate a three-cycle isotropic refinement. It has to be noted that the y-coordinate of one atom must be kept invariant. Since the origin can be chosen arbitrarily in the y-direction in space group $P2_1$, we have a non-defined origin in that direction, and refinement would result in unreasonable coordinate shifts. So we set the y-coordinate of C-1 invariant. The R-value decreases via 16.3%, 13.7% to 13.2% and we have standard deviations of bond lengths and angles of 0.015–0.020 Å and 1.0–1.5°. Before going to an anisotropic refinement, we checked whether we have the correct enantiomorphic form of the sucrose molecule. Note that the decision in favor of one form was made by the phase of one reflection, which defined the enantiomorphic form in the phase determination process. However, at that stage, the structure was still unknown, and although one of the two forms in question was selected, it could not be known if it was the D- or the L-form.

With the sucrose molecular structure now determined, we can check from the signs of the torsional angles in the pyranosyl ring of the molecule if we really have the D-form, which must be present for chemical reasons. The torsional angles calculated in the sequence given in Table 6.1 have the actual values −58.7, 62.2, −59.9, 57.3, −53.6, and 53.3°. A comparison with the signs listed in Table 6.1 shows that we have the correct enantiomorphic form, since this sign distribution corresponds to the D-form. If we had found the L-form, a simple transformation of the atomic positional parameters to their centrosymmetric equivalents would lead to the correct form of D-sucrose. It is recommended that this examination of the enantiomorphic form be done at the isotropic stage of refinement, since the transformation need only be done for the positional parameters. If it is done when the anisotropic thermal parameters are already included, they would also have to be transformed, and this is more complicated.

Having ensured that the atomic parameters for D-sucrose refer to the correct enantiomer, we continue refinement with anisotropic temperature factors for each non-hydrogen atom. We can no longer use the full-matrix mode, since we now have $23 \times 9 = 207$ parameters to be refined. The matrix size needed for this problem would be $(207 \times 208)/2 = 21528$ (see 6.2.1) which exceeds our version of the X-RAY system, for which the maximum core for the matrix is 15000 words. We therefore separate the structure into two blocks. For chemical reasons, it seems appropriate to have all atoms of the glucose moiety in one block and those of the fructose part in the second block. With this sub-division of parameters, we need two matrix blocks (glucose part: $13 \times 9 = 127$ parameters; fructose part: $10 \times 9 = 90$ parameters) of $(127 \times 128)/2 = 8128$ and $(90 \times 91)/2 = 4095$ words, giving a total of 12223 words core, almost one-half that amount needed for the full-matrix mode.

After two cycles of anisotropic refinement, the R-value reduces via 7.6% to 6.4%. The standard deviations of bond lengths and angles have been reduced to 0.008–0.010 Å and 0.6–0.8°. In a difference synthesis calculated at this stage, 14 of the 22 hydrogens can be located. They are introduced in a refinement cycle with isotropic temperature factors. All the remaining hydrogen positions are obtained from the next difference synthesis. Since the number of parameters has again increased, we now sub-divide the variables to be refined together in another way. First we refine all positional parameters together. We have 45 atoms, so we get $45 \times 3 = 135$ variables, needing a matrix of (135 × 136)/2 = 9180 words. In the next cycle, the thermal parameters are refined. We have 23 × 6 = 138 anisotropic temperature factors and 22 isotropic ones, hence a total of 160 variables, requiring (160 × 161)/2 = 12880 words core, so we are just at the upper limit of available core (a few words of core are needed for organization purposes). Refining our structure in that way (although each user may choose other reasonable sub-divisions of parameters), we get a final R-value of 4.4% and standard deviations of 0.004–0.007 Å for bond lengths and 0.4–0.7° for bond angles. We have not reached the precision of the last X-ray analysis of sucrose by Hanson, Sieker and Jensen [Acta Cryst. *B29*, 797 (1973)], which is about a factor of two or three times better. However, it was not our aim here to describe a high precision X-ray analysis, but rather the principles of how this method works.

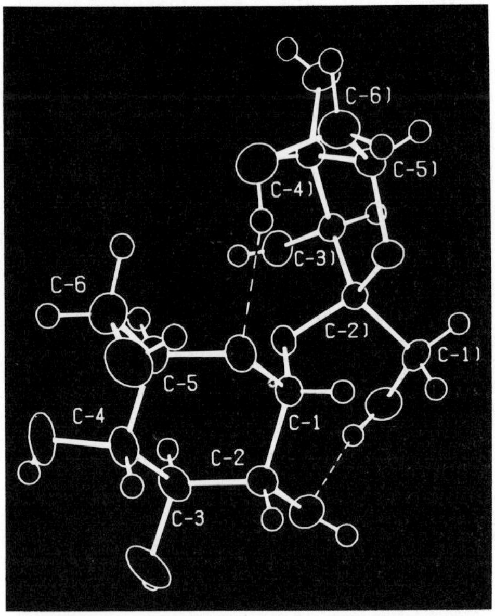

Fig. 6.17. ORTEP-representation of the sucrose molecule on a TV screen.

Having finished all refinement calculations, we can represent the sucrose model by an ORTEP plot. Since the molecule has rather complex geometry, getting a good view is not simple. We first use the representation of the model on a TV screen. The best picture obtained is represented in Fig. 6.17.

From all the atomic parameters we can now derive all the geometrical data needed for a discussion of the molecular and crystal structural geometry. We shall not enter in this discussion since sucrose has been discussed in a number of previous papers. We only state here that the structural problem of this compound was solved completely by the method of X-ray single crystal analysis as well as was done for the other structures and as has been done for a large number of structures in the last years.

Index

S. S. Augustithis

Atlas of the Textural Patterns of Basic and Ultrabasic Rocks and their Genetic Significance

Author: Professor Dr. rer. nat. S. S. Augustithis, Director, Chair of Mineralogy, Petrography, Geology, National Technical University, Athens, Greece.

1979. 21 cm x 29,7 cm. 352 pages. Numerous figures and tables.
Hardcover DM 255,–; $ 142.00 ISBN 3 11 006571 1

The book contains basic research in the geosciences and deals with the petrology, mineralogy, geochemistry and ore-mineralogy of the basic and ultrabasic rocks which are currently the focus of geoscience research.
Primarily, however, it is an Atlas of the microstructures of basic and ultrabasic rocks, from all over the world, showing the most common and important mineral phases and their intergrowths. The book contains over 750 photomicrographic illustrations, showing the great divergence of textural patterns and their genetic significance.
The petrofabric studies support interpretations for the genesis of basic and ultrabasic rocks which are in harmony with the prevailing new hypotheses of mantle-mobilization, ultra-metamorphism and volcanism.

Prices are subject to change